Rock Failure Mechanisms

Frontispiece

Seeing once is better than hearing for a hundred times

The Chinese proverb above is well known and continues to be used in modern-day China. During the Warring States Period in China, Duke Wen Hou of the State of Wei instructed his magistrate Ximen Bao to make decisions only after he had personally made investigation and analysis, instead of giving credence to official reports. He said that, "Seeing it for yourself is better than hearing it from others: on-the-spot investigation is better than just seeing it; and careful analysis is still better than investigation."

In this book, we illustrate rock failure—as it occurs in laboratory tests, in real rock masses, and in the output from computer simulations. So you will be seeing rock failure, albeit indirectly, many times in many different situations, perhaps a hundred times. We trust that the explanations and illustrations will provide a framework for understanding, thus enabling interpretation and appreciation of rock failure *in situ*. However, we also hope that you will be able to experience, investigate and analyse actual rock failure itself, although preferably not in adverse circumstances!

Rock Failure Mechanisms
Explained and Illustrated

Chun'an Tang
School of Civil and Hydraulic Engineering
Dalian University of Technology, Dalian, P.R. China

John A. Hudson
Department of Earth Science and Engineering
Imperial College of Science, Technology and Medicine
London, UK

CRC Press is an imprint of the
Taylor & Francis Group, an **informa** business

A BALKEMA BOOK

CRC Press/Balkema is an imprint of the Taylor & Francis Group, an informa business

© 2010 Taylor & Francis Group, London, UK

Typeset by Vikatan Publishing Solutions (P) Ltd, Chennai, India
Printed and bound in Great Britain by Antony Rowe (A CPI-group Company), Chippenham, Wiltshire

All rights reserved. No part of this publication or the information contained herein may be reproduced, stored in a retrieval system, or transmitted in any form or by any means, electronic, mechanical, by photocopying, recording or otherwise, without prior permission in writing from the publisher. Innovations reported here may not be used without the approval of the authors.

Although all care is taken to ensure integrity and the quality of this publication and the information herein, no responsibility is assumed by the publishers nor the author for any damage to the property or persons as a result of operation or use of this publication and/or the information contained herein.

Published by: CRC Press/Balkema
P.O. Box 447, 2300 AK Leiden, The Netherlands
e-mail: Pub.NL@taylorandfrancis.com
www.crcpress.com – www.taylorandfrancis.co.uk – www.balkema.nl

Library of Congress Cataloging-in-Publication Data

Rock failure mechanisms : explained and illustrated / editors, Chun'an Tang, John A. Hudson.
   p. cm.
Includes bibliographical references and index.
ISBN 978-0-415-49851-7 (hard cover : alk. paper) -- ISBN 978-0-203-84143-3 (ebook) 1. Rock mechanics.
I. Tang, Chun'an. II. Hudson, J. A. (John A.), 1940- III. Title.

TA706.R524 2010
624.1'5132--dc22

2010022873

ISBN: 978-0-415-49851-7 (Hbk)
ISBN: 978-0-203-84143-3 (eBook)

DISCLAIMER

No responsibility is assumed by the Authors or Publisher for any injury and/or damage to persons or property as a matter of products liability, negligence or otherwise, or from any use or operation of any methods, products, instructions or ideas contained in the material herein.

Dedication

To our wives, Juying Yang and Carol Hudson, for their unwavering support over many years.

Contents

Preface xiii
Acknowledgements xv
About the authors xvii
List of figures xix
List of tables xxxvii
Explanatory notes xxxix

1 Introduction 1

    1.1 The purpose of this book 1
    1.2 Why do things break? 1
    1.3 Rock failure in geological and recent history 2
    1.4 Rock failure in present day engineering 5
    1.5 The nature of rock—a natural material 7
        1.5.1 Discontinuities 7
        1.5.2 Inhomogeneity 7
        1.5.3 Anisotropy 8
        1.5.4 Inelasticity 10
    1.6 Numerical modelling of rock failure 10
    1.7 The content of this book 11

2 Rock failure in uniaxial tension 13

    2.1 Introduction 13
    2.2 Specimen simulation 15
    2.3 Numerical simulation results for the uniaxial tension case 16
    2.4 Further studies of simulated rock failure in uniaxial tension 22

3 Rock failure in indirect tension 29

    3.1 Generating a tensile stress through compressive loading 29
    3.2 Establishing the numerical simulation model for indirect tensile strength tests 30

viii Contents

3.3	Numerical simulations of rock failure in indirect tensile strength tests			32
	3.3.1	The disc test		32
		3.3.1.1	Stress distribution in the discs	32
		3.3.1.2	Effect of the material properties of the load-bearing strip on the disc test	35
		3.3.1.3	Effect of load-bearing strip width on disc test	36
		3.3.1.4	Effect of specimen size on the disc test	40
	3.3.2	The plate test		41
	3.3.3	The ring test		42
		3.3.3.1	Stress distribution along the loading diameter for ring specimens	43
		3.3.3.2	Effect of hole diameter on the failure pattern of ring specimens	44
		3.3.3.3	Effect of hole diameter on the ring test indirect tensile strength	49

4 Rock failure in uniaxial compression   **55**

4.1	Introduction			55
4.2	Numerical illustrations of rock failure in uniaxial compression			58
	4.2.1	Model description		58
	4.2.2	Numerical simulation results		59
	4.2.3	Summary of the numerical simulation observations		65
		4.2.3.1	The complete stress–strain curve	65
		4.2.3.2	Acoustic emission (AE) events and their locations	67
		4.2.3.3	Stress distribution and failure-induced stress redistribution	68
4.3	Rock failure modes in uniaxial compression			69
4.4	Factors affecting rock failure behaviour			72
	4.4.1	Model description		73
	4.4.2	Effect of end constraint in terms of the Young's modulus of the loading platens		73
	4.4.3	Effect of height to width ratio (slenderness) of the specimen		78
	4.4.4	Class I and Class II curves in uniaxial compression		82
	4.4.5	The size effect		85

5 Confinement and shear   **89**

5.1	The effect of confinement	89
5.2	Acoustic emission during shearing	95
5.3	Biaxial loading	97

6	**Effect of heterogeneity on rock failure**		**101**
	6.1 Introduction		101
	6.2 Heterogeneity-induced stress fluctuations		102
		6.2.1 Discs subjected to diametral loading	102
		6.2.2 Rock blocks under hydrostatic stress	103
	6.3 Heterogeneity-related seismic patterns		106
	6.4 Influence of heterogeneity on crack propagation modes		109
		6.4.1 Numerical specimen	109
		6.4.2 Numerical results and discussion	110
	6.5 The influence of heterogeneity on the meso-scale		114
		6.5.1 Digital image based modelling method	115
		6.5.2 Numerical model based on the digital image	115
		6.5.3 Simulation results for uniaxial compression	117
		6.5.4 Influence of interface strength	118
7	**The effect of rock anisotropy on rock failure**		**121**
	7.1 Introduction		121
	7.2 Numerical models		122
8	**Loading, unloading and the Kaiser Effect**		**129**
	8.1 Introduction		129
	8.2 Numerical simulation		130
9	**Time dependency of rock failure**		**135**
	9.1 Introduction		135
	9.2 A constitutive model for the time-dependent behaviour of rocks		137
	9.3 Illustrations of time-dependent micro-structural damage		138
		9.3.1 The creep test	138
		9.3.2 The relaxation test	141
	9.4 Degradation of building stones with time		144
10	**Coalescence of fractures**		**147**
	10.1 Introduction		147
	10.2 Modelling of crack growth from crack-like flaws in compression		147
		10.2.1 An angled crack-like flaw	148
		10.2.2 Crack growth from an array of crack-like flaws	158
		10.2.2.1 Wing crack growth from three flaw arrays	158
		10.2.2.2 Wing crack growth from randomly distributed multi-flaws	162
	10.3 Crack growth from a pore-like flaw in compression		165

10.3.1 Modelling crack growth from a single hole in specimens under compression — 166
    10.3.1.1 Crack growth from a single hole in specimens of different width — 166
    10.3.1.2 Crack growth from a single hole with different diameters — 167
    10.3.1.3 Modelling of crack growth from an array of holes in a specimen under compression — 172

11 Dynamic loading of rock — 181

11.1 Introduction — 181
11.2 The simulation models — 183
11.3 Simulation demonstration — 185
    11.3.1 Influence of heterogeneity on stress wave propagation — 185
    11.3.2 Influence of stress wave amplitude on the fracture process and failure pattern — 186

12 Rock failure and water flow — 189

12.1 Introduction — 189
12.2 Rock failure under hydraulic pressure — 190
12.3 Illustrations of fluid flow in heterogeneous initially intact rock — 193
    12.3.1 Evolution of flow paths — 196
12.4 Comparison with the rock degradation modelling by Yuan and Harrison (2005) — 200
12.5 Fluid flow in initially intact rock containing block inhomogeneities — 203

13 Rock failure induced by thermal stress — 209

13.1 Introduction — 209
13.2 Thermally-induced rock failure — 210
13.3 Thermal cracking of a disc–ring model — 212
13.4 Thermal cracking in models containing irregularly shaped inclusions — 215

14 Slope failure in rock masses — 219

14.1 Introduction — 219
14.2 Strength reduction rule and determination of safety factor — 220
14.3 A slope in a layered rock mass — 223
14.4 A slope in a jointed rock mass — 225
14.5 A slope in a jointed rock mass with differing joint persistence — 228

Contents xi

15 The fracture process when cutting inhomogeneous rocks — 229

15.1 Introduction — 229
15.2 Modelling rock cutting and the failure mechanism — 229
    15.2.1 Quasi-photoelastic fringe pattern — 230
    15.2.2 Fracture pattern — 231
    15.2.3 The chipping process — 233
15.3 The load–displacement response when cutting inhomogeneous rock — 234
15.4 The crushed zone during rock cutting — 236

16 Rock failure around tunnels in jointed rock — 237

16.1 Introduction — 237
16.2 Progressive failure around a tunnel in a jointed rock mass — 238
    16.2.1 Effect of dip angles on the stability of the tunnel — 239
    16.2.2 The effect of the lateral stress on the mode of tunnel failure — 243
    16.2.3 Displacements at the tunnel periphery — 247

17 Rock failure induced by longwall coal mining — 249

17.1 Introduction — 249
17.2 Illustrations of longwall mining simulations — 250
17.3 The Daliuta coal mine in China — 255
    17.3.1 The strata failure process — 256
    17.3.2 Pillar stresses — 258

18 Gas outbursts in coal mines — 263

18.1 Introduction — 263
18.2 Outbursts induced by cross-cutting from rock to coal seam — 266
18.3 Outburst as the working face approaches high methane pressure in the coal seam — 269

19 Particle breakage and comminution — 273

19.1 Introduction — 273
19.2 Single particle breakage — 274
    19.2.1 Breakage of single particle under diametral loading without confinement — 274
    19.2.2 Breakage of single particle under diametral loading with confinement — 276
19.3 Multiple particle breakage — 284
    19.3.1 Fragmentation process of a rock particle assemblage in a container — 285

	19.3.2 Force and displacement relation during the breakage process	287
	19.3.3 Energy considerations	290
	19.3.4 Size distribution	292
	19.3.5 Influence of particle shape	293

20 3-D modelling and 'turtle crack formation' in rock — 297

20.1 Introduction — 297
20.2 The three-layer model — 298
20.3 Fracture spacing measurements — 303

21 Concluding remarks — 307

References and bibliography — 309
Index — 321

Preface

In this book we explain and illustrate rock failure mechanisms. The subject of rock failure has been studied in a co-ordinated way since the 1960s: the International Society for Rock Mechanics was founded in 1962 and the first issue of the International Journal for Rock Mechanics and Mining Sciences was published in 1964. The way in which rock fails can be studied by examination of natural rock formations that have been stressed and strained over geological time, by laboratory experiments on rock samples, through *in situ* experiments, and by observing the results of rock excavation and loading during engineering construction.

Over the years, there have been three main developmental phases supporting rock engineering design: analysis based on elasticity theory; the use of rock mass classification systems; and computer modelling. The elasticity theory approach is useful because it enables the stresses around circular and elliptical holes to be determined, although the approach is most useful for deep excavations where the rock behaviour is essentially elastic. Rock mass classification is also useful because the variety of factors affecting rock behaviour can be accommodated in a mathematical expression, thus providing an index value for rock quality. Computer modelling started as a method of displaying analytical results and extending the analyses to more complex situations. However, in the last two decades, computer modelling has advanced by leaps and bounds so that it is now, not only the design tool of choice for rock engineering, but is also a research tool in its own right for exploring rock failure mechanisms. For example, a comprehensive knowledge of the state of stress throughout the micro-structure of a rock specimen or throughout a fractured rock mass several kilometres in size cannot be established by direct laboratory or *in situ* measurements but it can be studied through computer modelling using numerical techniques. For this reason, to illustrate rock failure mechanisms, many of the diagrams in this book are the output from numerical simulations. By many comparisons with the behaviour of real rocks, we have confidence that these simulations do indeed represent real rock failure behaviour.

When engineering on or in rock masses, one may wish to avoid failure (e.g. when excavating a cavern to host the turbines in a hydro-electric project) or one may wish to cause failure (e.g. in the block caving method of mining when a large rock block is undercut and breaks up as it descends). In both cases, wishing to avoid or to cause rock failure, it is important to understand the rock failure mechanisms and the many factors that can affect the mode of rock failure, in particular the nature of the applied stress state and the nature of the rock. The applied stress can be in the form of tension, compression or shear, and various combinations of these. The rock itself is generally discontinuous, inhomogeneous and anisotropic and occurs on a multiplicity of scales. This means that rock failure can be manifested in many ways. In the book we consider mainly brittle rock failure.

Thus, our intention in writing this book is to provide an overview of the physical manifestations of rock failure in the variety of circumstances that can occur. Accordingly, the Chapters follow the logic of an overall introduction explaining the geological background and engineering failure, then direct loading in tension, compression and shear, the effects of inhomogeneity, anisotropy, multiple loading and time dependency, the effects of water and heat flow, engineering projects, and finally the two concluding Chapters on 3-D modelling and concluding remarks. Five of the individual chapters are somewhat longer than the others because of the importance of their subject matter: Chapter 3 on indirect tension, Chapter 4 on uniaxial compression, Chapter 6 on the effect of rock heterogeneity, Chapter 10 on the coalescence of fractures and Chapter 19 on particle breakage.

We use photographs and computer simulation outputs to explain and illustrate the rock failure mechanisms. It is not our intention to provide detailed mathematical expressions characterising rock failure in the different circumstances, but rather to present illustrative examples of the rock failure mechanisms so that the overall spectrum of rock failure can be appreciated by all those concerned, including clients, consulting engineers, contractors, students, lecturers and researchers.

We hope that our objective will have been achieved and that you will enjoy reading this book and perusing the illustrations. If not, the fault is entirely ours.

Chun'an Tang, *Dalian, China*
John A. Hudson, *London, UK*
April, 2010

Acknowledgements

The first author, Chun'an Tang, wishes to gratefully acknowledge the following colleagues who have contributed to this book through the numerical simulations and discussion over the past fifteen years: Prof. K.T. Chau, Dr. Y.F. Fu, Dr. P. Jia, Prof. S.Q. Kou, Prof. P.K.K. Lee, Dr. L.C. Li, Dr. Z.Z. Liang, Dr. P. Lin, Prof. P-A. Lindqvist, Dr. H. Liu, Dr. H.Y. Liu, Dr. T.H. Ma, Dr. S.B. Tang, Prof. L.G. Tham, Dr. Y. Tsui, Dr. R.H.C. Wong, Dr. T. Xu, Prof. T.H. Yang, Dr. Q.L. Yu, Dr. Y.B. Zhang, Prof. W.C. Zhu.

The second author, John A. Hudson, acknowledges Prof. Charles Fairhurst of the University of Minnesota who guided his PhD work and instilled in him a life-long fascination with the phenomenon and manifold aspects of rock failure. He is also grateful for the academic companionship and many discussions over the last 20 years with Dr. John P. Harrison and Prof. John W Cosgrove in the Department of Earth Science and Engineering at Imperial College of Science, Technology and Medicine, London, UK.

The authors appreciate the valuable assistance provided by Dr. Xu Tao and Mr Yang Yue'feng in improving the presentation of the diagrams.

Some of the material contained in this book has been published in journal papers, as listed below. The authors are grateful to the journal publishers, Elsevier, Taylor and Francis, and Trans Tech Publications, for their kind permission to use the material in this book.

Tang, C.A., Liu, H., Lee, P.K.K., Tsui, Y. & Tham, L.G.: Numerical Tests on Micro-Macro Relationship of Rock Failure under Uniaxial Compression, Part I: Effect of Heterogeneity. *Int. J. Rock Mech. Min. Sci.* 37 (2000), pp. 555–569.

Tang, C.A., Tham, L.G., Lee, P.K.K., Tsui, Y. & Liu, H.: Numerical Tests on Micro-Macro Relationship of Rock Failure under Uniaxial Compression, Part II: Constraint, Slenderness and Size Effects. *Int. J. Rock Mech. Min. Sci.* 37 (2000), pp. 570–577.

Tang, C.A., Lin, P., Wong, R.H.C. & Chau, K.T.: Analysis of Crack Coalescence in Rock-Like Materials Containing Three Flaws—Part II: Numerical Approach. *Int. J. Rock Mech. Min. Sci.* 38 (2001), pp. 925–939.

Tang, C.A., Yang, T.H., Tham, L.G., Lee, P.K.K. & Li, L.C.: Coupled Analysis of Flow, Stress and Damage (FSD) in Rock Failure. *Int. J. Rock Mech. Min. Sci.* 39 (2002), pp. 477–489.

Wong, R.H.C., Tang, C.A., Chau, K.T. & Lin, P.: Splitting Failure in Brittle Rocks Containing Pre-Existing Flaws under Uniaxial Compression. *Eng. Fract. Mech.* 69 (2002), pp. 1853–1871.

Chau, K.T., Zhu, W.C., Tang, C.A. & Wu, S.Z.: Numerical Simulations of Failure of Brittle Solids under Dynamic Impact Using a New Computer Program—DIFAR. *Key Eng. Mat.* 261–263 (2004), pp. 239–244.

Tang, C.A., Liu, H.Y., Zhu, W.C., Yang, T.H., Li, W.H., Song, L. & Lin, P.: Numerical Approach to Particle Breakage under Different Loading Conditions. *Powder Technol.* 143–4 (2004), pp. 130–143.

Tang, C.A., Wong, R.H.C., Chau, K.T. & Lin, P.: Modeling of Compression-Induced Splitting Failure in Heterogeneous Brittle Porous Solids. *Eng. Fract. Mech.* 72 (2005), pp. 597–615.

Li, L.C., Tang, C.A., Li, C.W. & Zhu, W.C.: Slope Stability Analysis by SRM-Based Rock Failure Process Analysis (RFPA). *Geomech. Geoeng.* 1 (2006), pp. 1–12.

Tang, C.A., Zhang, Y.B, Liang, Z.Z., Xu, T., Tham, L.G., Lindqvist, P.-A., Kou, S.Q. & Liu, H.Y.: Fracture Spacing in Layered Materials and Pattern Transition from Parallel to Polygonal Fractures. *Phys. Rev. E*, 73, 5 (2006).

Tang, C.A., Tham, L.G., Wang, S.H., Liu, H. & Li, W.H.: A Numerical Study of The Influence of Heterogeneity on the Strength Characterization of Rock under Uniaxial Tension. *Mech. Mater.* 39 (2007), pp. 326–339.

About the authors

PROFESSOR CHUN-AN TANG

Chun'an Tang received his BSc from Central-South University and his MSc and PhD from Northeastern University in China. He was Vice Professor in the Mining Department, then Professor and Director of the Centre for Rock Instability and Seismicity Research in the Civil and Resources Engineering School, and then Chair Professor at the same University. Since 2006, he has been Professor in the School of Civil and Hydraulic Engineering at the Dalian University of Technology. He has authored and co-authored around 300 publications in the fields of rock mechanics, civil engineering and geophysics. He became Director of the Committee of Rock Fracture and Fragmentation of the Chinese Society for Rock Mechanics and Engineering in 2000 and Chairman of the China National Group of the International Society for Rock Mechanics in 2008.

EMERITUS PROFESSOR JOHN A. HUDSON

John A. Hudson graduated in 1965 from the Heriot-Watt University, UK, and obtained his PhD at the University of Minnesota, USA. He has spent his professional career in engineering rock mechanic—as it applies to civil, mining and environmental engineering—in consulting, research, teaching and publishing and was awarded the DSc. degree for his contributions to the subject. In addition to authoring many scientific papers, he edited the 1993 five-volume "Comprehensive Rock Engineering" compendium, and the International Journal of Rock Mechanics and Mining Sciences from 1983–2006. From 1983 to the present, he has been affiliated with Imperial College as Reader and Professor. In 1998, he was elected as a Fellow of the Royal Academy of Engineering in the U.K. and in 2007 he became President of the International Society for Rock Mechanics for the period 2007–2011.

List of figures

1	The generic complete force–displacement curve.	xl
2	The probability density values, $f(x)$ as a function of x for different values of the shape parameter, m, in the Weibull statistical distribution.	xli
3	Interpretation of the computer simulation illustrations.	xlii
1.1	Sets of rock fractures in a Carboniferous stratum, S. Wales, UK.	2
1.2	The distinction between structural geology and rock engineering in the context of interpreting past processes and predicting the future.	3
1.3	Runestone at Sigtuna, Sweden.	4
1.4	Failure occurring to the overhang at one of the Ellora Cave Temples excavated more than a millennium ago in the Deccan basalts, India.	4
1.5	Tunnel in hard crystalline rock at the Äspö Hard Rock Laboratory, Sweden.	5
1.6	Tunnel wall slabbing (spalling) due to high vertical stress caused by a large overburden in one of the access tunnels for the JinPing II hydroelectric project on the Yalong river in Sichuan province, China.	6
1.7	The discontinuity characteristics recommended by the International Society for Rock Mechanics for specification in a site investigation.	7
1.8	Computer simulation of a large rock mass containing many discontinuities.	8
1.9	The range of P-wave velocities measured on the site investigation cores at the Olkiluoto crystalline rock site in Finland.	9
1.10	Influence of anisotropy on the potential location for failure around an excavation in gneissic rock.	9
1.11	Complete stress–strain curve for a marble specimen loaded in uniaxial compression.	10
2.1	Section of a direct tension test specimen with a transverse crack across the specimen visible below the upper shoulder between the arrows—test stopped in the descending portion of the complete stress–strain curve. Specimen 20 mm in diameter, Tennessee marble.	14
2.2	Model geometry and mesh generation of the numerical specimen with two loading platens (the model is composed of $130 \times 130 = 16{,}900$ elements with dimensions equivalent to	

	130 × 130 mm. The variation of the gray colour represents the heterogeneity of Young's modulus, a lighter shade having a higher value).	15
2.3	Numerically obtained load and fracture event counts *vs.* deformation/displacement for specimen #1.	16
2.4	Numerically obtained cumulative fracture event counts and fracture event released energy accumulation *vs.* deformation for specimen #1 during the development of the stress–strain curve in Figure 2.3.	17
2.5	Numerically obtained maps of fracture event source locations at different loading steps for specimen #1 (the circles represent the source location of fracture events and the diameter of the circles represents the associated fracture event released energy value).	18
2.6	Failure process in different loading steps for specimen #1. (The gray colour represents the relative heterogeneity of the Young's modulus, and the black dots represent the micro-fractures).	19
2.7	Maximum shear stress distribution at the different loading steps for specimen #1 with $m = 1.5$. (The gray colour represents the relative shear stress (white being the high values), and the black dots represent the micro-fractures).	19
2.8	Variations in the shape of the jagged fracture surface for the four specimens #5 to #8 having different homogeneity indices, m (the larger the value of m, the more homogeneous is the specimen). The gray colour represents the relative heterogeneity in Young's modulus, and the black dots represent the micro-fractures.	21
2.9	Sensitivity of failure modes to the heterogeneity for the four specimens with homogeneity indices, $m = 1.5, 2, 3$ and 5 (left to right). a) the gray colour represents the relative shear stress; and b) the gray colour represents the relative heterogeneity of Young's modulus, with the black dots representing the micro-fractures.	21
2.10	Sensitivity of the load–deformation relations to the nature of the local heterogeneity.	22
2.11	Stress distribution and failure-induced stress redistribution for sections AA and BB before and after the macro-fracture develops in uniaxial tension ($m = 1.5$).	23
2.12	Failure mode of relatively homogeneous rock (specimen #7, $m = 3$). (Above: the circles represent the source location of fracture events. Below: the gray colour represents the relative shear stress).	24
2.13	Influence of heterogeneity on the shape of the load–deformation curves and the peak strengths.	25
2.14	Basic statistical model (Hudson and Fairhurst, 1969) illustrating the same trends as in Figure 2.13 and demonstrating the influence of the rock inhomogeneity.	26
2.15	Influence of heterogeneity on the shape of the fracture event count accumulation-deformation curves and their strength characterisation.	26

2.16	Six stages in the final rupture of the crack overlap (in the sample, the gray colour represents the relative shear stress, the lighter the shade, the higher the stress).	27
2.17	Displacement vector for a specimen which failed in the crack interface bridging manner.	28
3.1	The beam test and Brazilian test.	30
3.2	Geometries and loading conditions for disc, square plate and ring specimens. The co-ordinates and stress notations used are indicated.	31
3.3	Numerically produced 'fingerprint' contours of the maximum shear stress in diametrically loaded disc, plate and ring specimens of homogeneous and heterogeneous material.	33
3.4	Stress distribution across the loaded diameter for a disc specimen in the Brazilian test. Theoretical and numerical results for using a cardboard cushion when the Weibull shape parameter m is 5.0, E_s is 62.2 GPa, v_s is 0.28 and the bearing strip widths, b/D, are 0.04, 0.08 and 0.16.	34
3.5	Distribution of acoustic emission a) and maximum shear stress b) during the failure process of the numerical specimen. Fracturing is indicated in black. The specimen is cushioned with steel and b/D is 0.04 and its mechanical properties are: $m = 5.0$, $E_s = 62.2$ GPa, and $v_s = 0.28$.	36
3.6	Failure process of a disc specimen in the Brazilian test (here the specimens are cushioned with cardboard with b/D increasing for the illustrations a → d), with the maximum shear distributions at fracture initiation, peak load and post-peak given; for the numerical specimen, $m = 5.0$, $E_s = 62.2$ GPa, and $v_s - 0.28$.	37
3.7	Distribution of stresses when steel and cardboard are utilised as the cushion with different b/D values as given by the simulations. For the numerical specimen, $m = 5.0$, $E_s = 62.2$ GPa, and $v_s = 0.28$.	38
3.8	Numerically obtained load–displacement curves for specimens in Brazilian tests (the cushion material is cardboard, and for the numerical specimen, $m = 5.0$, $E_s = 62.2$ GPa, and $v_s = 0.28$).	39
3.9	Numerically obtained failure patterns for disc specimens having different sizes and load-bearing widths (For the numerical specimens, $m = 5.0$, $E_s = 62.2$ GPa, and $v_s = 0.28$).	41
3.10	Comparison of stress distributions across the vertical centreline of a square specimen and a disc specimen (for the numerical specimens, $m = 5.0$, $E_s = 62.2$ GPa, and $v_s = 0.28$).	42
3.11	Numerically obtained failure patterns for square plate specimens at three different loading levels. The three different relative load-bearing strip widths b/D are: a) $b/D = 0.04$, b) $b/D = 0.08$ and c) $b/D = 0.16$ (For the numerical specimen, $m = 5.0$, $E_s = 62.2$ GPa, and $v_s = 0.28$).	43
3.12	Numerically obtained distribution of horizontal stress along the loading diameter in the ring specimen for different relative	

	sizes of the central hole (for the numerical specimen, $m = 5.0$, $E_s = 62.2$ GPa, and $v_s = 0.28$).	44
3.13	Numerically obtained failure patterns of ring test specimens (specimen size D is 150 mm, and the cushioning material is cardboard).	45
3.14	Numerically obtained typical failure processes of two types (indicated by (a) and (b)) for ring specimens (specimen size D is 150 mm). In this Figure the gray degree indicates the magnitude of the shear stress. The numbers indicate the sequence of the failure process).	46
3.15	Numerically obtained load–displacement curves for ring specimens with different internal radii (the specimen size D is 150 mm and $b/D = 0.16$, with $m = 5.0$, $E_s = 62.2$ GPa and $v_s = 0.28$).	47
3.16	The typical failure patterns of ring specimens observed in experiments.	47
3.17	The different failure patterns for two ring specimens numerically simulated with the same geometry and loading conditions. (For these two specimens, D is 150 mm, b/D is 0.16 and r/R is 0.2).	48
3.18	Ring tests on Tennessee marble 50 mm diameter discs with a central hole ($r/R = 0.03$) illustrating the effect of direct loading via steel platens (for Specimen 52) or cushioning the loading points (for Specimen 53) on the location of failure.	49
3.19	Ring test on a plaster specimen, 150 mm in diameter, containing limestone chips to control the heterogeneity, with a central hole with $r/R = 0.022$, in which the failure path by-passed the hole.	50
3.20	Effect of a small hole ($r/R = 0.022$) located at various points along the 150 mm loaded diameter of a fine-grained plaster specimen on the formation of the secondary fractures.	50
3.21	Indirect tensile strength of numerical specimens with different internal radii compared with experimental results.	51
4.1a	Idealised complete stress–strain curve for a rock specimen loaded in uniaxial compression.	56
4.1b	Complete stress–strain curve for a marble specimen with unloading and re-loading hysteresis loops.	56
4.2	Sectional portion of a Tennessee marble specimen unloaded just after the peak of the complete stress–strain curve in uniaxial compression—illustrating the development of axial cracks which are the pre-cursor to the eventual coalescence of micro-fractures and the specimen collapse. (Specimen width 25 mm.)	57
4.3	Inhomogeneity of the initial five specimens used in the numerical studies.	58
4.4	Simulated complete stress–strain curve for the $m = 1.5$ specimen subjected to uniaxial compression with an axial displacement rate control of 0.002 mm/step.	59
4.5	Simulated acoustic emission (AE) results plotted vs. loading rate for the $m = 1.5$ specimen shown in Figure 4.3: a) nominal stress	

	(top graph); b) AE event rate and AE event accumulation (middle graph); and c) AE released energy and energy accumulation (lower graph).	60
4.6	Plots of AE locations for the specimen in Figure 4.5. Each circle represents one AE event and its relative magnitude. Stress states for each plot are indicated in Figure 4.5a. Event counts and the corresponding released energy are shown in Figure 4.5b and Figure 4.5c, respectively. The gray colour in plot 'a' shows all the AE locations which occurred before this step. The black colour represents the AE locations at each current step.	62
4.7	Plots of the simulated failure process corresponding to the same loading steps as in Figure 4.5. The gray colour in the images represents the Young's moduli of the elements.	63
4.8	Images of the simulated failure process corresponding to the same loading steps as in Figure 4.5. The gray colour in these images represents the maximum shear stress of the elements.	63
4.9	Plots of simulated displacement vectors corresponding to the same loading steps as in Figure 4.5. Stress states for each plot are indicated in Figure 4.5.	64
4.10	Influence of material heterogeneity on the stress–strain curves for five specimens with different homogeneity indices, m.	66
4.11	Influence of material heterogeneity on the cumulative AE events for five specimens with different homogeneity indices, m.	67
4.12	Influence of material heterogeneity on the failure modes for five specimens with different homogeneity indices; a) initial stress distributions, and b) final failure modes.	69
4.13	Variation of compressive strength for five rock types in the Finnish crystalline basement.	70
4.14	Sensitivity of failure modes to local micro-structural variation for five specimens with the same Weibull distribution generating parameter, $m = 1.5$. a) Initial stress distributions, and b) final failure modes.	71
4.15	Sensitivity of stress–strain curves to the local variation in micro-structure for five specimens with the same overall statistical distribution of local mechanical properties ($m = 1.5$ for all specimens).	72
4.16	Idealised deformation, specimen–platen interaction, stress states in the specimen, and failure modes within the specimen: left, the ratio of platen modulus to specimen modulus $E_p/E_s > 1$; and, right, $E_p/E_s < 1$.	74
4.17	Numerical simulation of the effects of end constraint for five specimens with different loading platen Young's moduli, a) initial stress distributions in the specimen and platens; and b) failure modes.	75
4.18	Numerical simulation of failure modes for specimens with different loading platens in terms of the relative Young's modulus of platen and specimen (the gray colour represents the shear stress):	

	a) stiffer constraint condition, $E_p/E_s = 10$; and b) softer constraint condition, $E_p/E_s = 0.1$.	76
4.19	Simulated stress–strain curves for five specimens with loading platens having different relative Young's moduli.	77
4.20	Simulated strength reduction with end constraint for five specimens with different loading platens in terms of Young's modulus.	77
4.21a	Specimens used in the numerical modelling to study the effect of specimen height: width ratio, illustrating the initial stress distributions in the specimens and platens.	78
4.21b	Failure modes for the five specimens with different height to width ratios shown in Figure 4.21a.	79
4.22a	Numerical simulation of failure modes for specimens with a height to width ratio of 0.5.	80
4.22b	Numerical simulation of failure modes for specimens with a height to width ratio of 3.	80
4.23	Stress–strain curves for five simulated specimens with different shapes in terms of the height to width ratio.	81
4.24	Strength reduction with specimen size for five simulated specimens.	81
4.25	Class I and Class II complete stress–strain curves—a Class II curve does not monotonically increase in strain.	82
4.26	Energy changes during micro-fracturing leading from stress–strain state A to stress–strain state B.	83
4.27	Thermally controlled stiff testing machine at the University of Minnesota in 1970.	83
4.28	Complete stress–strain curves obtained in a thermally controlled stiff testing machine.	84
4.29	The measurement of circumferential displacement as feedback in a closed-loop servo-controlled testing machine to obtain a Class II curve.	85
4.30	Use of the Particle Flow Code (PFC) to obtain the complete stress–strain curve for Lac du Bonnet granite.	85
4.31	Complete Class II stress–strain curves numerically simulated with an elastic-plastic cellular automation.	86
4.32	Five specimens used for numerical simulation of the size effect with the same height to width ratio but different sizes.	86
4.33	Stress–strain curves for five simulated specimens with different sizes but with the same height to width ratio tested in uniaxial compression.	86
4.34	Simulation of strength reduction with specimen size.	87
5.1	The numerical model for studying failure of rock in biaxial compression.	90
5.2	Complete stress–strain curves for numerical rock specimens with varying confining stress.	90
5.3	Relation between the compressive strength and confining pressure for the simulated rock specimens.	91

5.4	Failure envelope in normal stress–shear stress space for the simulated rock specimens.	91
5.5	Shear planes developed during a physical test on rock at a high confining pressure.	92
5.6	The formation of fractures and shear zones in rock over geological time.	92
5.7	Interaction matrix illustrating the effect of a second phase of fracturing or shearing on a rock that has already experienced an initial fracturing or shearing episode.	93
5.8	a) Rock material numerically modelled; b) Shear stresses in the simulated rock when subjected to a vertical to horizontal stress ratio of 3.	94
5.9	a) Fracture in the rock; b) Shear stresses around the fracture.	94
5.10	Complete stress–strain curves and AE characteristic curves for the simulated specimens subjected to different confining stresses.	96
5.11	AE curves and normalised cumulative AE energy (AEE) curves for model specimens under different confining stress.	96
5.12	Biaxial strength envelopes for rock (experimental and numerical results).	98
5.13a	Sketches of the loading cases for the specimens shown in Figure 5.13b.	99
5.13b	Failure patterns of simulated rock specimens under uniaxial and biaxial loading (b = the ratio of the applied displacement rates, u_x/u_y).	99
6.1	Simulated shear stress fields (equivalent to photoelastic fringe patterns) in discs subjected to vertical loading. Images a and b: finite element models with 40,000 elements with vertical loading conditions for homogeneous and heterogeneous discs respectively. The gray colour represents the variation of the elastic moduli of the individual elements; Images c and d: shear stress fields in the homogeneous and heterogeneous discs respectively.	103
6.2	Simulated shear stress changes along the line between the loading points of the discs shown in Figure 6.1.	104
6.3	Simulated stress fields in a heterogeneous rock block. a) Finite element model with 40,000 elements with boundary conditions, $\sigma_x = \sigma_y = 1$ MPa, and with the gray colour representing the variation of the elastic moduli of individual elements. b) Image of the resulting shear stress τ, with the gray colour representing the value of τ of the individual elements. The value from dark to light is 0–0.7 MPa. c) Image of the simulated principal stress σ_1. The gray colour represents the value of σ_1 of the individual elements. The value from dark to light is 0–1.7 MPa.	105
6.4	Shear stress fluctuations along the cross section A-A of the simulated rock block shown in Figure 6.3a.	106

6.5	Complete stress–displacement curves with the more homogenous specimens showing higher peak strength and more linearity during the failure process.	107
6.6	Acoustic emission (AE) *vs.* displacement of the specimen for $m = 1.1$ (significantly inhomogeneous).	108
6.7	Acoustic emission *vs.* displacement of the specimen with $m = 3$ (intermediate heterogeneity).	108
6.8	Acoustic emission *vs.* loading step/displacement for the specimen with $m = 10$ (strongly homogeneous).	108
6.9	Acoustic emissions of resin and three rock types with different variations in their mechanical properties subjected to stress loading. Colophony (pine resin) is the most homogenous, and pumice stone is the most heterogeneous among these four materials. They manifest different patterns of seismic activities: the main shock, pre-main shock, and swarm shock.	109
6.10	Numerical models for homogeneous and heterogeneous rock specimens containing a pre-existing flaw. a) relatively homogeneous model, $m = 20$ b) relatively heterogeneous model, $m = 2$.	110
6.11	Numerical simulation of the crack propagation path in a relatively homogeneous rock specimen ($m = 20$) containing a pre-existing flaw (The gray colour represents the value of the minimum principal stress).	111
6.12	Numerical simulation of the crack propagation path in a heterogeneous rock specimen ($m = 2$) containing a pre-existing flaw (The gray colour represents the value of the minimum principal stress).	111
6.13	a) Load–deformation and b) AE event rate–deformation curves obtained from the simulation of the heterogeneous rock specimen.	112
6.14	AE event locations in the numerical simulation of the heterogeneous rock specimen ($m = 2$) containing a pre-existing flaw.	113
6.15	Lemunda sandstone (mean grain size is 0.35 mm; minimum 0.05 mm, maximum 1.09 mm).	115
6.16	Process of obtaining the rock image (Q = Quartz, F = Feldspar, M = Mica).	116
6.17	Analysed image and an example of the transformation from pixel to grid.	116
6.18	Numerical model used for the simulation.	117
6.19	Force–displacement curve and associated AE events.	118
6.20	Distribution and redistribution of shear stress in the simulation (the lighter the gray colour, the higher the shear stress).	119
6.21	Progressive fracture process in the simulation.	119
6.22	Numerically obtained stress–strain curves for the rock with the interface strength being taken as 10%, 20%, 30%, 40%, and 50% of the mica strength.	120
6.23	Influence of interface strength on the overall strength of the samples.	120

6.24	Influence of interface strength on the sample failure modes (percentages indicate the interface strength *vs.* the mica strength).	120
7.1	Types of elastic anisotropy.	122
7.2	Seven transversely isotropic rock specimens with layers having different dip angles and composed of two materials, the lighter gray material being stiffer and stronger; β is the angle of the layers with respect to the axial direction of loading.	123
7.3	Complete stress–strain curves for seven rock specimens with different dip angles relative to the transverse anisotropy plane (*cf.* Figure 7.2), showing that the dip angles influence the shape of the curves during the failure process.	124
7.4	Plots of peak strength *vs.* dip angles for the seven rock specimens. The peak strength of the transversely isotropic rock decreases for dip angles 15° to 60°, and increases for dip angles 60° to 90°.	124
7.5	Plots of maximal axial strain *vs.* dip angles for the seven simulated rock specimens.	125
7.6	Plots of Young's modulus *vs.* dip angles for the seven simulated rock specimens.	125
7.7	Minor principal stress and fracturing during the failure process of the simulated rock specimen with a dip angle $\beta = 45°$.	126
7.8	Failure configuration and displacement vectors at the collapse point for the seven specimens with different dip angles.	127
7.9	Experimental results and numerical simulation for the validation exercise.	128
8.1	Numerical specimen with inhomogeneous elements, $m = 1.5$.	130
8.2	Numerical results of AE counts *vs.* percentage of strength.	131
8.3	Experimental results of AE counts under cyclical loading.	132
8.4	Curves of cumulative AE events *vs.* applied stress as a proportion of the compressive strength.	132
8.5	Complete stress–strain curve with loading and unloading cycles throughout.	133
8.6	Spatial distribution of AE events during the rock failure process: Black circles are acoustic emission; gray circles are tensile failure; white circles are shear failure.	133
9.1	Definitions of creep, stress relaxation and time-dependent unloading (dashed line) along the stiffness of the adjacent rock element (the spring above the rock specimen).	136
9.2	Strength degradation of an element following equation (9.2).	137
9.3	Numerically obtained strain *vs.* time and creep rate *vs.* time curves for the uniaxial creep test.	138
9.4	The shear stress field (gray shading) and fracturing (in black) during creep.	140
9.5	Numerically obtained AE locations in the rock specimen during the creep process.	140
9.6	Displacement vectors corresponding to the creep failure process.	141
9.7	Stepped load increase with time intervals.	142
9.8	Creep deformation at different stress levels.	142

9.9	Creep deformation in the stress intervals.	142
9.10	Acoustic emission sequence with stress intervals.	143
9.11	Stress relaxation curve.	143
9.12	Acoustic emission *vs.* time in a uniaxial relaxation test.	144
9.13	One face of the Finlandia Hall, Helsinki, Finland, showing convex bowing of the Carrara marble cladding panels due to large seasonal temperature changes.	145
9.14	Deterioration of the Caen stone at Norwich cathedral in the UK.	145
10.1	Schematic of numerical specimen geometry.	149
10.2a	Sequences of the wing crack growth under uniaxial compression for specimens containing crack-like flaws of lengths 10, 20 and 30 mm.	150
10.2b	Shear stress distribution during the wing crack growth under axial load for specimens containing crack-like flaws of length 10, 20 and 30 mm.	151
10.3	The influence of the initial flaw length (10, 20 and 30 mm) on the growth of wing crack under uniaxial compression.	152
10.4	Ultimate failure modes for specimens containing pre-existing crack-like flaw lengths of 10, 20 and 30 mm (in these images the flaw is not so visible but is present between the main developed cracks).	152
10.5	Stress distribution during stages of the growth of the wing cracks during uniaxial compression for specimens containing a crack-like flaw at ψ angles of 25°, 45° and 60°.	154
10.6	The influence of the varied initial flaw angle ψ (25°, 45° and 60° from the axial load direction) on wing crack growth in specimens under uniaxial compression.	155
10.7	Ultimate failure modes for specimens containing crack-like flaws at angles ψ of 25°, 45° and 60° to the applied load.	155
10.8	Stress distribution during crack growth under uniaxial compression for specimens with widths 25, 50 and 100 mm.	156
10.9	The influence of the specimen width (25, 50 and 100 mm) on the growth of wing cracks under uniaxial compression.	157
10.10	Ultimate failure modes for specimens with widths 25 mm, 50 mm and 100 mm.	157
10.11	Stress fields showing the interaction between flaws for the three types of three-flaw arrays: Diagonal (D), Vertical (V) and Horizontal (H). The flaw numbers are I, II and III read from the top down or left to right.	159
10.12	Stress distribution of crack growth under uniaxial compression for specimen containing three crack-like flaws (D Model, V Model and H Model).	160
10.13	Crack growth under uniaxial compression in the specimen containing three pre-existing crack-like flaws arranged in the D Model configuration. For the arrangement of flaws I, II and III, refer to	

	Figure 10.11, and the data for the isolated crack are obtained from Figure 10.9.	161
10.14	The influence of the flaw arrangement (D model, V model and H model) on the growth of cracks under uniaxial compression. The normalised crack length is taken as the mean value from the three flaws I, II and III. The growth of the isolated crack is plotted for comparison.	161
10.15	The failure modes of the specimens containing three crack-like flaws with the different arrangements of D model, V model and H model.	162
10.16a	Sequence of progressive failure during growth of cracks under uniaxial compression for a specimen containing multi-flaws (background is variation in elemental stiffnesses).	163
10.16b	Stress distribution during the growth of cracks under uniaxial compression for specimen containing multi-flaws (background is shaded to show the shear stress variation).	164
10.17	Stress–strain curve for specimen containing multi-flaws with the cumulative crack length value. The points a to j correspond to the stages in Figure 10.16 with the same letters.	165
10.18	Crack growth from single hole a) and an array of holes b) in PMMA plates, after Sammis and Ashby (1986).	166
10.19	Sequence of crack growth from single holes under uniaxial compression for numerical specimens with widths 25, 50 and 100 mm.	168
10.20	Crack growth in specimens under uniaxial compression showing the influence of the specimen width.	169
10.21	The failure modes of the specimens with different widths of 25, 50 and 100 mm.	169
10.22	Stress distribution during crack growth under uniaxial compression for specimens containing a hole of diameter 10, 15 and 20 mm.	170
10.23	Increase in normalised crack length in specimens under uniaxial compression showing the influence of the hole radius.	171
10.24	The failure modes of the specimens containing a single hole but with different diameters of 10, 15 and 20 mm.	171
10.25	Stress field showing the interaction between holes. The fringe pattern in the central images has the same appearance as a photoelastic pattern because both are generated by the magnitudes of the shear stresses.	173
10.26	The stress σ_z along the longitudinal axis of the central hole in the specimens under compression showing the influence of the three-hole orientation.	174
10.27	Stress distribution during crack growth from three holes under uniaxial compression, showing the influence of the nearby holes on the central hole (from top to bottom, D, V and H models).	175
10.28	Crack growth for the specimen containing three holes arranged in a diagonal line (D Model) showing the influence of nearby holes on the central hole.	176

10.29	Crack growth in specimens containing three holes, showing the influence of hole arrangement.	176
10.30	The failure modes of the specimens containing three holes in different arrangements.	177
10.31	Sequence and stress distribution of crack growth from multiple holes uniformly arranged in a specimen under uniaxial compression.	177
10.32	Complete stress–strain plot and cumulative crack length plot for specimen containing uniformly distributed holes.	178
10.33	Sequence of crack growth from multi-holes randomly arranged in a specimen under uniaxial compression.	178
10.34	Uniaxial compression of a plaster specimen containing cylindrical holes.	179
11.1	Two dynamic natural fracture events as the fracture, with its curved front, moves to the left. Note the surface irregularities on the fracture surfaces which are typical of dynamic fracturing.	182
11.2a	Laboratory experiment to evaluate the influence of fractures on pre-split blasting.	182
11.2b	Example of the half-barrels of pre-split blastholes in a fractured rock mass in Scotland.	183
11.3	Numerical model of the sample (left centre) and transmitter bars, (400 × 15 elements with 5 mm length scale for the elements in the sample).	184
11.4	The numerical Brazilian disc sample and the diametral loading conditions (the sample with 160 × 160 elements).	184
11.5	Stress wave propagation along the bars and the sample for $m = 2$.	185
11.6	Geometry of the stress wave *vs.* stress–time curves at the A (black curve) and B points (gray curves) along the two bars.	186
11.7	Applied stress waveforms with three peak values of stress.	187
11.8	Stress history obtained in the transmitter platen during the simulation of a sample with input peak stress of 150 MPa.	187
11.9	Fracture sequence and failure patterns for different stress wave amplitudes, the peak stresses being 75 MPa, 100 MPa and 150 MPa.	188
12.1	Experimental configuration for the FSD tests.	191
12.2	Experimentally obtained relations among stress, strain and permeability for sandstone.	192
12.3	Numerically obtained relations between load, permeability and loading step and the associated acoustic emissions (AE events, relative counts) of the specimen with homogeneity index $m = 1.5$.	193
12.4	Failure process of specimen ($m = 1.5$) containing hydraulic pressure. Plot shows the inhomogeneity and failure of the rock elements.	195
12.5	Shear stress evolution during failure of the specimen ($m = 1.5$) containing hydraulic pressure.	196
12.6	Sequential plots showing the maps of AE source locations during failure of specimen ($m = 1.5$) containing hydraulic pressure.	197
12.7	Flow velocity field with arrows indicating the velocity vectors during failure of the specimen ($m = 1.5$).	197

12.8	The change of flow paths with transmissivity evolution for overall flow parallel with the shear direction at different displacements of 2, 3, 5, 10 and 15 mm with constant normal loading. The legend is the order of magnitude of transmissivity (m^2/sec).	198
12.9	Numerically obtained load–displacement (loading step) curves for the four specimens with different homogeneity indices, m.	199
12.10	Permeability–displacement (loading step) curves for the four specimens with different homogeneity indices, m.	200
12.11	Complete axial stress–axial strain curve, volumetric strain–axial strain curve and permeability evolution curve for simulated uniaxial compression.	201
12.12	Fracture patterns, permeability plots, and flow vectors for each labelled loading stage shown in Figure 12.11.	202
12.13	Histograms showing evolution of elemental permeability under uniaxial compression. The letters on the right-hand horizontal axis correspond to the loading stages in Figures 12.11 and 12.12.	203
12.14	Specimen configuration with grains introduced in addition to the background inhomogeneity.	204
12.15	Shear stress field in the rock specimen shown in Figure 12.14.	204
12.16	Relations between load, loading step and the associated acoustic emission.	205
12.17	Plots of the failure process of the rock sample subjected external loading and hydraulic pressure.	205
12.18	Pore pressure gradient in the rock specimen (lighter colour is higher pressure).	206
12.19	Fluid migration in the rock sample (the arrows indicating the velocity vectors).	206
13.1	Natural spalling of granite in the Gobi desert due to hot days and cold nights.	210
13.2	Artificial spalling of granite induced by excavation and rock heating in the Äspö Hard Rock Laboratory, Sweden.	210
13.3a	Carrara marble quarry in Italy.	211
13.3b	Portion of Finlandia Hall in Helsinki, Finland, showing the convex bowing of the Carrara marble cladding slabs which were originally flat.	211
13.4	Numerical model—initial state before differential thermal expansion.	212
13.5	Shear stresses with different homogeneity indices.	212
13.6	The disc–ring model.	213
13.7	Stress distribution in the discring geometry (stiff inclusion and matrix) at $\Delta T = 100°C$.	214
13.8	Failure pattern of the disc–ring model obtained by experiment.	214
13.9	Plots of the model failure process for a uniform temperature increase. The gray levels represent the magnitudes of a) elastic modulus, b) shear stress and c) maximum principal stress—the lighter the gray, the higher the value. The black points/lines represent failed elements.	215

13.10	The failure process for the irregular inclusions model with a uniform temperature change (upper images for uniform temperature increase; lower images for uniform temperature decrease). The gray levels represent the magnitude of the minimum principal stress—the lighter the gray, the higher the stress. The black points represent failure elements. Cracks can be seen where the failure elements have accumulated.	216
13.11	Flowchart illustrating an approach for thermal conductivity modelling of a heterogeneous rock domain.	217
14.1	Small wedge failure caused by the adverse conjunction of two joints, Loch Lomond, Scotland.	220
14.2	Large wedge failure at the Tectonic Bore Mine, Western Australia caused by two major fractures creating the wedge which slipped into the open-pit mine. Left: the wedge beginning to form; right: the wedge accelerating and creating dust which escapes along other pre-existing fractures.	220
14.3	Fracture orientation data and slope stability stereographic analysis for Rubha Mor rock slope, Loch Lomond, Scotland.	221
14.4	The simulation model.	222
14.5	Numerical results for the stability of a soil slope.	222
14.6	Failure pattern comparison from the numerical simulation and limiting equilibrium solution.	222
14.7	AE counts with trial safety factor.	223
14.8	Safety factors with different slope angles for the numerical simulation and limit equilibrium analysis.	223
14.9	Simulation results for the stability analysis of a slope in a layered rock mass.	224
14.10	Failure mode of a rock slope with two sets of joints.	226
14.11	Slope failure mode for a rock mass with 80% joint persistence.	227
15.1	Model for the numerical simulation of inhomogeneous rock cutting.	230
15.2	Quasi-photoelastic stress fringe pattern in cutter and rock (contours of shear stress).	230
15.3	The fracturing process in cutting (maximum principal stress distribution).	231
15.4	Modelling the chipping process in rock cutting (Young's modulus distribution).	233
15.5	Load–displacement curve and associated AE.	235
16.1	Schematic illustration of stress-induced rock failure around a deep level South African mine tunnel.	238
16.2	Models with different dip angles of the joints (these models consist of 40,000 elements; the horizontal stress is 1 MPa and the vertical stress is 5 MPa).	239
16.3	Failure modes of a tunnel in a layered rock mass after excavation and the induced stress redistribution. For the model types, see Figure 16.2.	240
16.4	Maximum principal stress vector diagram around an excavated tunnel in a layered rock mass.	241

		List of figures xxxiii
16.5	Key points on the tunnel perimeter.	242
16.6	Displacements of the key points in the different models. The horizontal ordinate is the model number.	242
16.7	The failure process of a rock mass with inclined joints (dip angle is 30°) under different lateral stress coefficients. The vertical load is 1.0 MPa and the lateral stress coefficients are 0.5, 0.7, 1.0, 1.2, and 1.5.	244
16.8	Graphical method to distinguish the sliding zones and bending zones.	245
16.9	Failure zone when the lateral stress coefficient is 0.5.	245
16.10	Horizontal displacement of the left sidewall after excavation. The horizontal co-ordinates indicate evenly spaced points on the two sidewalls, point 1 being at the top and point 27 at the bottom.	245
16.11	Horizontal displacements of the right sidewall after excavation. The horizontal co-ordinates indicate evenly spaced points on the two sidewalls, point 1 being at the top and point 27 at the bottom.	246
16.12	Vertical displacement of the tunnel floor after excavation. The horizontal co-ordinates indicate evenly spaced points on the tunnel floor.	246
16.13	Vertical displacement of the crown-midpoint after excavation as a function of the lateral stress.	246
17.1	Physical model test illustrating the strata collapse behind an advancing longwall face.	250
17.2	Numerical simulation model of mining and overburden strata (shading represents the strata stiffnesses). The longwall face is in a direction into the page and it advances to the right of the page.	251
17.3	Overburden strata failure induced by longwall mining (left: elastic modulus distribution; right: shear stress distribution). The longwall face is in a direction into the page and it advances to the right of the page.	252
17.4	Numerically simulated result of overburden strata failure induced by mining (with a thicker immediate roof).	254
17.5	Schematic diagram of the overburden strata layout and lithology (Daliuta coal mine, Shenhua Corporation in China).	255
17.6	Numerical simulation of the overburden failure process when a multi-coal stratum is mined.	257
17.7	Support stress within the pillars of the boundary coal pillar.	258
17.8	Stresses (MPa) in the coal pillar before and after the large weighting.	259
17.9	Horizontal displacements (mm) of the surface with the working face advancing. The y-axis is the subsidence in mm and the different curves represent the amount of horizontal displacement corresponding to the different advance distances of the working face.	260
17.10	Vertical subsidence (mm) of the surface with the working face advancing. The y-axis is the subsidence in mm and the different curves represent the amount of subsidence corresponding to the different advance distances of the working face.	260

18.1	Coal and gas outburst induced by underground mining.	264
18.2	'Spherical shell losing stability' model during outbursts.	265
18.3	Numerical mechanical and seepage model of coal and gas outbursts induced by 'cross-cutting' penetration.	266
18.4	Numerically simulated coal and gas instantaneous outbursts process and shear stress distributions during instantaneous outbursts. Legend: e.g. Step 1–9 stands for the ninth calculated failure step in the first time step for modelling the outbursts process.	268
18.5	Simulation model for a coal seam with gas-containing inclusions. (The grayscale represents the relative Young's moduli for the elements, brighter elements having a higher modulus.)	269
18.6	Simulated outburst when the gas-containing inclusions are approached during progressive mining (Shading indicates relative stiffnesses of the elements).	270
18.7	Associated evolution of the stress field (Figure 18.6) in the coal seam and the roof and floor rock strata. Shading indicates the relative shear stresses.	271
19.1	2-D irregularly shaped particle model and the loading conditions. Shading indicates the variation in the Young's moduli of the elements.	274
19.2	The shear stress fringe contours in a homogeneous particle subjected to diametral loading without confinement.	275
19.3	The normalised stress distributions, σ_x (compression positive, acting horizontally and to the left in the Figure) and σ_y (right), along the loading axis inside the irregularly shaped particle (both homogeneous and heterogeneous) under diametral loading without confinement.	275
19.4	Load–displacement and energy–displacement curves for the particle subjected to diametral loading without confinement, see Figure 19.5.	276
19.5	Failure mode and associated *major principal stress* field in the particle under diametral loading without confinement.	277
19.6	Failure mode and associated *minor principal stress* field in the particle under diametral loading without confinement.	278
19.7	2-D irregularly shaped particle model and the loading and confining conditions.	279
19.8	The fringe (shear stress) contours in the particle under diametral loading with confinement.	279
19.9	Normalised stress distributions, σ_x and σ_y, along the loading axis inside the irregularly shaped particle under diametral loading with confinement.	280
19.10	Load–displacement and energy–displacement curves for the particle under diametral loading with confinement.	280
19.11	Failure mode and associated *major principal stress* field in the particle under diametral loading with confinement.	281
19.12	Failure mode and associated *minimum principal stress* field in the particle under diametral loading with confinement.	282
19.13	Cumulative energy release and the confinement index.	283

19.14	Failure modes of the three cases with different geometry and loading conditions.	284
19.15	Model of a crushing chamber containing 27 rock particles.	285
19.16	Elastic stress distributions in homogeneous particles in a crushing chamber before breakage.	286
19.17	Progressive fragmentation process for rock particles inside a crushing chamber.	287
19.18	The major principal stress distribution within rock particles inside a crushing chamber during the inter-particle breakage process.	288
19.19	The minor principal stress distribution in rock particles inside a crushing chamber during the inter-particle breakage process.	289
19.20	Resultant force–displacement curve during the inter-particle breakage process. (Points on the curve labelled alphabetically A, B, etc. correspond to the images in Figures 19.17, 19.18 and 19.19.)	290
19.21	Relation between the failure event rate and the cumulative failure event rate during the inter-particle breakage process.	291
19.22	Relation between the elastic energy release (ENR) and the cumulative ENR during the inter-particle breakage process.	291
19.23	The fragment size distributions corresponding to the images in Figure 19.17.	292
19.24	Elastic stress distributions of irregular 'mono-dispersed' particles in a crushing chamber before breakage.	293
19.25	Simulated progressive fragmentation process for irregularly shaped but mono-dispersed rock particles inside a crushing chamber.	294
20.1	Examples of a) a parallel fracture pattern in a road surface and b) a polygonal rock fracture pattern in 'turtle cracking' mode, from the Chinese Sinian period (600 million years ago).	298
20.2	FEM model with a heterogeneous central embedded layer bonded to the top and bottom layers and consisting of 1.6 million elements ($200 \times 200 \times 40$). The horizontal plane is defined as the x-y plane and z is the vertical direction. The whole lower boundary is fixed in the z-direction, and the top boundary is free to displace as required. A constant displacement increment is implemented in the x-direction along the left and right boundaries, and another increment in the y-direction along front and back boundaries. λ is the loading ratio of the displacements in the x and y directions.	299
20.3	Fracture evolution (left column, with top layer not shown) and induced stress redistribution (right column) for model with principal stress ratio $\lambda = 1$ (isotropic stretch). The cross-sections are taken from the central plane in the embedded layer. The stress is expressed as minimum principal stress. The stages a to e represent the different characteristic stages of the fracture process.	300
20.4	Models with different principal stress ratios varying from uniaxial tension, $\lambda = 0$, to equal biaxial tension, $\lambda = 1$.	301

20.5	Interface fracturing induced stress redistribution for model with principal stress ratio $\lambda = 1$. The cross-sections are taken from the interface plane: a) the interface debonding; b) minimum principal stress redistribution.	302
20.6	Measurement of fracture spacing using scanlines across the central point of the numerical model and parallel to the two principal stress directions. The points along the scanlines are the intersection points with fractures. The distance L between two neighbouring points represents its fracture spacing.	303
20.7	Modelling results for fracture spacing, L, vs. principal stress ratio, λ. a) and b) Spacing L_x and L_y vs. principal stress ratio λ; c) Spacing L_y/L_x vs. the principal stress ratio, λ.	304
21.1	Numerical compilation and analysis of a synthetic fractured rock mass.	308

List of tables

3.1	Material properties used in the indirect tensile test simulations.	31
3.2	The maximum loads and tensile strengths obtained from the numerical simulation of disc specimens with different sizes and relative load-bearing widths, b/D, and cushioned with cardboard.	40
3.3	The ratio of σ_r/σ_t obtained from the numerical ring tests.	51
3.4	The ratio of σ_{rc}/σ_t obtained from numerical ring tests.	52
6.1	Mechanical properties of minerals and interface for the reference granite.	117
17.1	Physical and mechanical parameters of the overburden rock strata.	256
18.1	Mechanical and seepage parameters used in the numerical model.	267

Explanatory notes

Engineering orientation of the book The discussion and illustrations of rock failure presented in this book are orientated towards engineering applications. There is another major potential application to the formation of joints and faults and other structures in structural geology to which we have occasionally referred, but our main theme is brittle rock failure in the context of engineering. For structural geology books, see Section 1.3.

Subject descriptions 'Rock Mechanics' is the study of the statics and dynamics of rocks and rock masses. 'Rock Engineering' concerns the design and construction of structures on or in rock masses. 'Engineering Rock Mechanics' is the study of the statics and dynamics of rocks and rock masses in anticipation of the results being applied to engineering.

Background rock mechanics knowledge A certain level of reader rock mechanics knowledge has been assumed when we wrote the book—knowledge at about undergraduate level. If you wish to improve your background knowledge in the subject area, we recommend the two books: "Engineering Rock Mechanics: An Introduction to the Principles" by J.A. Hudson and J.P. Harrison (1997) and "Engineering Rock Mechanics: Illustrative Worked Examples" (2000) by J.P. Harrison and J.A. Hudson, both published by Elsevier in English and by Science Press of Beijing in Chinese (2009). The 4th edition of "Fundamentals of Rock Mechanics" by J.C. Jaeger, N.G.W. Cook and R.W. Zimmerman, published in 2007 is also helpful.

Generic force–displacement curve In the book, we discuss a variety of specific test loading conditions on a rock specimen. All these examples are illustrated by the generic force–displacement curve shown in Figure 1. As the specimen is loaded, some damage occurs in the pre-peak region, the maximum load-bearing capability occurs at the peak, and the specimen continues to disintegrate in the post-peak region. Often the curve in the post-peak region is jagged because of the repeated coalescence of micro-cracks during the failure process. In order to obtain the complete curve, it is necessary to consider the displacement as the independent variable (i.e. the cause) and force as the dependent variable (i.e. the effect) as plotted in Figure 1.

However, and in line with traditional usage, when referring to the application of force or displacement (or stress and strain) to a rock or numerical specimen, we have

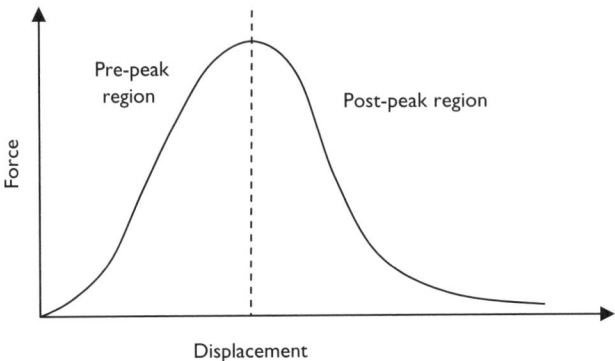

Figure 1 The generic complete force–displacement curve.

used the term 'loading', even though usually this may actually refer to the application of displacement or strain.

Synonyms The words 'displacement' and 'deformation' have been used interchangeably throughout the book, as have 'inhomogeneous' and 'heterogeneous'.

Also, in the book we present many examples of numerical computer models simulating specific rock loading configurations. There is, however, a difference between the word 'model' and the word 'simulation', Eberhardt (2003), in that 'modelling' refers more to the overall computer process and generic studies, whereas 'simulating' refers to the computer replication of a specific geometry and loading configuration. However, it is not always easy or necessary to be so pedantic in describing the modelling/simulation in this book, so in some cases we have used the words more or less interchangeably.

On the other hand, the terms 'pressure' and 'stress' are not synonyms. The quantity 'pressure' is a scalar value and requires only one value for its specification, e.g. 10 MPa. The quantity 'stress' is a second order tensor and requires six pieces of information for its specification in 3-D, e.g. the principal stresses are 25 MPa, 10 MPa and 8 MPa in the directions North, West and vertically.

Experimental results In line with the title of the book, we are keen to illustrate the many facets of rock failure mechanisms. However, there is a limit to the type of diagrams based on experimental work using physical specimens that can illustrate some key aspects, e.g. we cannot illustrate the *real* distribution of the stresses within a granite micro-structure when it is loaded, nor indeed the stresses in a large rock mass containing joints and faults. For this reason, we have used many illustrations from computer simulations to demonstrate the points. For a wide-ranging presentation of laboratory rock testing results, we recommend "Experimental Rock Deformation—The Brittle Field" (2005) by M.S. Paterson and T.-f. Wong, published by Springer and "Experimental Rock Mechanics" (2007) by K. Mogi, published by Taylor and Francis. Both these books deal with laboratory scale experiments.

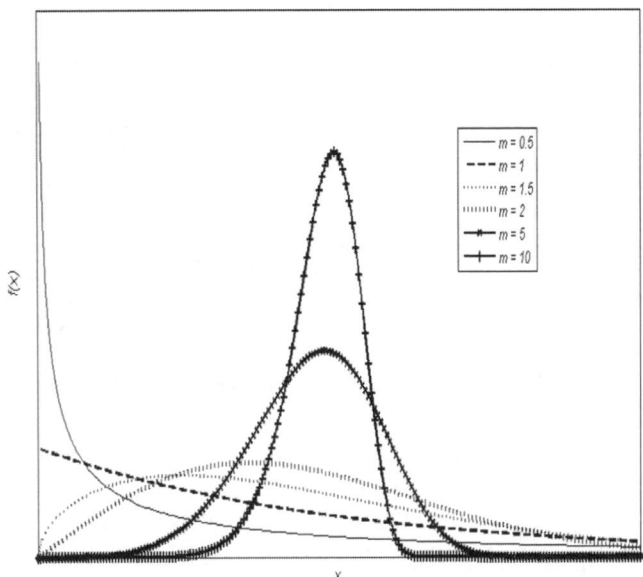

Figure 2 The probability density values, f(x), as a function of x for different values of the shape parameter, m, in the Weibull statistical distribution.

Weibull distribution In all the computer simulations in the book, the Weibull probability distribution is used to characterise the variation in the properties of the simulated rock elements. The Weibull cumulative distribution function is expressed as $F(x) = 1 - e^{-((x-x_u)/x_0)^m}$ where x_u is the lowest possible value of x (assumed to be zero in the simulations here) and x_0 and m are the scale and shape parameters respectively. This distribution is particularly useful for the simulations because changing the value of the shape parameter, m, provides a wide variety of distribution types, as illustrated in Figure 2.

Computer simulation outputs There are two main types of computer simulation output in the book, as indicated in Figure 3.

Each element in the finite element simulation using the RFPA (Realistic Failure Process Analysis) code is separately assigned elastic modulus and strength values from the Weibull distribution function for a given value of m, i.e. the assigned modulus and strength value for each element are uncorrelated.

References and bibliography This book is intended to explain and illustrate rock failure mechanisms. A vast amount of material, based on analytical, experimental and computer-based research, has already been published on this subject but it has not been our intention to provide a comprehensive survey of all previous work in the subject area. Accordingly, we have highlighted the main subject areas and concentrated

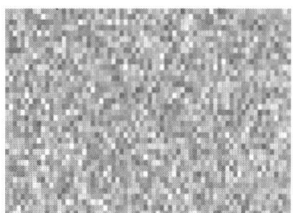 

| The grayscale variation in the local stiffness and strength properties of the rock micro-structure (lighter gray indicates higher values) | The stresses in the rock micro-structure will be indicated in the relevant Figure caption as the major or minor principal stress, or the shear stress, as above (lighter gray indicates higher values) |

Figure 3 Interpretation of the computer simulation illustrations.

on key historical references, the relevant textbooks, those references with testing results enabling comparison with the computer simulations, and references reporting on recent related numerical modelling work. If you are an author who considers that your book or papers have been overlooked, please accept our apologies: our emphasis here is only on explaining and illustrating the rock failure mechanisms.

Chapter 1

Introduction

1.1 THE PURPOSE OF THIS BOOK

As noted in the Preface, the purpose of this book is to enable readers to become familiar with the way in which rock fails in a variety of engineering circumstances—in laboratory tests, on or near the surface in foundations and slopes, and underground in tunnels, caverns and mines. In contrast to other engineering disciplines where materials are manufactured to certain specifications, rock masses are pre-existing natural substances and so we recognise the contribution that structural geology brings to the understanding of pre-existing rock mass features. Our objective is to provide sufficient illustrations and supporting explanation to provide readers with an overall understanding of the modes of rock failure when different rock geometries are subjected to a variety of stress states. It is not our intention to provide detailed formulae characterising the rock failure nor to provide a comprehensive literature survey, but to provide an understanding of why the rock breaks in different ways and hence provide the background for a reader to delve more deeply into any one of the applications, whether as an academic study or more directly as related to a rock engineering problem.

1.2 WHY DO THINGS BREAK?

In 2003, Mark Eberhardt, professor of chemistry and geochemistry at the Colorado School of Mines in the USA, published a book with the title "Why things break" (Eberhardt, 2003). In this book and with reference to the history of tool making, the author notes the antiquity of using shaped rock pieces as tools and that only certain types of rock, i.e. flint, obsidian and petrified wood, fractured to produce a sharp cutting edge. He also explains the distinction between bending (when planes of atoms slide across one another) and breaking (when the planes of atoms are pulled apart). He explains that the tendency to break or bend in a polycrystalline material is dependent on a dislocation's ability to move across grain boundaries: when this movement is suppressed, the material fractures. As with metals, rock fracture begins at the tip of a pre-existing crack although, as we shall see, the initiation of such crack growth does not necessarily lead to total structural collapse of the rock volume in question.

Professor Eberhardt notes that, through our understanding of the fracturing of materials in general, in less than a century we have moved from a society where breakage was considered the norm to the current time when we can make tough

2 Rock failure mechanisms

containers, strong sheets of glass and polymers that absorb the energy of bullets. However, these advances have been achieved by varying the manufacturing processes of the materials. In rock mechanics, we do not have the same advantage with the pre-existing rock. Nevertheless, despite the fact that the study of rock failure does not have such a long history as for other materials, we do now understand rock failure and we are able to simulate such failure using numerical computer modelling. In the book, we will illustrate both real rock failure and simulated rock failure.

1.3 ROCK FAILURE IN GEOLOGICAL AND RECENT HISTORY

During the history of the Earth, there have been many tumultuous events involving the deformation and fracture of rock formations. Thus, there is considerable forensic opportunity to study the nature of earlier rock failure and to take advantage of structural geology knowledge. Because brittle rock failure occurs through the application of stress, in particular tensile stress and shear stress, a primary feature of the structural geology analysis of natural rock failure considers extensional and shear failure.

Also, because the rock fractures are created by stress within the rock mass and stress has three orthogonal principal components, the fractures generally form in sets, as is clearly evident in the Carboniferous stratum in Figure 1.1. This orthogonality of the three principal stress components also leads to the Anderson (1942) classification of fault types as thrust, transcurrent and reverse. Even though rock fractures can be caused by a variety of natural and anthropogenic means, e.g. meteorite impacts and

Figure 1.1 Sets of rock fractures in a Carboniferous stratum, S. Wales, UK.

underground blasting (Bäckström, 2008), it is the application of stress to the rock that causes the fractures. For this reason, the existence and measurement of rock stress has received considerable attention (e.g. Amadei and Stephansson, 1997; Zhang and Stephansson, 2010).

Furthermore, during geological history, there has been a series of stress changes, resulting in a succession of superimposed rock failure events. Thus, rock masses, as viewed today represent the cumulative effect of many such episodes; the geologist or engineer has to decode the current geometry to understand the succession of failure events and hence to understand the discontinuous nature of the rock mass. In some cases, the geometry of the fractures is clear; in other cases, it can be much more complex. However, because the subject of structural geology is now so well developed, there is a considerable body of knowledge to support understanding of the discontinuous nature of rock masses (e.g. Hobbs *et al.*, 1975; Price and Cosgrove, 1990; Bahat, 1991; Davis and Reynolds, 1996; Mandl, 2000; Bahat *et al.*, 2005).

As indicated in Figure 1.2, the geologist interprets the past geological processes through study of the rock structures observed today and is able to make predictions regarding natural events such as volcanic eruptions and landslides. Similarly, civil, mining and petroleum engineers are able to study past engineering construction located on and in rock masses in order to understand how rock failure has affected the engineering—whether the intention was to avoid rock failure or to cause it. Armed with this knowledge, the engineer can then predict the consequences of a particular design, e.g. the result of locating a cavern with certain dimensions at a certain depth and orientation in a rock mass containing given stresses. This predictive capability is especially important in rock engineering: if one cannot predict the consequences of a particular rock engineering approach, there is no basis for coherent design; and, for this reason, we need to be able to understand rock failure.

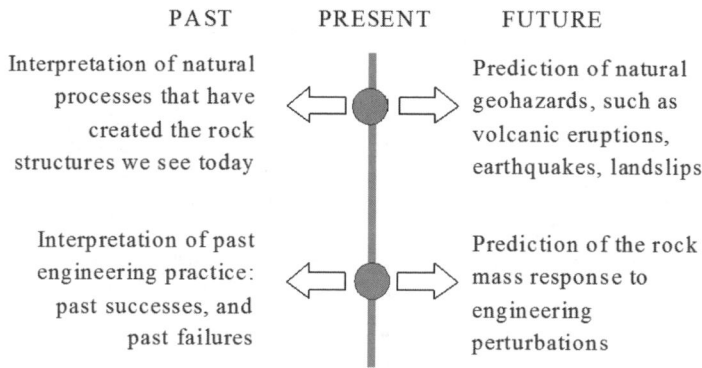

Figure 1.2 The distinction between structural geology and rock engineering in the context of interpreting past processes and predicting the future.

4 Rock failure mechanisms

Once, the rock or rock mass has been used for a particular purpose, it can deteriorate significantly such that its engineering purpose is prejudiced. The runes engraved on the hard crystalline rock shown in Figure 1.3 are still visible after many hundreds of years of exposure to the weather but the structural integrity of the Ellora excavations in India, shown in Figure 1.4 and created at least a millennium ago, has been seriously affected by the presence of the pre-existing fractures in the Deccan basalt.

Figure 1.3 Runestone at Sigtuna, Sweden.

Figure 1.4 Failure occurring to the overhang at one of the Ellora Cave Temples excavated more than a millennium ago in the Deccan basalts, India.

Introduction 5

1.4 ROCK FAILURE IN PRESENT DAY ENGINEERING

The word 'failure' has negative connotations, implying something that is not intended or desired. This is true in civil engineering for projects such as foundations, slopes, dams, transport tunnels, storage caverns, etc. for which the function of the project can easily be invalidated by rock failure. However, in mining engineering, many techniques rely on the intentional stimulation of rock failure, as in longwall coal mining where the overlying strata are intended to collapse once the longwall face has moved on, and the block caving method of metal mining in which a large block of the rock mass is undercut, causing the rock mass to fracture and the broken rock to flow through draw points.

Consider the excavation of the tunnel shown in Figure 1.5. We are interested in rock failure because:

a   it is necessary to break the rock in order to excavate the tunnel (using a tunnel boring machine in this case); but
b   it is then necessary for the rock surrounding the tunnel to remain relatively unbroken or to be supported so that it does not collapse as a result of the concentrated *in situ* stress in the rock around the tunnel periphery. In fact, the excavation periphery is the interface between inducing rock failure during construction and avoiding rock failure during the tunnel's use. The tunnel in Figure 1.5 has remained stable and retained its circular profile cut by the tunnel boring machine. By contrast and as shown in Figure 1.6, there is an example of high rock stresses causing slabbing in the wall of a tunnel excavated by blasting.

Figure 1.5 Tunnel in hard crystalline rock at the Äspö Hard Rock Laboratory, Sweden.

6 Rock failure mechanisms

Figure 1.6 Tunnel wall slabbing (spalling) due to high vertical stress caused by a large overburden in one of the access tunnels for the JinPing II hydroelectric project on the Yalong river in Sichuan province, China.

In more extreme cases, severe damage can be caused to underground excavations and Ortlepp (1997) provides many photographs of rock failure in the deep South African mines.

Because the engineer designing structures to be located on or in rock masses is interested in both causing and avoiding rock failure, this naturally leads to the need for rock mechanics research in the laboratory and in the field. Many rock engineering problems involve potential and actual unstable rock failure, such as spalling, rockbursts, coal and gas outbursts and crack development in hydraulic fracturing. For this reason, the brittle failure of rock has received considerable attention since the seminal paper "The Failure of Rock" by Cook (1965). Various models and fracture criteria have been invoked in attempts to capture the essential features of the mechanisms that lead to brittle fracture in intact rock and in the rock mass. Although much progress has been made and theories and models, such as fracture mechanics and damage mechanics, have provided techniques to solve fracture problems in rock, it is only relatively recently that computer-based approaches have been developed that are capable of capturing the complete process of fracture initiation, propagation and coalescence and hence of investigating the fracture-induced progressive failure of rock. In preparing this book, we have drawn upon our experience of engineering projects, site investigation, laboratory testing and numerical codes to illustrate the manifold aspects of rock failure.

1.5 THE NATURE OF ROCK—A NATURAL MATERIAL

A major difficulty in characterising and modelling the failure mechanisms for rock subjected to various loads is the fact that rock, a natural material, is Discontinuous, Inhomogeneous, Anisotropic and Not Elastic (in short, a DIANE material). This is in contrast to the ideal material for analytical solutions which is Continuous, Homogeneous, Isotropic and Linearly Elastic (a CHILE material).

1.5.1 Discontinuities

The discontinuous nature of rock and rock masses is caused by the many fractures present on all scales, from the minute cracks within crystal grains to faults which can extend over many kilometres. The International Society for Rock Mechanics (ISRM) provides guidance on rock testing (Ulusay and Hudson, 2007) including the discontinuity features recommended for specification in a site investigation, as illustrated in Figure 1.7.

There is a hierarchy of such pre-existing rock fractures present which form a complex array as discussed in the structural geology books referenced in Section 1.3. However, modern computer methods are now beginning to be capable of representing such arrays, as shown by the example in Figure 1.8.

1.5.2 Inhomogeneity

The intact rock and rock mass inhomogeneity are caused by many factors, depending on whether the rock is of igneous, sedimentary or metamorphic origin. An igneous rock mass can have different properties at different locations because of the processes occurring in the fluid magma motions during the formation of the rock. A sedimentary rock mass, such as a sandstone, limestone, mudstone sequence, is composed of strata having different properties. A metamorphic rock can be inhomogeneous because the

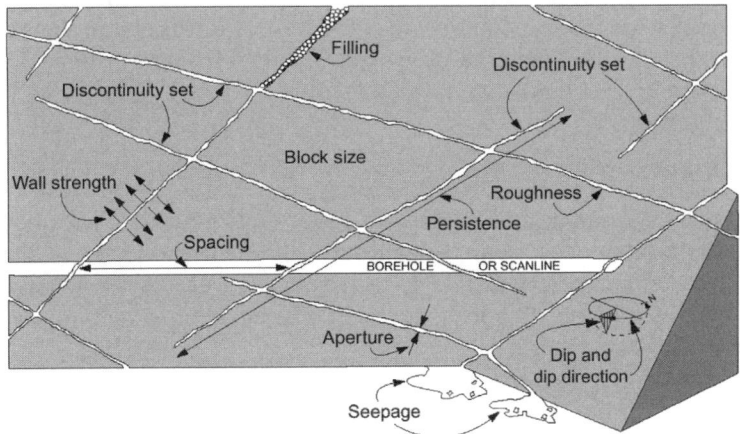

Figure 1.7 The discontinuity characteristics recommended by the International Society for Rock Mechanics for specification in a site investigation.

8 Rock failure mechanisms

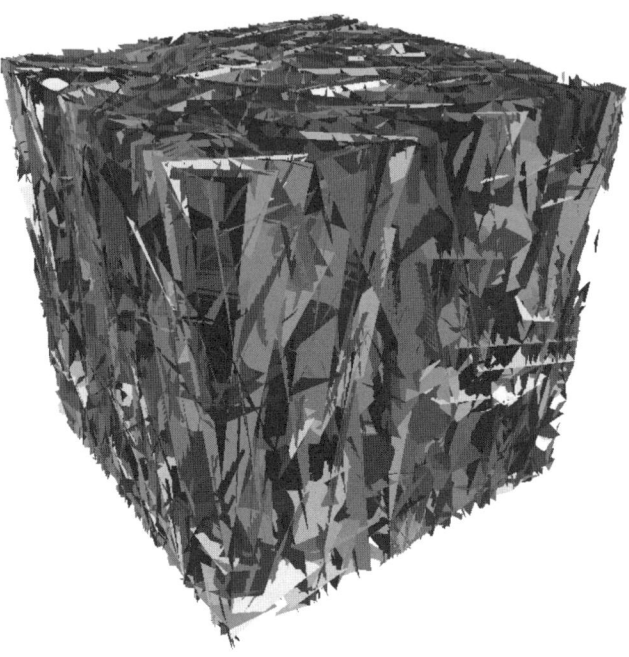

Figure 1.8 Computer simulation of a large rock mass containing many discontinuities (from Mas Ivars, 2010).

metamorphic processes were different at different locations or because the original rock was inhomogeneous.

During the site investigation for a potential radioactive waste repository in Finland, the P-wave (compressional wave) velocity was measured on a large number of site investigation cores. For the rocks in this Fennoscandian bedrock, comprising mainly diatexitic gneiss, pegmatitic granite and veined gneiss, the P-wave velocities of the rock samples vary from 3929 m/s to 7134 m/s, as illustrated in Figure 1.9. This is typical of the spread in material properties obtained from measurements on many rock samples from a rock mass.

1.5.3 Anisotropy

A rock or rock mass can be anisotropic as a result of its inhomogeneity. For example, the strata in a sedimentary sequence will have different properties in the different directions parallel and perpendicular to the strata sequence. Alternatively, the intact rock may contain an inherent anisotropy due to its mode of formation, such as slate, or the gneiss shown in Figure 1.10.

As intimated in Figure 1.10 and as we will show in Chapter 7, the anisotropy has a pronounced effect on the strength of the rock as a function of the direction of the applied stress—and hence on the location where failure might initially occur around a tunnel periphery. To consider the potential for rock failure around the opening, the stress (i.e. the pre-existing stress concentrated by the circular opening) is compared to

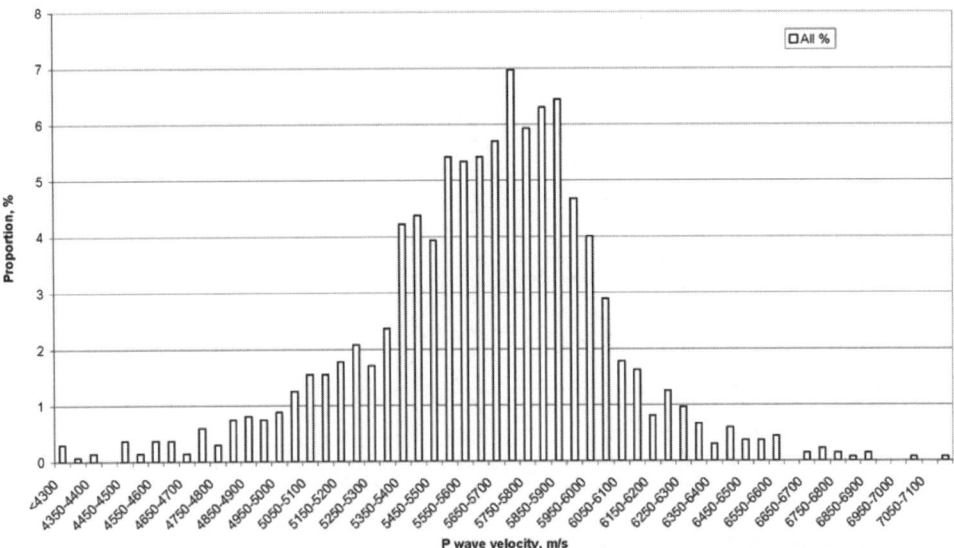

Figure 1.9 The range of P-wave velocities measured on the site investigation cores at the Olkiluoto crystalline rock site in Finland (from Aaltonen et al., 2009).

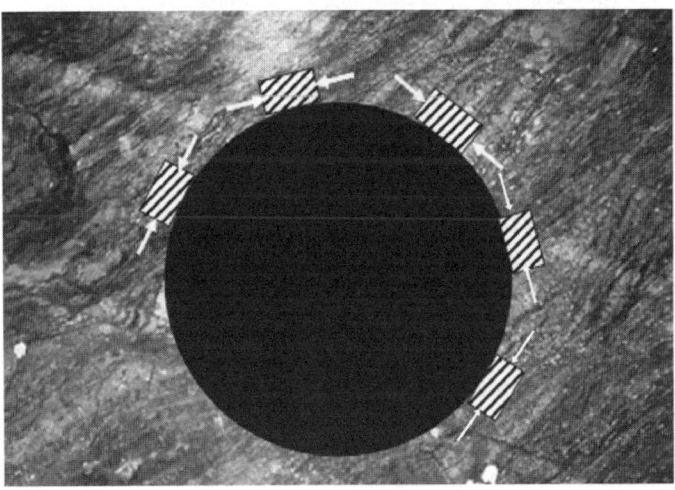

Figure 1.10 Influence of anisotropy on the potential location for failure around an excavation in gneissic rock.

the local rock strength. This can be envisaged by the loading of each of the rectangles placed around the opening in Figure 1.10 with the shading representing the direction of the foliation. Each of these cases must be studied separately due to the different local rock strengths, the different local rock stress, and the different local effect of the angle of the foliation with respect to the tangential stress around the opening.

10 Rock failure mechanisms

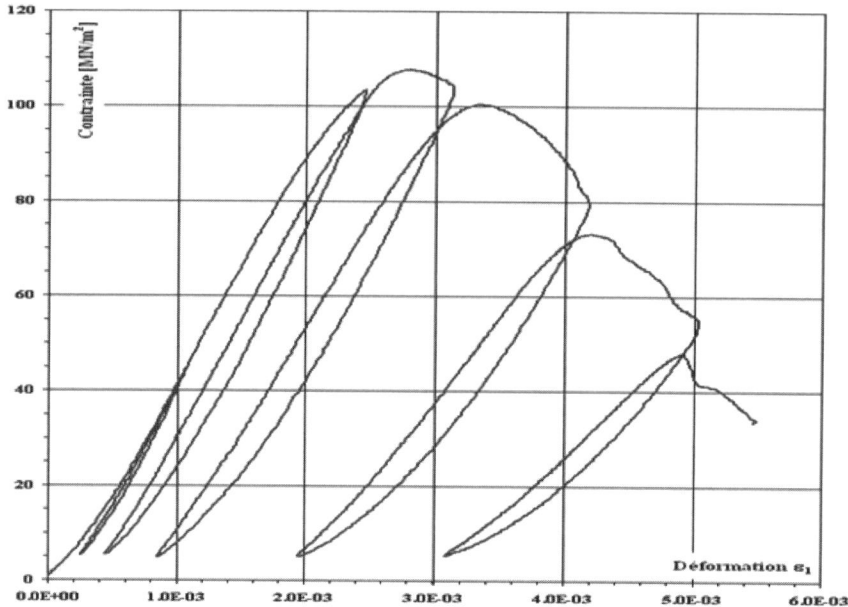

Figure 1.11 Complete stress–strain curve for a marble specimen loaded in uniaxial compression (with thanks to EPFL, Lausanne, Switzerland).

1.5.4 Inelasticity

For a rock to be elastic, all the energy input to a rock volume when it is loaded must be recoverable when it is unloaded, i.e. there should be no energy loss through hysteresis. Although some rocks do show very little hysteresis when loaded up to and close to the peak load, there is considerable energy lost in others when they are loaded and unloaded. We will be studying the complete stress–strain curve for rock in the following chapters and demonstrating that the generation of micro-cracks during the failure process causes the rock to be inelastic both before and after the peak load.

The complete stress–strain curve for a marble specimen is shown in Figure 1.11. The hysteresis effect is shown not only by the energy lost within the unloading-loading loops but, more importantly, by the loss of energy during the progressive failure process.

1.6 NUMERICAL MODELLING OF ROCK FAILURE

As a consequence of the DIANE nature of rock and rock masses, analytical models cannot be relied upon in general to provide the necessary predictive capability for rock engineering design because, by definition, they cannot fully account for rock failure This is because analytical models are generally two dimensional models based on the assumption of a CHILE material and having a closed form solution. In the seminal paper on the failure of rock by Cook (1965) in which an elastic analysis is presented representing the effect of a single crack, he notes that, "These equations are obviously only exact for a single crack in an infinite plate, but evaluation of the effect of finite crack density would make the problem formidably complex." So, given the inelastic nature of the *in situ* rock, together with its discontinuous, inhomogeneous, anisotropic and three-dimensional nature, it is not possible to analytically examine and evaluate the complete mechanical behaviour of a DIANE rock, except as an approximate approach. The problem becomes even more intractable if gas or fluid, as in coal and gas outbursts, hydraulic fracturing, etc., is involved.

In an attempt to supplement analytical modelling and to incorporate some of the rock mass idiosyncrasies, albeit indirectly, rock mass classification systems have been developed mainly to support rock engineering design, the most well known ones being the Rock Mass Rating (RMR), the Quality (Q), and the Geological Strength Index (GSI). These have proved to be most useful for rock engineering, especially for estimating rock support. Although the originators of these systems have used their knowledge of rock failure to develop the appropriate rock classification indices and have linked their systems to numerical models, the systems do not allow us to directly study the mechanisms of rock failure.

This leads us to numerical models which can be used to simulate the detailed rock fracturing sequence and are thus useful for understanding rock failure mechanisms on both the small and large scales (e.g., Jing and Stephansson, 2007). Many computer codes have been developed that are either applicable to rock masses or have been directly written for that purpose, e.g., the Itasca codes (see www.itascacg.com). In this book, most of the computer generated illustrations are outputs from the RFPA (Realistic Failure Process Analysis) code developed by the first author (see www.mechsoft.com).

1.7 THE CONTENT OF THIS BOOK

In this book, we explain and illustrate many of the ways in which rock can fail. When dealing with rock in civil and mining engineering and other types of engineering, it is important to understand the processes by which the rock fails under load so that safe structures can be built on and in the rock. As has already been noted, it is crucial for the rock engineering designer to have a predictive capability—it must be possible for the designer to determine the consequences of, for example, the different geometries, different depths and different orientations of proposed excavations. Moreover, there are many ways in which the rock can be loaded to failure, so it is important for geologists, engineers and researchers to have a clear understanding of the failure processes under different conditions. For this reason, the chapters in this book explain, simulate and illustrate rock failure in different circumstances in the laboratory and in the field.

Thus, it is hoped that the book will serve as an illustrated guide and explanation of the many aspects of rock failure—for students, teachers, researchers, clients, consulting engineers and contractors—with applications in geology plus civil, mining, petroleum and environmental engineering. It is also intended to provide readers with fluency and competence in understanding the rock failure concepts so that they can be applied in engineering practice.

Chapter 2

Rock failure in uniaxial tension

2.1 INTRODUCTION

Prediction of rock failure is one of the central problems in the mechanics of rocks, and tensile failure is one of the most important failure modes. Materials can fail only in tension or shear and so, even in direct compression, it is found that, on the micro- and meso-level, the fracturing of rock is essentially a tensile phenomenon (see Chapters 3 and 4). Also, it is recognised that the tensile strength of rock is among the most important parameters governing engineering aspects such as rock blasting. Thus, and since the tensile strength of rock is much lower than its compressive strength, the consideration of the tensile strength of rock is required in designs of underground structures.

Figure 2.1 shows a 'dogbone' specimen which has been tested in direct tension. As might be expected, failure occurs by a through-going crack, which in the photograph is just visible traversing the specimen below the upper shoulder. The geometry of the crack has clearly been influenced by the irregularity of the micro-structure as it found its way through and around the individual marble crystals.

However, complete stress–strain curve results from experimental direct tensile tests are rare. Although some experiments on the failure mode and the complete stress–strain curve for rocks subjected to uniaxial tension with un-notched specimens have been conducted, a detailed study on the tensile failure behaviour of rock is still incomplete due to the difficulty in carrying out such tests. Stress concentration at the grip location of the specimen is one of the obstacles that is hard to overcome; another is the occurrence of a bending moment because of the potential for non-coaxial gripping and hence curvature of the specimen. However, even if the apparatus is essentially perfect, slight imperfections in specimen preparation or inhomogeneity in the material itself can lead to non-uniform tensile stresses across the section where failure occurs, as demonstrated in Figure 2.1.

During the past decades, many combinations of specimen preparation and grip/pull systems have been used. However, little attention has been given to the investigation of the influence of the heterogeneity of rock on the progressive failure leading to collapse in uniaxial tension. This lack of fundamental knowledge of tensile failure under uniaxial loading led to the simulations reported here. In this Chapter, uniaxial tensile tests with inhomogeneous simulated specimens of brittle rock material are numerically studied using the 2-D finite element code RFPA^{2D}.

In Chapter 1, Section 1.5, we discussed the nature of rock as a natural material, i.e. a DIANE material, and particularly in the current context the fact that the intact rock is inhomogeneous. Sometimes, this inhomogeneity is small, as is the case with a fine-grained

14 Rock failure mechanisms

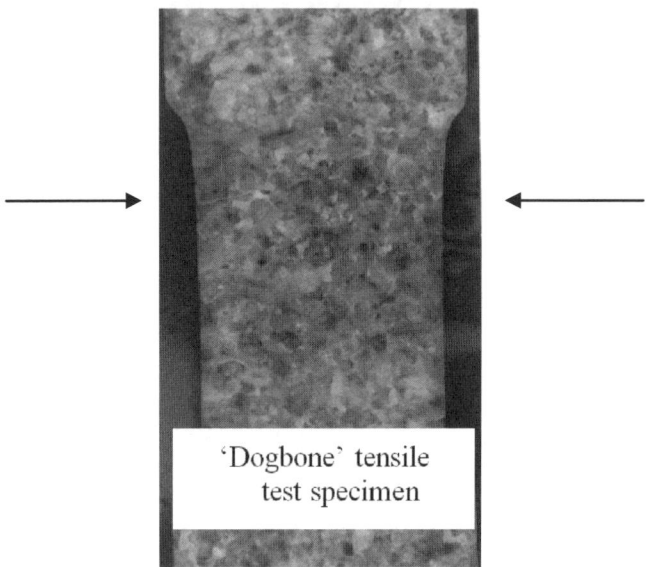

Figure 2.1 Section of a direct tension test specimen with a transverse crack across the specimen visible below the upper shoulder between the arrows—test stopped in the descending portion of the complete stress–strain curve. Specimen 20 mm in diameter, Tennessee marble (from J A Hudson, personal collection).

uniform rock such as Solenhofen limestone (see Figure 3.1 in the next Chapter); in other cases, it can be large, for example in a coarse-grained granite. The capability of the RFPA^{2D} code to simulate different degrees of inhomogeneity, plus the localised fracture events and their source locations, is helpful in studying the failure mechanisms of rock. When a rock specimen is loaded in tension, this material inhomogeneity produces locally inhomogeneous stresses through the specimen. As a consequence, because of the heterogeneous inclusions (stiff or soft), the magnitude of the local stress is significantly altered from that of the average or applied stress, especially when the material is strongly heterogeneous. Of course, the integrated stress across any specimen section perpendicular to the loading axis must equal the applied load. Since the local stress and the local strength vary in an essentially random fashion, the site of failure initiation in the specimen also varies randomly and does not necessarily coincide with the maximum stress location.

In this Chapter, we illustrate the application of the RFPA^{2D} code to model the fracturing of intact rock subjected to uniaxial tension. In the next Chapter, we do the same for the case of indirect tension, i.e. when the tensile stress is generated by a compressive load. We emphasise that many numerical computer codes have been developed which are capable of such simulations (e.g. Fang and Harrison, 2002): the RFPA finite element code is used here for convenience and because of its capability to output diagrams illustrating the gradual breakdown of the rock micro-structure. Also, the discussion in this Chapter is restricted to the behaviour observed at the laboratory scale and is focused on deriving a sound physical understanding of the fracture phenomena.

2.2 SPECIMEN SIMULATION

The specimen to be numerically loaded is shown in Figure 2.2. The model mesh contains $130 \times 130 = 16900$ elements equivalent to a dimensional size of 130×130 mm. In the RFPA code, a random number generator is used to generate a distribution of elastic moduli and strengths for the elements representing the rock. The Weibull distribution was used as the probability density distribution of these parameters (see the description of the Weibull distribution in the Explanatory Notes at the beginning of the book) because of its flexibility in being able to characterise a wide range of heterogeneity modes though variation of the shape parameter, m.

The mechanical properties for the simulated specimens used in this investigation are the following: homogeneity index (m), 1.5, 2, 3, 5; mean elemental elastic modulus (E_0), 60 GPa; mean elemental strength (σ_0), 200 MPa; Poisson's ratio (v), 0.25; tension cut-off (λ), 10%; friction angle (ϕ), 30°. Although the simulated rock is heterogeneous, it is assumed through the application of the Weibull probability density function that the rock is statistically homogeneous.

Two groups of specimens were numerically tested: (1) four specimens with the same heterogeneity index, $m = 1.5$, representing a relatively heterogeneous material (see the graph of the Weibull probability density function for various values of m in the Explanatory Notes at the beginning of the book), and (2) four specimens with different heterogeneity indices, $m = 1.5, 2, 3, 5$, representing a spectrum of heterogeneity, from relatively heterogeneous to relatively homogeneous. For specimens #1, #2, #3, #4, $m = 1.5$; for specimens #5, #6, #7, #8, $m = 1.5, 2, 3, 5$.

In all cases, the specimens undergo a plane stress tension, imposed by a relative motion of the upper and lower rigid platens with a constant rate of 0.0002 mm/step. We use a failure strength approach so that micro-fracturing occurs when the stress of an element satisfies the strength criterion. A Mohr-Coulomb criterion envelope with a tensile cut-off is used so that the failure of the elements may be either in shear or in tension. As load is applied to the specimen, the fractures will grow, interact, and

Figure 2.2 Model geometry and mesh generation of the numerical specimen with two loading platens (the model is composed of $130 \times 130 = 16{,}900$ elements with dimensions equivalent to 130×130 mm. The variation of the gray colour represents the heterogeneity of Young's modulus, a lighter shade having a higher value).

coalesce, resulting in non-linear rock behaviour and in the formation of macroscopic fractures and cracks. The stress and deformation distribution throughout the specimen is then adjusted instantaneously after each failure event to reach the equilibrium state. At locations with increased stress due to stress redistribution, the stress may again reach the critical value and further ruptures will occur. The process is repeated until no further elements are stressed to failure. Further external displacement is then increased. In this way, the system develops a macroscopic fracture. Owing to the stress redistribution and the long-range deformation-induced interactions, a single important elemental failure may induce an avalanche of additional failures in neighbouring elements—leading to a chain reaction and the release of further energy.

Energy is stored in the elements during the loading process and this elastic strain energy, W_p, is released as acoustic emission (AE) when an element fails:

$$W_i = \frac{1}{2E}(\sigma_1^2 + \sigma_3^2 - 2v\sigma_1\sigma_3)V \qquad (2.1)$$

where i is the elemental number, E is the elemental elastic modulus, σ_1 and σ_3 are the major principal stress and minor principal stress respectively, v is the Poisson's ratio, and V is the elemental volume. In a brittle or quasi-brittle material such as rock, the acoustic emission occurrences are thus related to the elemental failure events. As an approximation, it is reasonable to assume that the energy releases can be established from the strain energy releases of the failed elements.

2.3 NUMERICAL SIMULATION RESULTS FOR THE UNIAXIAL TENSION CASE

Figure 2.3 shows the simulated load–displacement relation for specimen #1 with homogeneity index $m = 1.5$. The model predicts a non-linear load–displacement curve

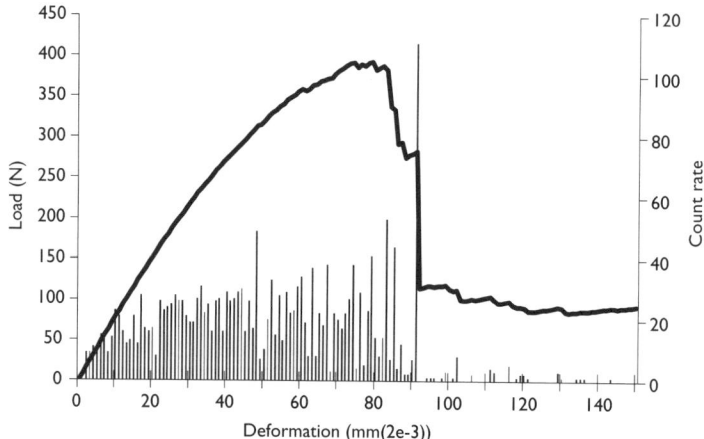

Figure 2.3 Numerically obtained load and fracture event counts vs. deformation/displacement for specimen #1.

similar to the typical curve for inhomogeneous material observed in laboratory tests (e.g. Peng, 1975; Okubo and Fukui, 1996), and, as we shall see later in Chapter 4, similar to the shape of the compressive stress–strain curves. Note also, that the complete curve is shown in Figure 2.3, i.e. including the portion after the peak load. In a load-controlled test, the specimen will break rapidly at the peak stress; however, in this computer simulation it is the deformation that is the control variable and so the complete curve can be obtained. This can also be achieved in physical tests using a servo-controlled testing machine in which deformation is the controlled variable.

It is found that, although the individual elements in the numerical model are elastic and brittle (with 20% residual strength), a sizeable non-linearity exists before the maximum load is reached, and the curve has a clear post-peak region (strain softening) as the specimen degrades. Figure 2.3 also shows the numerically obtained fracture event counts (event rate) as a function of deformation. For this relatively heterogeneous material, $m = 1.5$, many fracture events occurred before the peak load was reached, manifested by the steadily reducing tangent stiffness of the specimen. After the peak load, the breaking elements coalesce to form through-going fractures evident from the sharp drops in the descending portion of the curve and the associated large fracture event counts.

Figure 2.4 shows the fracture event accumulation and fracture event released energy accumulation as functions of the deformation. It is important to note that, as shown in Figure 2.4, although more than 80% of the fracture *events* occurred before reaching the peak load, more than 80% of the fracture event released *energy* is dissipated after the peak load. It is found that there are at least two occasions when the rate of the released fracture event energy accumulation shows a clear increase: the first corresponds to the load drop immediately after the peak load, as shown in Figure 2.3; and the second corresponds to the maximum fracture event count, where the largest load drop is induced.

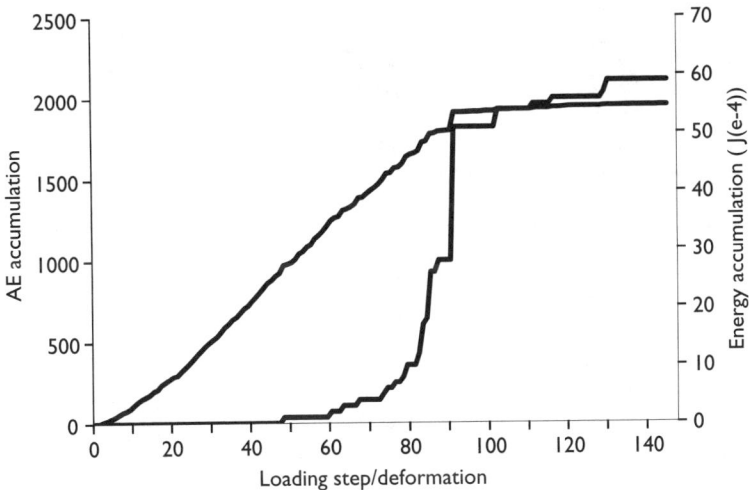

Figure 2.4 Numerically obtained cumulative fracture event counts and fracture event released energy accumulation vs. deformation for specimen #1 during the development of the stress–strain curve in Figure 2.3.

18 Rock failure mechanisms

In Figure 2.5, we present the locations of the fracture events that occurred during the loading stages. Each circle represents one fracture event and the diameter of the circle represents the magnitude of the fracture event released energy. The individual plots of the simulation shown in Figure 2.5a–h are also shown in Figure 2.6 for the elastic modulus and the fractures and Figure 2.7 for the shear stress. The various stages in the microstructural breakdown of the specimen are described in the following bullet points.

- Facture event locations for events occurring during loading to 75% peak load are shown in Figure 2.5a (step 48). Note that the local failure events are distributed throughout the specimen, reflecting the statistically uniform degradation during this portion of the simulation. It is difficult to predict where the macro-crack will initiate at this stage.
- In step 62, however, some fracture events have occurred around a certain cluster that appears to be the nucleation site for the future fracture plane. While a few events are still occurring throughout the specimen, most events are now clustered near the nucleation zone on the left side of the specimen (step 64 to step 79).
- From step 79 to step 86, a distinct fracture event active zone has developed along the left zone of the specimen and the events decrease in intensity throughout the rest of the specimen. It is important to note that the maximum load is not reached at step 92 where the highest count of fracture events is recorded, but at step 79 where the fracture events are less than half of the maximum count. This is attributed to the heterogeneity of the specimen.
- The final fracture events (Figure 2.5h) show the fracture event active zones are connected together to form a large fracture event active zone that is found to be consistent with the site of macro-cracks. Again, the fracture event maps show

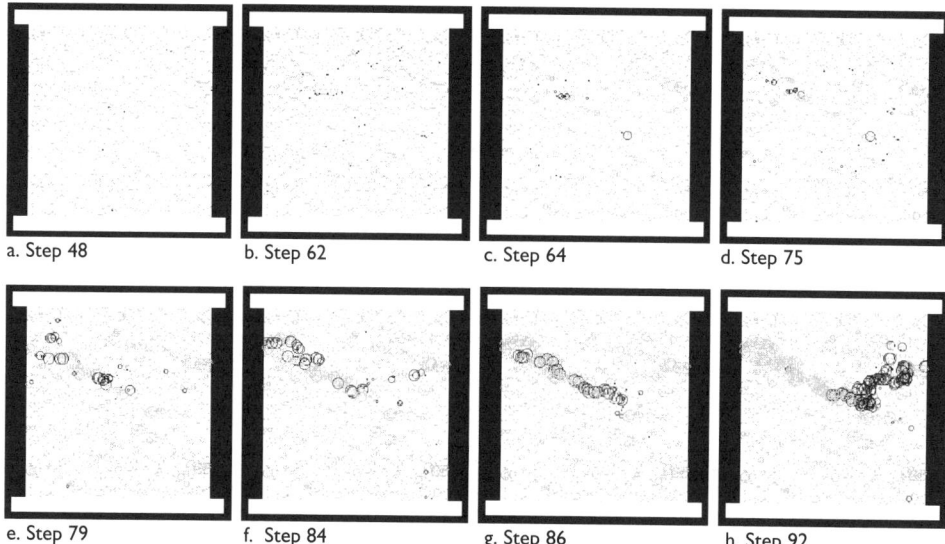

Figure 2.5 Numerically obtained maps of fracture event source locations at different loading steps for specimen #1 (the circles represent the source location of fracture events and the diameter of the circles represents the associated fracture event released energy value).

Rock failure in uniaxial tension 19

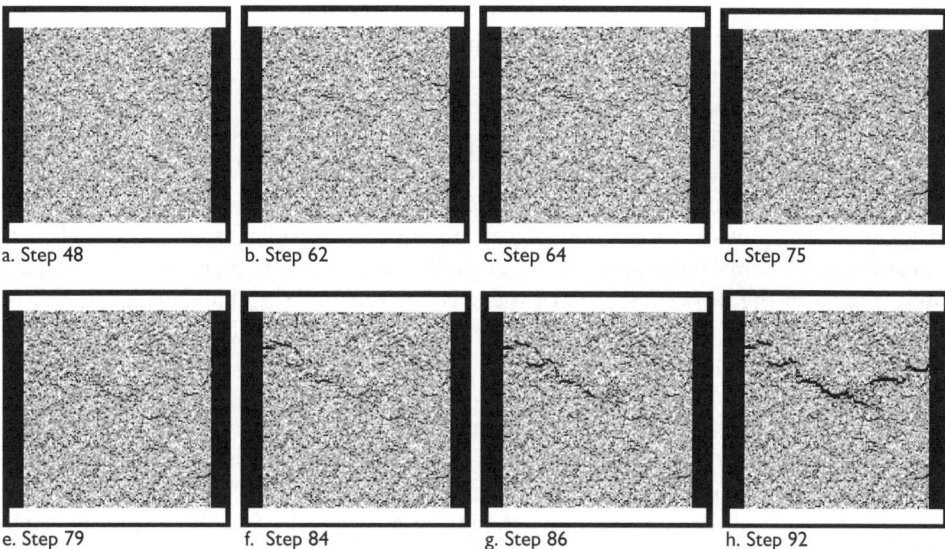

Figure 2.6 Failure process in different loading steps for specimen #1. (The gray colour represents the relative heterogeneity of the Young's modulus, and the black dots represent the micro-fractures).

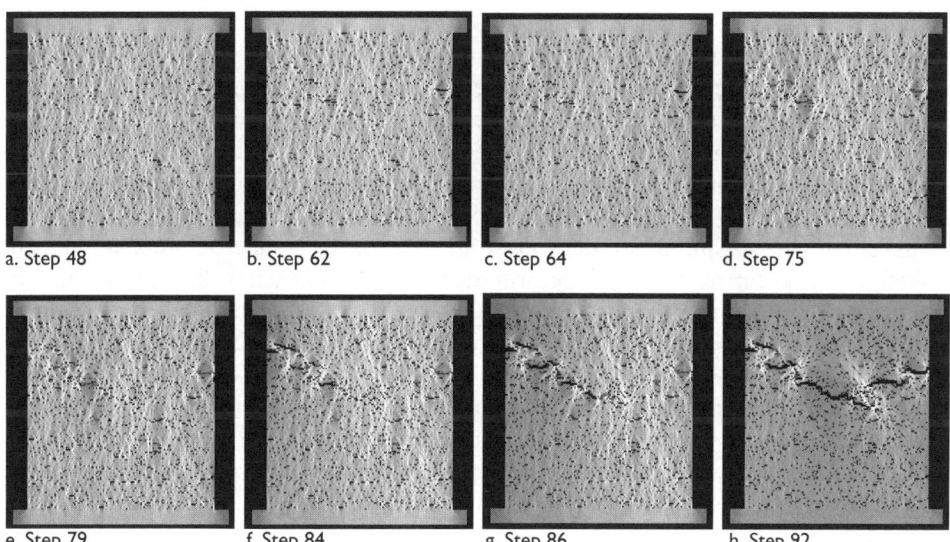

Figure 2.7 Maximum shear stress distribution at the different loading steps for specimen #1 with $m = 1.5$. (The gray colour represents the relative shear stress (white being the high values), and the black dots represent the micro-fractures).

that, although most of the fracture events occurred prior to the maximum load and are distributed throughout the specimen, the fracture events along the narrow active zone dissipated most of the acoustic energy (as can be seen by the diameters of the fracture events).

Generally, the failure process of the specimen can be divided into three stages. Initially, and as shown in Figures 2.6a and 2.7a, randomly distributed micro-fractures are formed due to the heterogeneity of the specimen. The stress distribution shown in Figure 2.7a shows a totally disordered picture on the micro-scale, although it is statistically uniform on the meso-scale. Once the crack nucleation is initiated at some sites, mostly in the upper left side of the specimen in this simulation (Figure 2.6b–c and Figure 2.7b–c), the micro-structure proceeds relatively quickly to establish the embryonic macro-cracks, but these are not connected (as shown in Figure 2.6d–f and Figure 2.7d–f).

Following this phase of macro-crack development is the growth phase, in which a band of micro-fractures develop, indicating a zone of intense stress concentration at the front area of the crack process zone (Figure 2.7g). It can be seen that the cracks extend from the left side of the specimen into the interior. As opposed to the beginning stage of loading, the stress field at this stage shows on the macro-scale a strongly heterogeneous picture (Figure 2.7f–h). At the left side of the specimen, a relatively lower stress region forms behind the advancing fracture front. On the other hand, a relatively high stress region forms on the right side, indicating that this portion of the specimen now bears most of the load from the loading frame. Finally, due to the strong interactions between the cracks under this high stress field, the cracks propagate in an unstable manner and the isolated cracks coalesce to form a macro-crack (Figure 2.6g–h). The crack system shown in this simulation, which is similar in principle in all such simulations, is not as simple as assumed in fracture mechanics; indeed, it is difficult to define a crack process zone or the crack tip in such a heterogeneous material.

From this simulation for a relatively heterogeneous rock with $m = 1.5$, we develop a clearer understanding of material strength. Comparing the load–deformation curve shown in Figure 2.3 and the failure processes shown in Figures 2.6 and 2.7, it is evident that the maximum strength does not necessarily indicate immediate abrupt failure of the specimen. The load–deformation curve becomes non-linear when microfractures occur in the specimen (Figure 2.7a–c). At point e (Step 79), the specimen reaches its maximum strength. Although the load bearing capacity drops dramatically from point f (Step 84) to g (Step 86), this does not indicate the collapse of the specimen under the deformation controlled simulation. The specimen rapidly lost its load bearing capacity, yet the crack does not propagate thoroughly until point h (Step 92). Subsequently, a distinct transition occurs and the load capacity begins to decrease at a much slower rate until it reaches its residual strength, at about 20% of its maximum strength.

In order to study the influence of the degree of heterogeneity on the failure mode, four specimens (#5 to #8) with different homogeneity indices, $m = 1.5, 2, 3$ and 5, are numerically simulated under uniaxial tension conditions (noting that specimen #1 ≡ #5). Figure 2.8 represents example variations in the shape of the jagged fracture surfaces for the four specimens. Note that the relatively heterogeneous specimen ($m = 1.5$) manifests a fracture surface with irregularities more than twice as large as

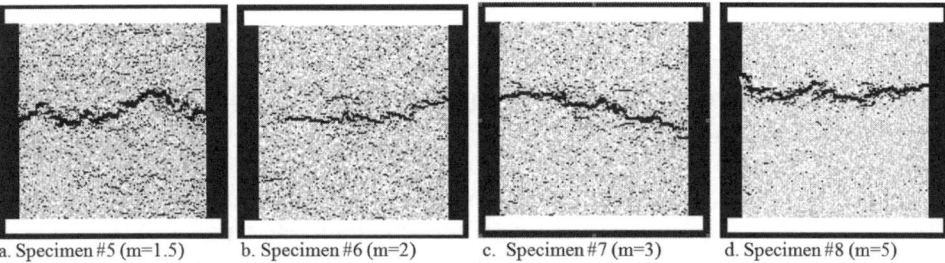

a. Specimen #5 (m=1.5)　　b. Specimen #6 (m=2)　　c. Specimen #7 (m=3)　　d. Specimen #8 (m=5)

Figure 2.8 Variations in the shape of the jagged fracture surface for the four specimens #5 to #8 having different homogeneity indices, m (the larger the value of m, the more homogeneous is the specimen). The gray colour represents the relative heterogeneity in Young's modulus, and the black dots represent the micro-fractures.

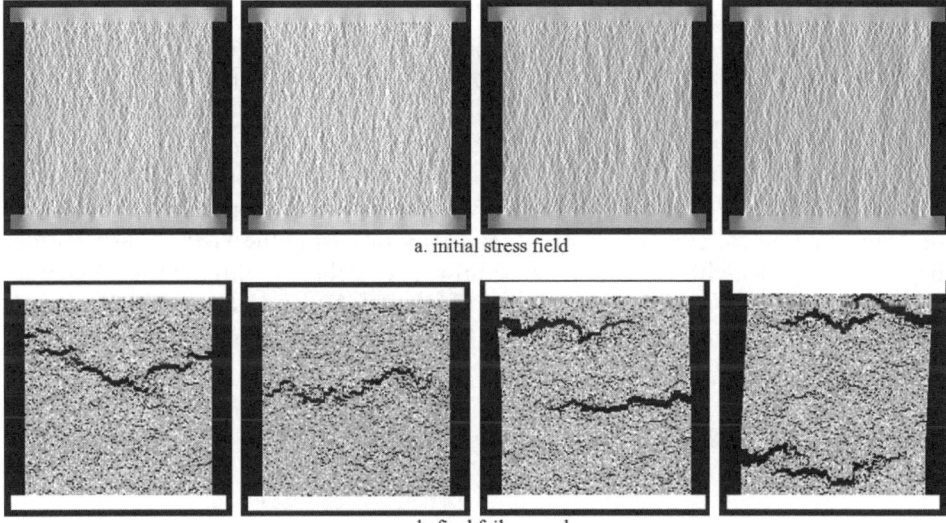

a. initial stress field

b. final failure mode

Figure 2.9 Sensitivity of failure modes to the heterogeneity for the four specimens with homogeneity indices, $m = 1.5, 2, 3$ and 5 (left to right). a) the gray colour represents the relative shear stress; and b) the gray colour represents the relative heterogeneity of Young's modulus, with the black dots representing the micro-fractures.

that of a relatively homogeneous specimen ($m = 5$), thereby suggesting that the degree of a rock's brittleness is expressed by such variation, which in turn indicates the heterogeneity of the specimen.

Figure 2.9 shows the results of the final failure modes of the four specimens having different homogeneity indices. The results indicate that the variation of the failure mode is strongly sensitive to the local inhomogeneous features of the specimen. Although the mechanical properties and the initial stress distribution for the four specimens are statistically homogeneous on the macro-scale, the localised

22 Rock failure mechanisms

zones or major crack positions for the four specimens are different, which results in different final failure modes.

2.4 FURTHER STUDIES OF SIMULATED ROCK FAILURE IN UNIAXIAL TENSION

Local failure in the rock occurs when the local stress reaches the local strength. A corollary is that we can consider two types of failure: one type when the stress is high; and one type when the strength is low. It is possible that a misalignment of the loading frame may cause bending of the specimen and hence a higher-stress region in the specimen, which may induce different fracture modes. On the other hand, failure may start at a point where the stresses are not highest but the local strength is lower— due to pores, micro-cracks, grain boundaries, etc. Moreover, the observation from simulations that the macro-cracks nucleated in different specimens with the same parameters in different ways suggests that nucleation is strongly controlled by the nature of the heterogeneity in a group of adjacent elements. Indeed, it is found in experiments that the localised fracture zones or major crack positions occur differently for different specimens due to the randomly distributed micro-defects in the specimens and cannot therefore be predicted.

Here we explore this phenomenon with numerical simulations. These simulations were conducted with four specimens having the same seed parameters ($m = 1.5$ for all four specimens, #1–#4) as used in the above study. Since the computer generates the mechanical parameters of the elements randomly following the Weibull distribution, the four specimens with the same macro-properties will have different local characteristics for individual elements. As shown in Figure 2.10, although the failure modes differ, the load–deformation curves for the four specimens demonstrate a more or less identical shape.

A detailed study of Figure 2.10 indicates that the load–deformation curves can be divided into three stages. In stage A, the initial portions of the four curves are similar,

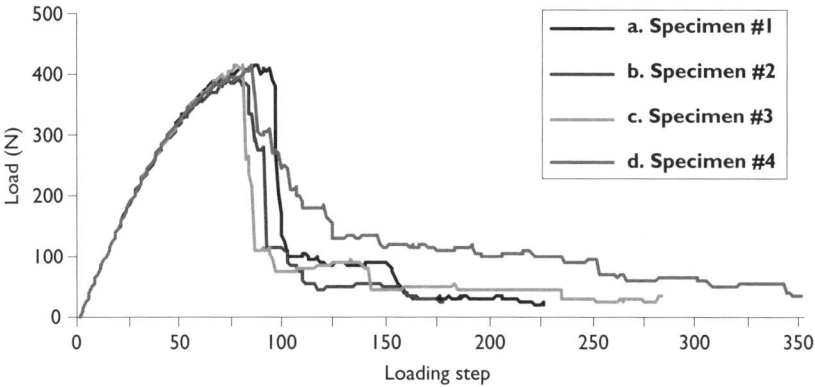

Figure 2.10 Sensitivity of the load–deformation relations to the nature of the local heterogeneity.

indicating that the randomly distributed micro-fractures have little influence on the overall stress field and the deformation in the specimen is statistically uniform. This is a common phenomenon for all four specimens modelled. A slight difference among the slopes of the load–deformation curves can be observed when the load reaches about 90% peak load. This implies that the specimen tends to deform non-uniformly. Around the maximum load, specimen #2 undergoes a larger deformation than the other specimens, as shown in Figure 2.10. This difference is attributed to the crack initiation and propagation geometry. The biggest difference in the curves, however, is found to be at the stage of relatively unstable failure immediately after the peak load and beyond. Also, when the failure process returns to stability, the difference in the residual strength stage tends to become smaller, with the exception of specimen #4 which has a larger residual strength and deformation capability (as shown in Figure 2.10). Again, the higher residual strength of specimen #4 is also attributed to its specific failure mode.

In a conventional physical direct tension experiment on rock, the stress distribution throughout the specimen cannot be obtained. The advantage of numerical simulation is that we can obtain detailed information about this stress distribution, including the failure-induced stress redistribution, as illustrated in Figure 2.11. Although the micro-fractures shown in Figure 2.7 provide similar information regarding the failure process as in Figures 2.5 and 2.6, the shear stress field displayed in Figure 2.7 sheds some new light on the failure mechanism of the specimen. Although the stress distribution at the beginning

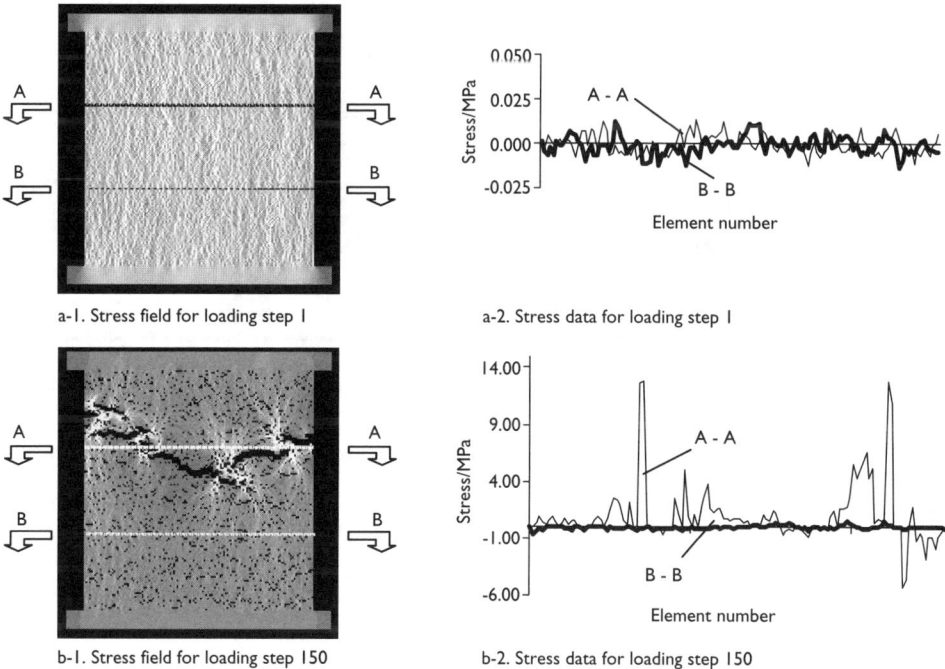

Figure 2.11 Stress distribution and failure-induced stress redistribution for sections AA and BB before and after the macro-fracture develops in uniaxial tension ($m = 1.5$).

of loading is statistically homogeneous on the macro-scale, it is highly disordered on the micro-scale—due to the micro-scale heterogeneity of the specimen. The failure-induced stress redistribution, and, particularly, the high stress concentration due to crack initiation, propagation and coalescence is demonstrated clearly in these Figures.

Figure 2.11 shows the stress variation along the two sections A-A and B-B as shown by the lines in Figure 2.11a. The state of the stress field in the specimen is initially statistically uniform across the specimen, but only before macro-fracturing or strong interaction between micro-fractures occurs. As soon as large fracture zones develop, the situation changes dramatically, and significantly non-uniform stress distributions develop, especially when the fracture zone is not immediately stress-free (as shown in Figure 2.11b).

In Figure 2.5, the simulation results for the relatively heterogeneous specimen #1 with $m = 1.5$ show that the macro-fracture nucleation involves the occurrence of a large number of small-magnitude energy events. In contrast to this, few localised events occur during the early stage of loading in the more homogeneous specimen #7 having $m = 3$, as shown in Figure 2.12. These results suggest interesting new avenues of research for failure prediction. The nucleation sites for strongly homogeneous rock, such as specimen #8 with $m = 5$, seldom occur prior to the peak load. Once the crack nucleation occurs, further growth of the crack is rapid and uncontrolled. Such a failure process would be difficult to predict. On the other hand, the results for specimen #1 show that a relatively heterogeneous rock generates localised zones of intense fracture events well before collapse, which implies that this type of unstable failure can be predicted in advance (Figures 2.5–2.7).

Also, the heterogeneity of the rock has an important influence on the shape of the load–deformation curves and the strength characterisation. Figure 2.13 shows the

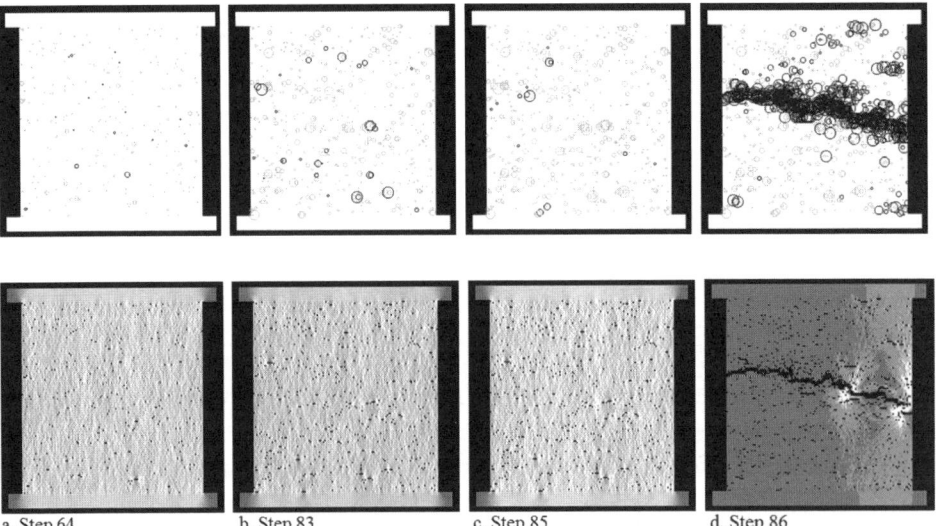

Figure 2.12 Failure mode of relatively homogeneous rock (specimen #7, m = 3). (Above: the circles represent the source location of fracture events. Below: the gray colour represents the relative shear stress).

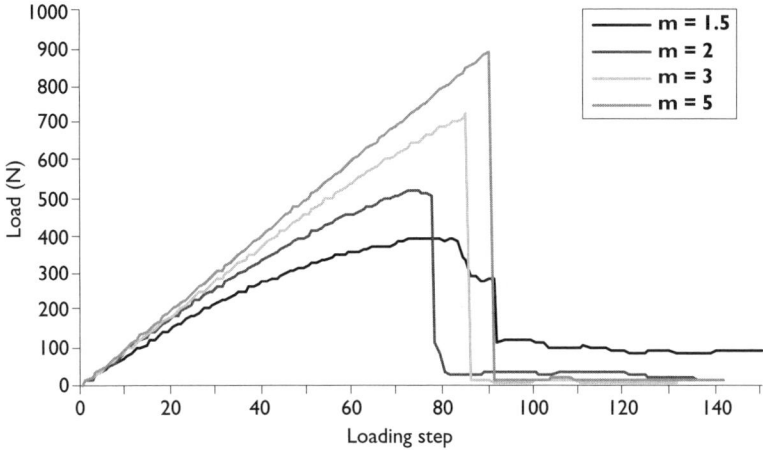

Figure 2.13 Influence of heterogeneity on the shape of the load–deformation curves and the peak strengths.

simulated load–deformation curves for various homogeneity indices, $m = 1.5, 2, 3,$ and 5. It can be seen that, under uniaxial tension, the shape of the more heterogeneous rock has a gentler post-peak behaviour. The maximum strength of the specimens is correlated with the homogeneity index: the higher the value of the homogeneity index, the higher the strength of the specimen. As a result, the load deformation curve becomes more linear and the strength loss is also sharper in this case.

The differing shapes of the complete stress–strain curves in Figure 2.13 are caused by the different degrees of rock inhomogeneity involved. Indeed, this can be demonstrated via an even simpler statistical model (Hudson and Fairhurst, 1969) in which one establishes the survival fraction of many elements with random strengths subjected to random strains, with the mean strain gradually increasing. The continuously reducing survival fraction is then used as a measure of the elastic modulus so that complete stress–strain curves can be generated, as illustrated in Figure 2.14.

The numerical results thus demonstrate that the fracture event patterns are influenced greatly by the degree of heterogeneity of the materials. Figure 2.15 shows the event count accumulation as a function of the deformation for the four specimens with their different homogeneity indices. The results demonstrate that the relatively heterogeneous specimen emits more fracture events as precursors of macro-fracture than those of the relatively homogeneous specimen. However, in physical experiments, even for a heterogeneous rock specimen, few fracture events can be detected at the beginning of the loading stage unless the stress state reaches a certain level—ensuring that the emitted fracture energy can exceed the preset threshold level for the recording equipment. The reason for the lower energy release of fracture events in the initial loading stage is simple: most of the elements failing at this stage had a lower strength than those which failed later, particularly for the more heterogeneous materials.

It is generally considered that the microscopic failure mechanisms in tension and compression are substantially different from each other. In compression, a failed por-

26 Rock failure mechanisms

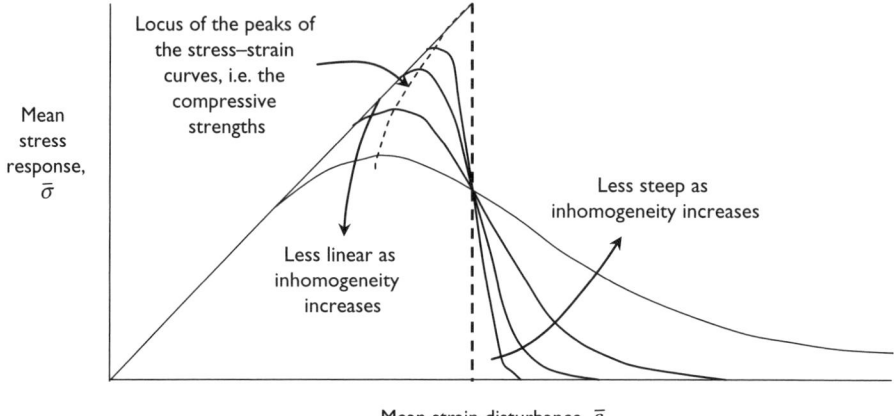

Figure 2.14 Basic statistical model (Hudson and Fairhurst, 1969) illustrating the same trends as in Figure 2.13 and demonstrating the influence of the rock inhomogeneity.

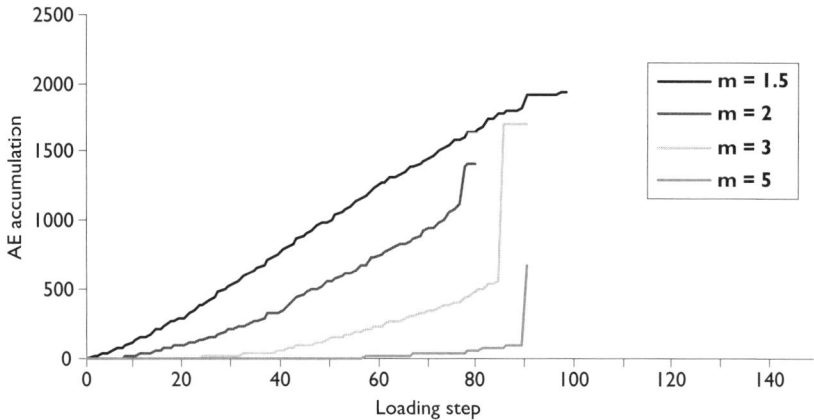

Figure 2.15 Influence of heterogeneity on the shape of the fracture event count accumulation-deformation curves and their strength characterisation.

tion does not completely lose its loading bearing capacity because of friction with the surrounding rock (see Chapter 4). However, in tension the loading capacity of the failed portion decreases more rapidly than in compression. This consideration suggests a comparatively gentle decrease of load bearing capacity of a specimen in compression; however, the shapes of the stress–strain curves do not differ much from each other in tension and compression. This has also been observed by Okubo and Fukui (1996) in their experiments.

In Okubo and Fukui's (1996) experimental study, one result of particular interest was obtained in uniaxial tension testing: a large amount of residual strength remains in the post-failure region, i.e. before total failure. Peng (1975) conducted uniaxial tension tests and successfully obtained complete stress–strain curves for four rocks. He also found large residual strength for all four rocks. The main difference compared to our results is that in the simulation the first visible crack appears comparatively earlier, for example, at 90% of the peak strength in the pre-peak region, while in the physical experiments cracks could not be observed until the stress decreased to less than 20% of its peak value in the post-peak region. The presence of such residual strength in the post-peak region before total failure may be attributable to several reasons. The first is the heterogeneity or strength variation from location to location. If the local strength variation is large, the specimen failure initiates in the weak portions. Even in the post-peak region, strong portions are still intact and so some amount of load-bearing capacity remains before final separation. Conversely, a macro-crack extends rapidly from boundary to boundary in a homogeneous specimen.

In the final failure modes shown in Figure 2.9, the cracks have almost crossed the sections of the specimens but still have a non-zero residual stress bearing capability. The cracks are not continuous, but rather overlap and branches exist. Six illustrations of the development of the crack interface bridging in specimen #8 are shown in Figure 2.16 for the post-peak region. In order to indicate the loading stages corresponding to the pictures of the failure process, the load–deformation curve is also shown in Figure 2.16. The specimen was loaded in 350 steps—which is more than four times the step number corresponding to the maximum load. However, the specimen still can bear 10% of the maximum load at the end of the process. Figure 2.16e (step 344) shows that three main crack branches are initiated: two start from the upper right corner and bottom left corner of the specimen; and another initiates in the

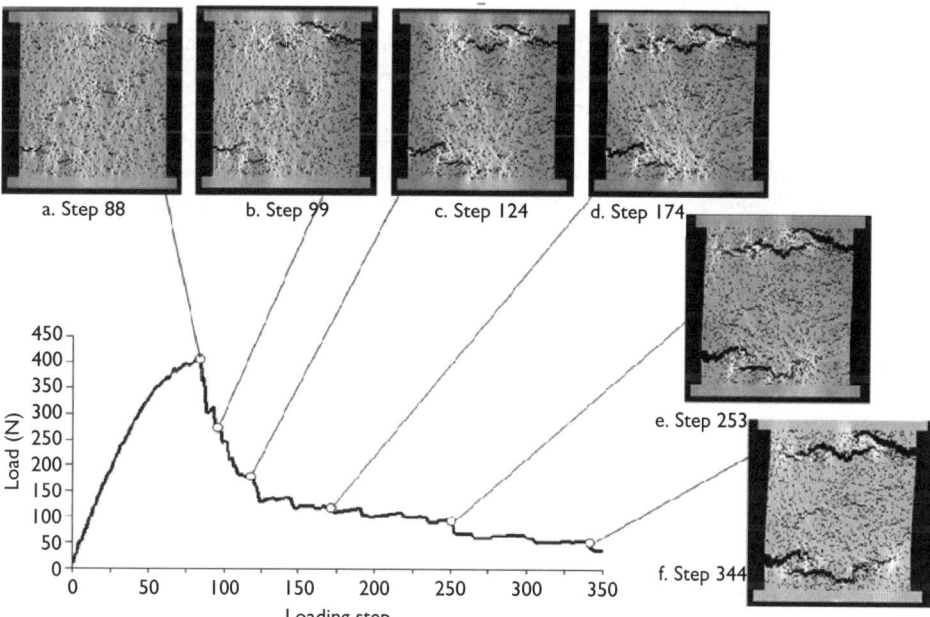

Figure 2.16 Six stages in the final rupture of the crack overlap (in the sample, the gray colour represents the relative shear stress, the lighter the shade, the higher the stress).

28 Rock failure mechanisms

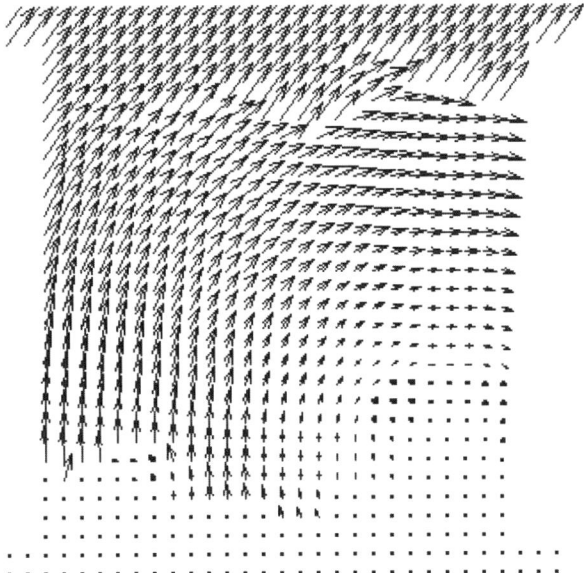

Figure 2.17 Displacement vector for a specimen which failed in the crack interface bridging manner.

central part of the specimen. Note that the crack in the centre becomes inactive later, while the two cracks which start from the two corners become active afterwards.

Instead of propagating directly, two other isolated cracks nucleate and propagate in an overlapping manner (see Figure 2.16c at step 124) with the crack branches separated by an intact ligament. Defining a crack tip for such crack systems, as in fracture mechanics, is not straightforward. Upon further loading, one of the cracks, starting from the left boundary, is almost arrested, while the other one propagates with a pathway like the Chinese traditional symbol sign of the dragon, 龙. In Figure 2.16d–e, the tip of the lower crack has almost reached the upper crack and the failure of the ligament still does not occur, and the stress can still be transferred through the ligament.

Figure 2.17 shows the displacement vectors for the same specimen as shown in Figure 2.16 at step 344. Clearly visible is the way in which the crack interface bridges develop, and how the ligament rotates.

Chapter 3

Rock failure in indirect tension

3.1 GENERATING A TENSILE STRESS THROUGH COMPRESSIVE LOADING

In Chapter 2, we discussed the failure of rock in direct tension. However, it is difficult to conduct direct tensile tests in the laboratory because of the problem of obtaining a true tensile stress field in the rock specimen. The preparation of specially shaped specimens, like the dog-bone specimen in Figure 2.1 in Chapter 2, is time consuming and not suited to field work at a construction site. Thus, methods have been developed for testing rock specimens in tension—but with overall compressive loading, Figure 3.1.

It can be seen in Figure 3.1a that the overall compression of a beam in three-point loading produces a tensile stress at the base of the beam. Knowing the load at beam failure and the dimensions of the beam, the tensile stress at failure can be calculated, assuming that the rock is elastic. Similarly, in Figure 3.1b, the overall compression of the disc produces a tensile stress perpendicular to the loaded diameter. Figure 3.1c shows a Solenhofen limestone disc after it has been tested to failure in this way. The straight line tensile crack between the loading points reflects the extremely fine grain of the Solenhofen limestone. The more irregular lines are cracks formed by the immediately subsequent loading of the half discs. The Brazilian test is so named (perhaps apocryphally) because a church was being moved in Brazil on concrete rollers—which split under the load of the church.

In fact, there are many such specimen geometries whereby a tensile stress can be generated by the overall application of a compressive load. However, the Brazilian test, as an indirect tensile test, is commonly and widely used to determine the tensile strength of rock thanks to its easy sample preparation and simplicity of testing. The International Society for Rock Mechanics (ISRM) suggests the Brazilian test as the standard test for determining the tensile strength of rock (Ulusay and Hudson, 2007).

Although splitting of the specimen into two halves is the expected failure mode, as in Figure 3.1c, other rupture modes can occur. Also, the splitting strength of the samples can vary significantly with the specimen size and loading conditions. Moreover, in addition to elasticity theory being used to calculate the tensile strength existing in the sample at failure, the stress field at the centre of the disc is biaxial. For these reasons, the Brazilian tensile strength should be regarded as an index property rather than a material property—because a material property does not depend on the specimen geometry nor the loading conditions of the test.

30 Rock failure mechanisms

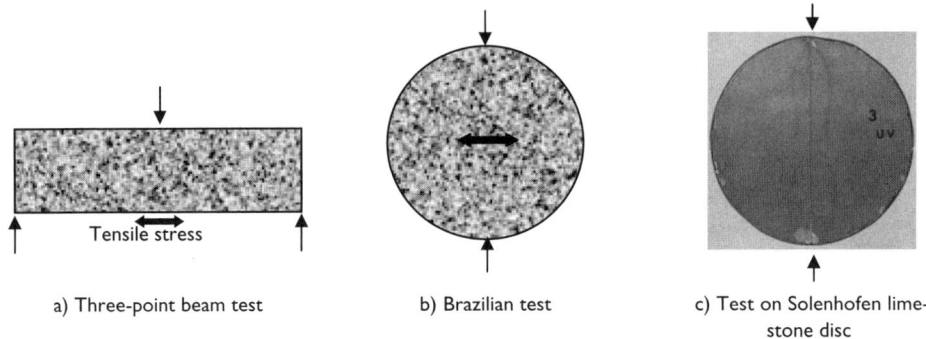

a) Three-point beam test   b) Brazilian test   c) Test on Solenhofen limestone disc

Figure 3.1 The beam test and Brazilian test.

However, it is useful for rock mechanics applications to numerically capture and hence understand the indirect tensile failure mechanism during the complete failure process for rock samples subjected to the Brazilian test with different boundary conditions. Apart from splitting tests on discs (or cylinders) and square plates, ring specimens have also been developed in which a disc with a central hole is subjected to diametral compression, with failure occurring at the central hole (Addinall and Hackett, 1964; Hudson, 1969). Numerical simulations of both the Brazilian test and ring test are presented in this Chapter to illustrate the failure patterns and tensile strength of disc and ring specimens under different geometry and loading conditions.

Based on the theory of elasticity, the stress distribution in the disc, square plate or ring subjected to diametral compression can be calculated and combined with a strength criterion (such as the Griffith criterion) to enable the location of failure initiation to be determined (e.g. Fairhurst, 1964); and the stress distributions, as well as the damage or failure zone in the Brazilian test have been obtained. However, there are conflicts between the theoretical solutions based on the theory of elasticity and experimental results. This is due, *inter alia*, to the fact that both the theoretical analysis using the theory of elasticity and the numerical analyses using the FEM and BEM methods neglect the inhomogeneity of rock, i.e. the I in the DIANE nature of rock as discussed in Chapter 1, Section 1.5. So, in the computer simulations presented here, we simulate the failure of rock specimens and incorporate inhomogeneity.

3.2 ESTABLISHING THE NUMERICAL SIMULATION MODEL FOR INDIRECT TENSILE STRENGTH TESTS

Disc (or cylinder), square plate (or prism) and ring specimens of rock with different sizes loaded diametrically using bearing strips of various widths are numerically evaluated here. In the numerical simulations, the three kinds of specimens with diameters, $D = 30, 76, 90, 120, 150$ and 180 mm (as shown in Figure 3.2) are studied. In order to study the effect of the width of the load-bearing strip, three different load-bearing strip widths, b, equal to 4%, 8% and 16% of the specimen size ($b/D = 0.04$, 0.08 and 0.16) are considered in the numerical simulation. For the ring test, the internal diameter of the ring is also a variable to be numerically studied.

Here the numerical models of the disc and ring specimens are established according to the mechanical properties of the granite used by Mellor and Hawkes (1971). The *direct* tensile strength of rock is also taken to vary from 12.0 to 15.0 MPa with the other mechanical properties of the rock specified as Young's modulus, 62.2 GPa, uniaxial compressive strength, 185 MPa, and Poisson's ratio, 0.28. As in the direct tension simulations in Chapter 2, the Weibull distribution is used to introduce the inhomogeneity, with the parameters listed in Table 3.1.

The parameters are specified according to the so-called 'standard numerical specimen' which is 200 mm × 100 mm in dimensions and composed of 200 × 100 elements and has the above-mentioned mechanical properties for this rock. For convenience of comparison of numerical results from the different shaped specimens, the same mechanical parameters are specified for the square plate and ring specimens. Since, in the numerical simulations, we assume that the cushion materials, including steel and cardboard, will not fail, they are specified to be homogeneous and to have a high strength. It is expected that failure will usually initiate at a weak element in the numerical specimen, and this failure will lead to a stress concentration, which in

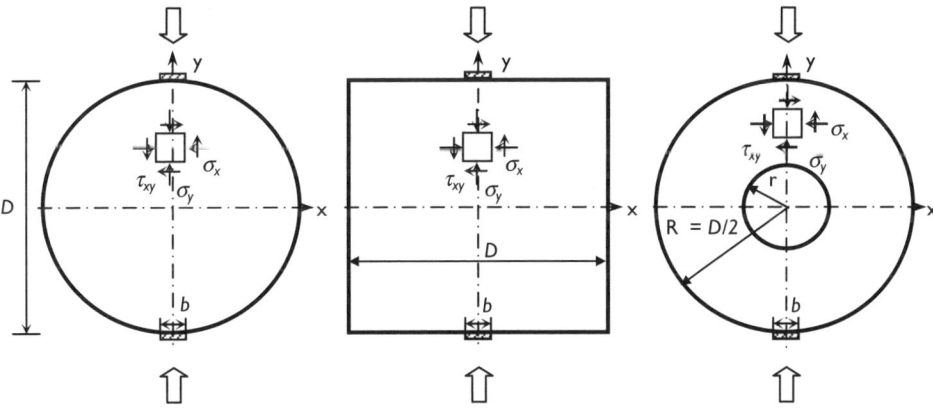

Figure 3.2 Geometries and loading conditions for disc, square plate and ring specimens. The co-ordinates and stress notations used are indicated.

Table 3.1 Material properties used in the indirect tensile test simulations.

	Rock	Steel	Cardboard
Homogeneity index (m):	5	homogeneous	homogeneous
Mean compressive strength (σ_0)	510 MPa	800 MPa	500 MPa
Mean (Young's) elastic modulus (E_0)	70 GPa	21 GPa	10 MPa
Poisson's ratio (v)	0.28	0.3	0.3
Ratio of compressive to tensile strength	12	1	1
Frictional angle (ϕ)	30°	–	–

turn leads to continued propagation of the fracture. Experimental observations have indicated that failure may initiate under the loading points in the Brazilian test, rather than at or near the centre of the disc, Hudson et al. (1972).

In all cases, the numerical specimens are simulated as a plane stress problem. The displacement is imposed on the upper loading platen step by step and the lower platen is fixed in the vertical direction, which can thus simulate compressive loading with displacement control (mirroring the use of displacement control when a closed-loop servo-controlled testing machine is used for laboratory studies of indirect tensile testing). The equation used to calculate the splitting tensile strength, S, can be summarised via the following Equation (3.1) assuming that the specimen is actually fractured due to the tensile stress generated in the plane of fracture.

$$S = k\frac{2P}{Dt} \tag{3.1}$$

where k is a coefficient depending on the specimen geometry and experimental conditions (see below), P is the load required for fracture, D is the diameter of the disc or ring (alternatively the edge length of the square plate), t is the thickness (in the numerical simulation t is the unit thickness of an element because the element is simulated as a plane stress problem).

$k = 1/\pi$ for the disc test; $k = 0.30 - 0.315$, for the square plate test when b/D is $0.04 - 0.16$; $k = [6 + 38(r/R)^2]/\pi$, for the ring test, where r and R are the internal and external radii of the ring, i.e. $R - D/2$.

The splitting tensile strengths obtained from the disc, square and ring specimens are denoted as σ_d, σ_{sq} and σ_r respectively. In the numerical simulations, the maximum load P can be directly obtained, and then the tensile strength of the specimen calculated according to Equation (3.1). In this work, the direct uniaxial tensile strength, σ_t, obtained from the numerical simulations is considered as the material property of the rock. The splitting tensile strength obtained from the Brazilian, plate or ring test can then be compared to the σ_t value.

3.3 NUMERICAL SIMULATIONS OF ROCK FAILURE IN INDIRECT TENSILE STRENGTH TESTS

In this section, the numerical results are analysed and compared with corresponding experimental observations on strength and failure patterns.

3.3.1 The disc test

3.3.1.1 Stress distribution in the discs

In Figure 3.3, the fringe contours of maximum shear stress in the disc, plate and ring specimens when they are simulated both as homogeneous and heterogeneous materials are illustrated. It can be seen that the heterogeneity of the rock has a strong local effect on the stress fields within the specimens. This demonstrates the usefulness

Rock failure in indirect tension 33

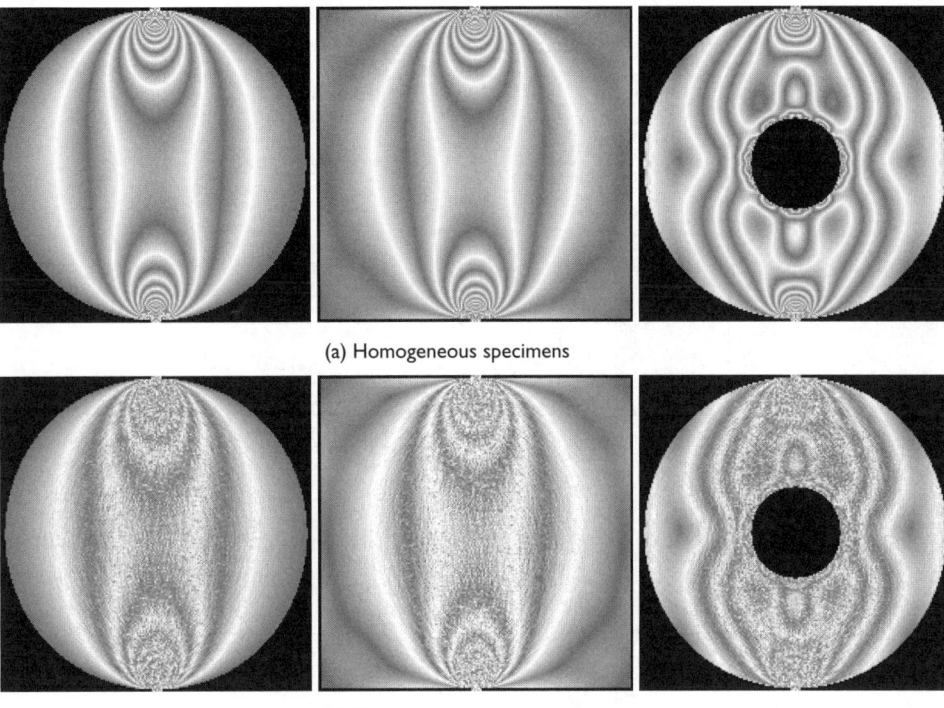

(a) Homogeneous specimens

(b) Heterogeneous specimens

Figure 3.3 Numerically produced 'fingerprint' contours of the maximum shear stress in diametrically loaded disc, plate and ring specimens of homogeneous and heterogeneous material.

of the numerical simulation approach because there is currently no experimental method available to obtain such detailed 'fingerprint' patterns for heterogeneous materials.

It can be seen from Figure 3.4 that the tensile stress distribution at the central area of the specimen shows good agreement with the theoretical results, and the stresses at the centre of the disc are not much affected by the load-bearing strip width. Because the Young's moduli of the elements in the numerical specimen vary, the stress distribution across the loaded diameter is also heterogeneous and fluctuates around the theoretical solution. Some deviations exist near the loading points when the load-bearing strip is relatively narrow. In the theoretical analysis, the compressive load is simplified by applying uniform pressure radially over a short strip of the circumference at each end of the diameter but, in the numerical simulations, a uniform displacement is applied on the loading platen in order to simulate the real loading conditions applied in the laboratory testing. The high compressive stress near the loading points when b/D is 0.04 may cause the failure to firstly originate at this position (as has been postulated for laboratory tests without the load distributer, Hudson *et al.*, 1972). With the increase of the load-bearing strip width from 0.04 to 0.16, the stress concentration near the loading points is relieved greatly, so the specimen failure tends to initiate from the centre of the disc specimen.

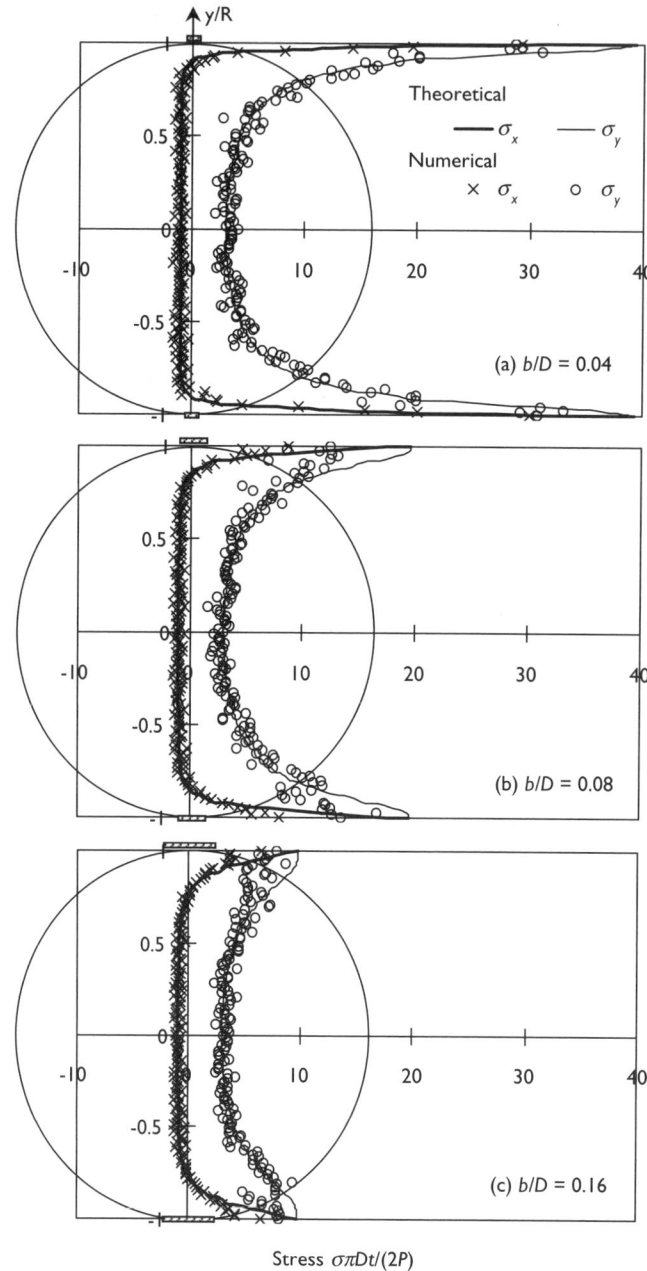

Stress $\sigma \pi Dt/(2P)$

Figure 3.4 Stress distribution across the loaded diameter for a disc specimen in the Brazilian test. Theoretical and numerical results for using a cardboard cushion when the Weibull shape parameter m is 5.0, E_s is 62.2 GPa, v_s is 0.28 and the bearing strip widths, b/D, are 0.04, 0.08 and 0.16.

3.3.1.2 Effect of the material properties of the load-bearing strip on the disc test

With regard to the cushioning at the load application region, various materials including blotting paper, araldite, cardboard, hardboard, polyethylene, timber and copper have been used. If flat steel platens are used to load the specimen, the contact width and the contact stresses can be calculated according to the theory of elasticity (Yanagidani et al., 1978) as follows, Equation (3.2).

$$\frac{b}{D} = 2\left(\frac{2P}{\pi Dt}\right)^{1/2}\left[\frac{1-v_p^2}{E_p} + \frac{1-v_s^2}{E_s}\right]^{1/2} \tag{3.2}$$

where P is the applied load, D is the diameter of the disc, t is the thickness of the specimen, and the subscripts p and s refer to platen and specimen respectively, E_p and v_p are the Young's modulus and Poisson's ratio of the platen, while E_s and v_s are the Young's modulus and Poisson's ratio of the rock specimen. The width of the contact strip can be calculated if the rock is assumed to be linearly elastic.

For a steel cushion material with Young's modulus E_p = 210 GPa, and v_p = 0.3, E_s = 62.2 GPa, and v_s = 0.28, it can be calculated from Equation (3.2) that the relative load-bearing strip width b/D is 0.008752 $\sqrt{2P/(\pi Dt)}$. Assuming the rock is elastic before failure and $2P/(\pi Dt)$ is close to its uniaxial tensile strength (15.0 MPa), we obtain b/D as 0.034. Similarly, if cardboard (with a Young's modulus of 10 GPa and a Poisson's ratio of 0.3) is used as cushioning material, b/D is 0.08. However, when the soft cushion is used, it can lose its shape along the entire width or even be crushed and fit tightly along the specimen's entire width, meaning that a wide load-bearing width is achieved between the specimen and loading platens.

When stiffer cushion materials (such as a steel cushion or no cushion) are used, the fracture initiates near the upper load point (as shown in Figure 3.5). Failures at the centre of the specimen are also found, but it is the fracture initiation near the upper load point which causes the specimen to lose most of its loading capacity. The evident crack propagation path is from the upper loading point to the centre of the specimen and to the lower loading point. Under this loading condition, some elements near and below the upper loading point are fractured in shear mode; however, the dominant failure is caused by tensile failure of many elements (as shown by the black elements in Figure 3.5b).

In the case of the softer cushion material (cardboard, for example) with relative width b/D of 0.04, the fracture initiates at the centre of the disc (as shown in Figure 3.6a). Finally, unstable failure occurs along the diameter of the specimen and a significant loss of load takes place. The elements in the numerical specimen are dominantly damaged in tensile mode, and no shear damage of elements near the loading points is found. Comparing Figure 3.5 and Figure 3.6a, we find that the failure initiation location is changed due to the different cushion materials: the tensile strengths obtained from the test cushioned with cardboard and steel are 15.08 MPa and 13.76 respectively. It can be seen that the results with the cardboard cushion are closer to the direct tensile strength of this numerical specimen—as might be expected intuitively.

36 Rock failure mechanisms

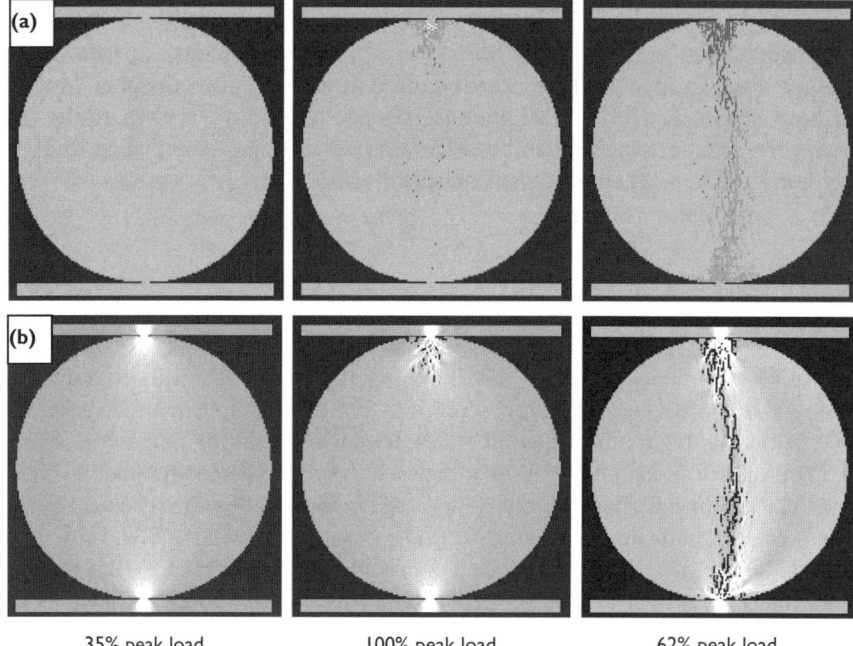

Figure 3.5 Distribution of acoustic emission a) and maximum shear stress b) during the failure process of the numerical specimen. Fracturing is indicated in black. The specimen is cushioned with steel and b/D is 0.04 and its mechanical properties are: $m = 5.0$, $E_s = 62.2$ GPa, and $v_s = 0.28$.

The different failure processes due to the different cushion materials can be explained from the stress distributions as shown in Figure 3.7, noting that the stresses σ_x and σ_y in the cushion are normalised by $2P/(\pi Dt)$. It can be seen that the soft cushion (cardboard) has the ability to uniformly distribute the applied load P, with the stress concentration at the edge of the cushion being relatively weak compared with the conditions for a specimen directly loaded by a steel platen (or cushioned with a steel plate). When the load-bearing strip is narrow (for example, b/D is 0.04), the failure initiation location is much more sensitive to the cushion materials. In line with results from the theoretical analysis of Fairhurst (1964), the strength obtained from this numerical specimen with a steel cushion is lower than those specimens where fracture initiates near the centre of the specimen.

3.3.1.3 Effect of load-bearing strip width on disc test

So, to ensure that the Brazilian test is valid, cushion materials are used between the loading platens and specimen to increase the contact area, and therefore relieve the stress concentration. Additionally, radiused end-caps between the specimen and

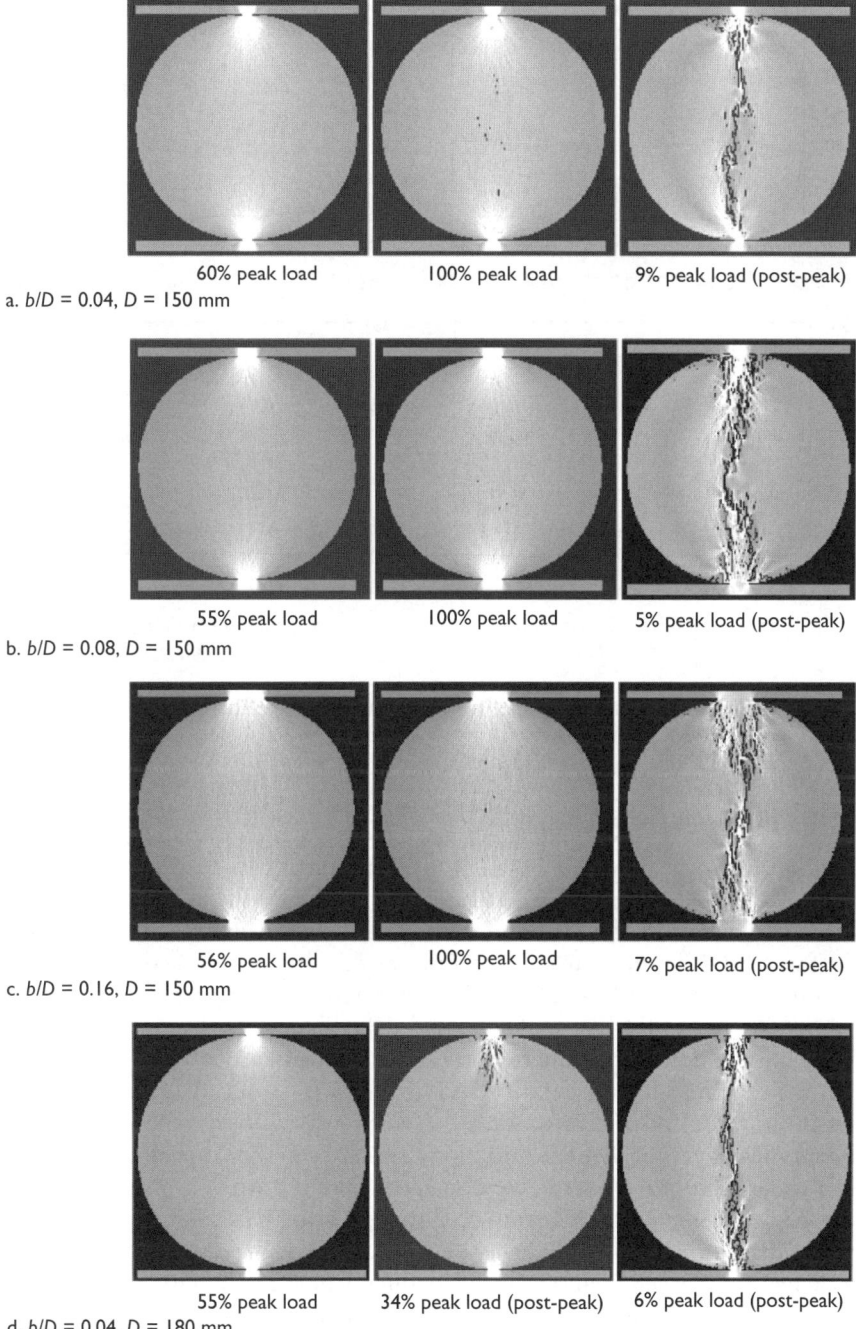

Figure 3.6 Failure process of a disc specimen in the Brazilian test (here the specimens are cushioned with cardboard with *b/D* increasing for the illustrations a → d), with the maximum shear distributions at fracture initiation, peak load and post-peak given; for the numerical specimen, $m = 5.0$, $E_s = 62.2$ GPa, and $v_s = 0.28$.

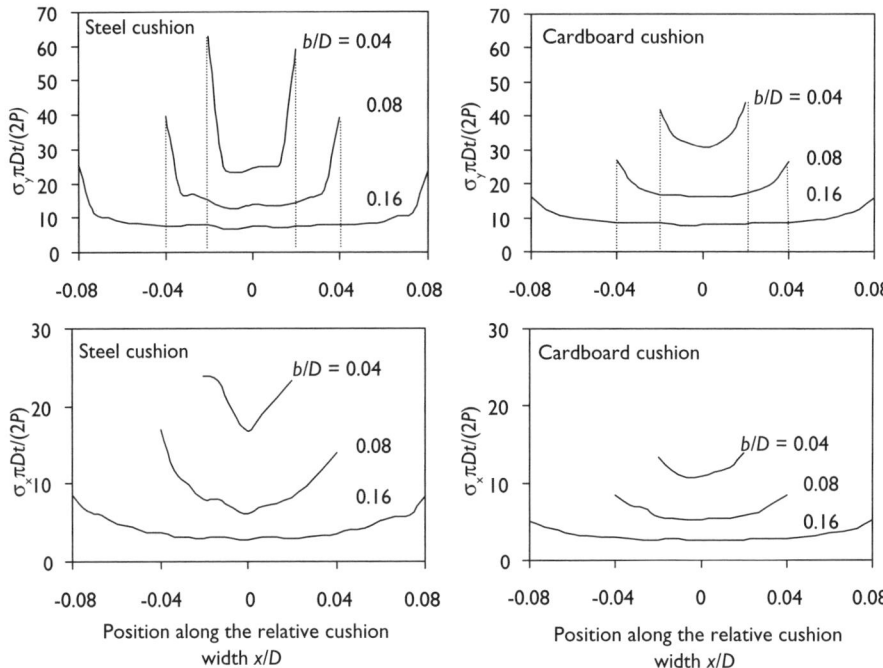

Figure 3.7 Distribution of stresses when steel and cardboard are utilised as the cushion with different b/D values as given by the simulations. For the numerical specimen, $m = 5.0$, $E_s = 62.2$ GPa, and $v_s = 0.28$.

the platen or a curved-jaw loading jig can be designed to distribute the load over a known arc. Also, a flattened Brazilian disc specimen can be used to increase the contact area with loading platens. In any case, a sufficient cushion width must be guaranteed to distribute the stress concentration between the specimen and loading platens for a valid Brazilian test. Therefore, in this investigation, the contact width between the specimen and the platens is considered as an important factor affecting the failure patterns and strength of rock in the Brazilian test.

The failure patterns of numerical specimens, when cardboard is selected as the cushion material and for the relative widths $b/D = 0.08$ and 0.16, are presented in Figures 3.6b and c, respectively. With the cushion conditions, the distributions of maximum shear stress at damage initiation, peak load and post peak load are shown. Under a wider load-bearing strip, the damage is initiated at the centre of the disc and no damaged elements near the loading point are found. The failure (or damaged elements) are distributed in a relatively wide zone at the centre of the specimen. Finally, many elements at the centre and near the loading platens fail in tensile mode almost at the same loading step, and a sudden loss of loading capacity of the specimen occurs.

The primary crack at the centre propagates along the diameter of the specimen and secondary diagonal cracks near the loading platens are also developed. The specimen substantially loses its load bearing capability (as shown in Figure 3.8). For the specimens, fracture initiation does not mean the collapse of the specimen. Generally,

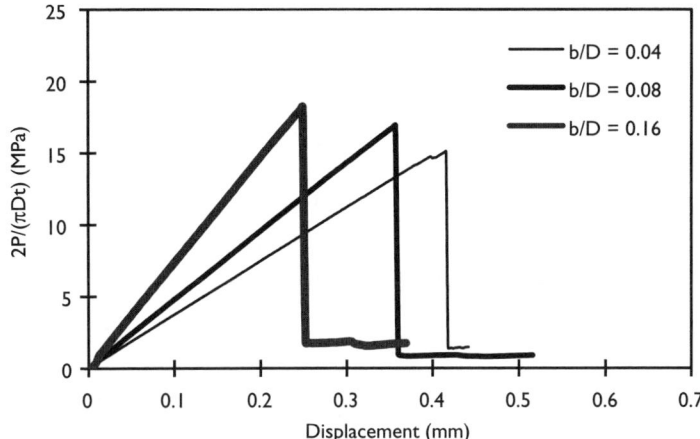

Figure 3.8 Numerically obtained load–displacement curves for specimens in Brazilian tests (the cushion material is cardboard, and for the numerical specimen, $m = 5.0$, $E_s = 62.2$ GPa, and $v_s = 0.28$).

the initiation of fractures usually occurs when the load is 40–60% of the ultimate value of final failure. This result agrees with the acoustic emission observation of Yanagidani *et al.* (1978). Also, due to the heterogeneity, the path of the primary crack in the specimen is not a straight line—as compared to that in Figure 3.1c for the super-fine grained Solenhofen limestone. This is a common phenomenon that can be observed in many experiments on rocks.

Despite the fact that the stress distributions at the centre of the specimen are similar with the increase of b/D from 0.04 to 0.16, an increase in load-bearing capacity of the specimens is found (as shown in Figure 3.8). When Equation (3.1) is used to calculate the splitting test strength for the specimens, the calculated tensile strengths from the specimens ranged from 15.08 MPa to 18.21 MPa with the increase of relative load-bearing strip width b/D from 0.04 to 0.16. The ratios of Brazilian tensile strength to direct testing tensile strength are seen to vary from 1.005 to 1.214.

We find that sometimes the evident failure near the loading platen makes the Brazilian test invalidated when b/D is 0.04. For example, when the specimen diameter $D = 180$ mm and $b/D = 0.04$, a few elements become damaged at the centre of the disc, but in the end the specimen's main loss of its loading capability is because of the elements near the loading platen—damaged in shear or tensile mode, as shown in Figure 3.6d. From this Figure, we find that failure near the loading platen is observed before the main fracture of the specimen, although the final failure pattern is quite similar to other numerical specimens. Based on these numerical results, a relative load-bearing strip width b/D that is equal or greater than 0.08 is suggested in order to guarantee the fracture initiates at the centre of the disc.

Furthermore, in some cases, failure may not initiate at the theoretically predicted points only, but at other points where a critical combination of stress and strength is reached: many elements with low strengths may be damaged initially even at the mean applied low stress, but then this failure may change the stress distribution in the specimen, which then influences the final failure patterns of the specimens. When the cardboard cushion is used and b/D is larger than 0.04, not only the elements at the centre along the loaded diameter are damaged, but other elements near the centre are also damaged in tensile mode. Conversely, when the division of the specimen is caused mainly by failure of elements near the loading platens, the damage of many elements can also be found at the centre of the specimen. This phenomenon is confirmed by fracture event observations of rock subjected to physical Brazilian tests.

3.3.1.4 Effect of specimen size on the disc test

The maximum loads and tensile strengths obtained from the numerical simulation of disc specimens with different sizes and relative load-bearing width b/D are listed in Table 3.2. In this Table, only the results for cardboard as a cushioning material are given. It can be seen that the tensile strength of these numerical specimens decreases with an increase in the specimen size. The tensile strengths obtained from the numerical simulations lead to a steady value for diameters above about 76–90 mm. It is also found that the size effect is slightly stronger for the highest relative width of bearing strip ($b/D = 0.16$). For the specimen 180 mm in diameter and with a relative load-bearing strip width b/D of 0.04, its tensile strength is much lower because the failure initiates near the loading platen and thus makes the Brazilian test invalid.

Note that, as a consequence of the effect of the bearing strip width, the tensile strength obtained from numerical simulations of the same size specimens can be considerably different, particularly for a small specimen (for example $D = 30$ mm). From Table 3.2, we find that the maximum and minimum tensile strengths are 21.2 MPa and 15.0 MPa respectively when the invalid Brazilian test values are excluded, a difference of 41%. A similar result was found in experiments with differences of up to 40% for granite with different specimen sizes and load-bearing strips (Hobbs, 1965). For the failure patterns of differently sized specimens as shown in Figure 3.9, it can be seen that these patterns are closely related to the load-bearing strip value and are independent of the specimen size. This indicates that the failure mechanism is the same and not dependent on the specimen size.

Table 3.2 The maximum loads and tensile strengths obtained from the numerical simulation of disc specimens with different sizes and relative load-bearing widths, b/D, and cushioned with cardboard.

D (mm)	Load P (N)			Tensile strength $S_d = \frac{2P}{\pi DL}$ (MPa)		
	$b/D = 0.04$	$b/D = 0.08$	$b/D = 0.16$	$b/D = 0.04$	$b/D = 0.08$	$b/D = 0.16$
30	745	874	999	15.9	18.5	21.2
76	1816	1971	2125	15.2	16.5	17.8
90	2123	2291	2454	15.0	16.2	17.4
120	2813	3126	3257	14.9	16.6	17.3
150	3545	3953	4112	15.0	16.8	17.4
180	3857	4668	4963	13.6	16.5	17.6

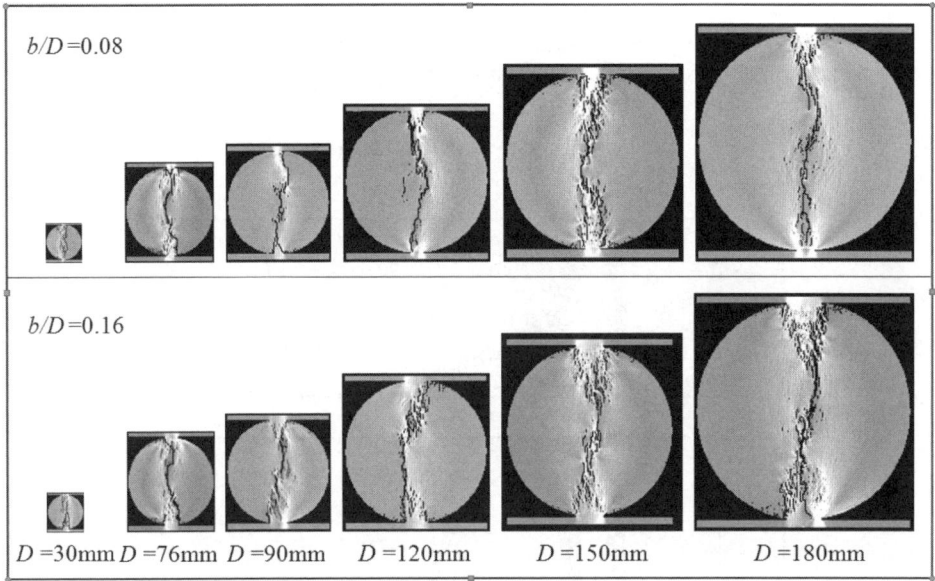

Figure 3.9 Numerically obtained failure patterns for disc specimens having different sizes and load-bearing widths (For the numerical specimens, $m = 5.0$, $E_s = 62.2$ GPa, and $v_s = 0.28$).

3.3.2 The plate test

When a square plate is compressed between a pair of symmetrically placed flattened indenters, tensile stresses also develop in its central region. The distribution of horizontal and vertical stresses across the vertical centreline of specimens is given in Figure 3.10 and is compared with those for the disc specimen. In this Figure, the side length of the square specimen and the diameter of the disc specimen are the same, and the values of stresses are normalised with $2P/(\pi Dt)$, this being the formula for the tensile strength as determined with the Brazilian test. It is concluded that the stress distributions across the vertical centrelines of the specimens are similar.

The failure patterns of the square plate specimens as shown in Figure 3.11 are also similar to those of the disc (or cylinder) specimens. The failure process and strength of the plate specimen are also affected by the relative load-bearing strip width b/D. When the cardboard is cushioned between specimen and loading platen, and the value of b/D is 0.04, fracture initiation near the loading platens is also found causing the indirect strength test to be invalidated (Figure 3.11a). When the b/D is 0.08 or 0.16, fracture initiates at the centre of the specimen and propagates towards the loading platens, and finally the secondary cracks are also observed near the loading platens (Figure 3.11b and c).

The indirect tensile strengths σ_{sq} calculated from the peak loads of this specimen are 14.1, 17.1, and 18.1 MPa when the relative load-bearing strip widths are 0.04, 0.08 and 0.16, respectively. The numerical result for $b/D = 0.04$ is invalid (because

42 Rock failure mechanisms

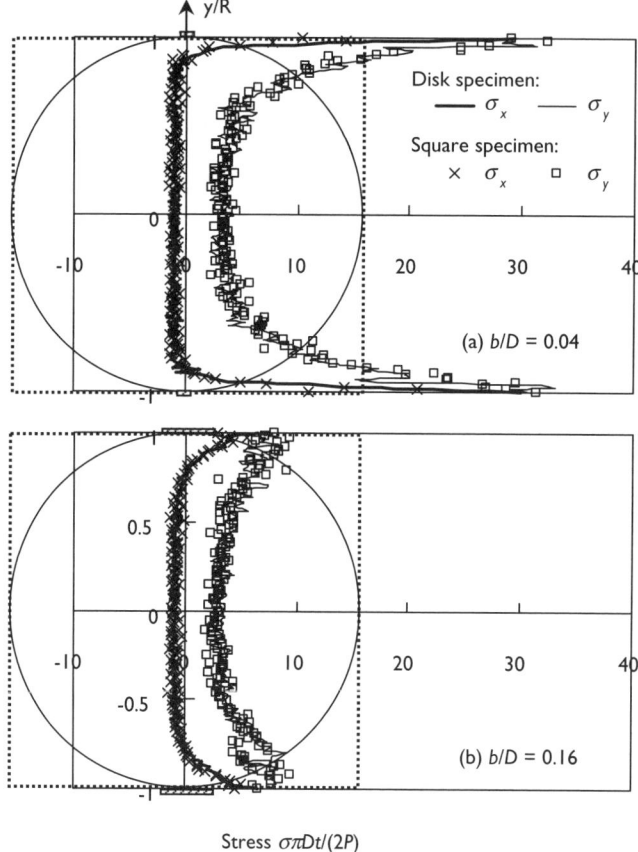

Stress $\sigma \pi Dt/(2P)$

Figure 3.10 Comparison of stress distributions across the vertical centreline of a square specimen and a disc specimen (for the numerical specimens, $m = 5.0$, $E_s = 62.2$ GPa, and $v_s = 0.28$).

the fracture initiates near the loading platens); its calculated tensile strength is much lower than the uniaxial tensile strength σ_t. If the plate test is valid, the tensile strengths obtained from these specimens are found to be higher than those for the disc specimen with the same dimensional parameter D.

3.3.3 The ring test

The geometry of ring specimens and the loading condition have been given in Figure 3.2. In the numerical simulations for the ring test, the ratios of internal to external radius used are 0.02, 0.1, 0.2, 0.3, 0.4, 0.5, 0.6 and 0.7. The Weibull distribution parameters for the numerical specimens are the same as those for the Brazilian test specimens as listed in Table 3.1. The effects of the internal radius of ring and relative load-bearing strip width b/D on the failure modes and loading capacity are numerically simulated.

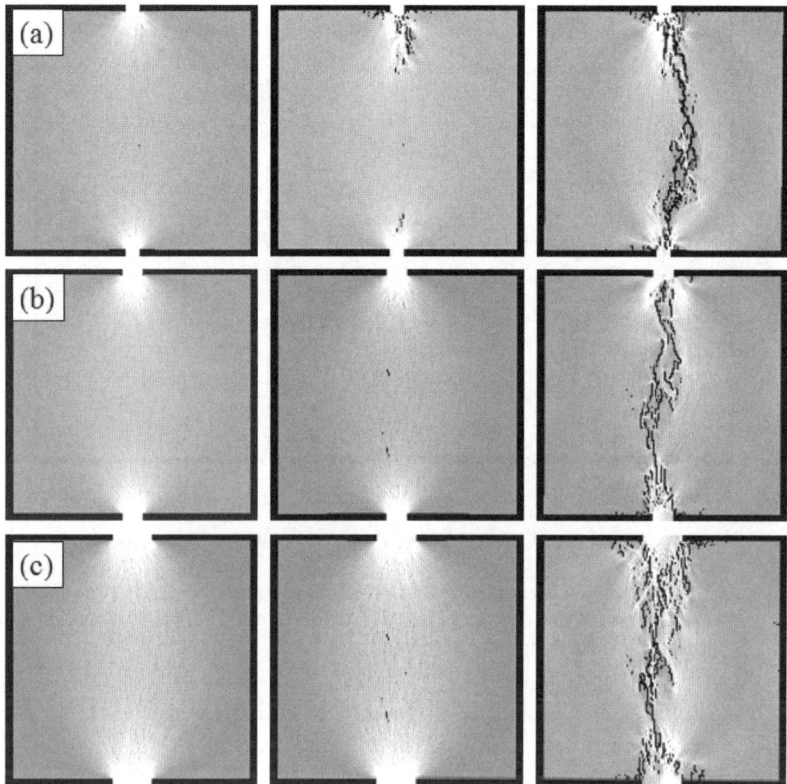

Figure 3.11 Numerically obtained failure patterns for square plate specimens at three different loading levels. The three different relative load-bearing strip widths b/D are: a) b/D = 0.04, b) b/D = 0.08 and c) b/D = 0.16. (For the numerical specimen, $m = 5.0$, $E_s = 62.2$ GPa, and $v_s = 0.28$.).

3.3.3.1 Stress distribution along the loading diameter for ring specimens

Due to the stress concentration, the critical tensile stress normal to the loading diameter at the intersection of the loading diameter with the hole can be expressed, Equation (3.3), as

$$\sigma_x = -k \frac{P}{\pi R t} \qquad (3.3)$$

where P is the applied load, and k is the stress concentration factor reflecting the degree of stress concentration induced by the hole. In this investigation, the distribution of initial horizontal stress along the loading diameter is given from numerical simulation, as shown in Figure 3.12. The values of the stresses are again

44 Rock failure mechanisms

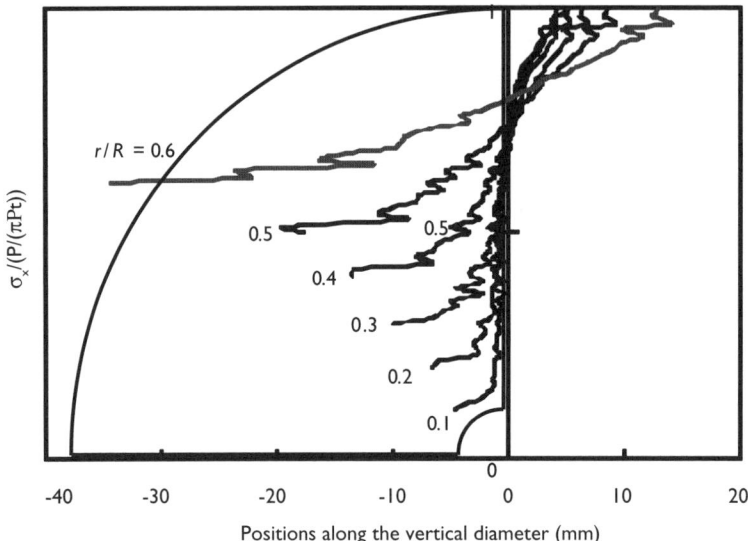

Figure 3.12 Numerically obtained distribution of horizontal stress along the loading diameter in the ring specimen for different relative sizes of the central hole (for the numerical specimen, $m = 5.0$, $E_s = 62.2$ GPa, and $v_s = 0.28$).

normalised by $P/(\pi Rt)$. In the ring test, the introduced stress raiser at the surface of the hole promotes the failure, and also the contact stresses are lower than those in the Brazilian test.

3.3.3.2 Effect of hole diameter on the failure pattern of ring specimens

Figure 3.13 shows the failure patterns of ring specimens when the r/R varies from 0.1 to 0.6 and the relative load-bearing strip width b/D is 0.04, 0.08 and 0.16. In this Figure, each numerical specimen is referred to with a lower case letter (from a to f) and a column number (from 1 to 3). For example, the specimen with an r/R of 0.1 and a relative load-bearing strip width of 0.08 is referred to as Figure 3.13a–2.

When r/R is 0.1, the cardboard is employed as a cushion and sufficient load-bearing width is guaranteed to distribute the applied compressive load; due to the existence of the hole, the stress concentrating effect of the hole alters the stress distribution sufficiently (when its relative diameter r/R is not smaller than 0.1) to cause the failure to initiate at the hole. Then the crack propagation is upwards and downwards to the loading platens and it ceases to propagate on reaching the cushion. Finally, wedges near the loading platens are also formed. Under this condition, the ring specimen generally splits into two halves. The width of the wedge formed is related to the relative load-bearing strip width b/D.

The failure process for this typical failure pattern is shown in Figure 3.14a. In these specimens the crack initiation occurs at 73% peak load, the crack then has a stable propagation process, and finally the specimen fails after the peak load, when

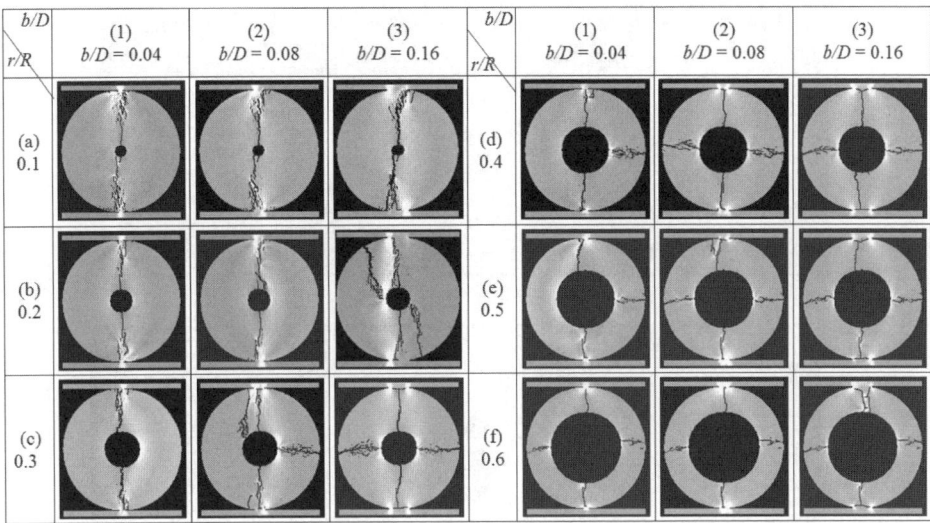

Figure 3.13 Numerically obtained failure patterns of ring test specimens (specimen size D is 150 mm, and the cushioning material is cardboard).

the primary crack across the loaded diameter and the secondary cracks are formed. When b/D is 0.04 and the value of r/R is 0.1 and 0.2, similar failure patterns are also found from numerical simulations (see also Figure 3.13).

In Figure. 3.14(b) another typical failure process of a ring test specimen is presented. With the increased size of the hole in the specimen, the primary fracture unavoidably occurs along the loading diameter at or near the surface of the hole, and propagates with the increase of external load. The primary cracks cease to propagate when they reach the contact zone between the specimen and cushion, and high tensile stresses are induced at the left and right edges of the specimen. The secondary cracks started at the right and left edges, and propagated horizontally towards the surface of the central hole (see also Figure 3.13). It is also due to the heterogeneity of the specimen material that the secondary crack initiates at one edge (left or right edge) and propagates towards the surface of the internal hole of the ring specimen. When the crack reaches the internal surface of the hole, the tensile stress concentration at this edge is released and is transferred to the opposite edge; therefore, at the other edge, the crack also initiates and propagates internally to the hole surface. Finally, the ring specimen splits into four pieces with almost the same size along the horizontal and vertical diameters (as shown in Figure 3.14b). This failure pattern occurs commonly in the ring specimens with a larger hole, for example, when r/R is 0.6, all the specimens fail in this pattern. For every crack initiation and propagation, the stress redistribution is clearly observed from Figure 3.14b and the corresponding load drop is seen in the load–displacement of the loading platen as shown in Figure 3.15.

Sometimes, the secondary cracks initiate only at one edge (as shown in Figure 3.13c–2, Figure 3.13d–1 and Figure 3.13e–1) when the load-bearing strip is narrow. This is because the primary crack deviates from the central diameter, which

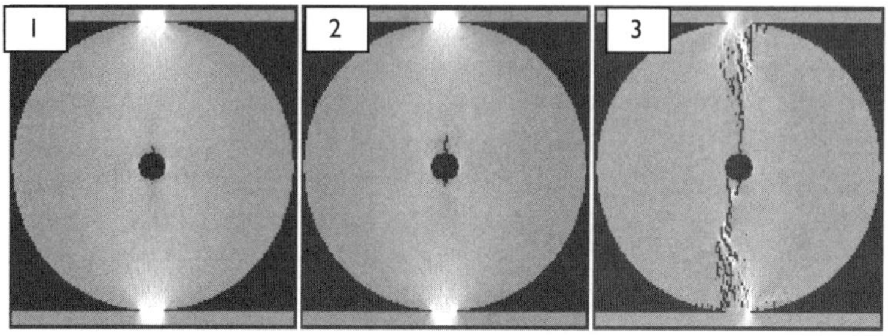

46 Rock failure mechanisms

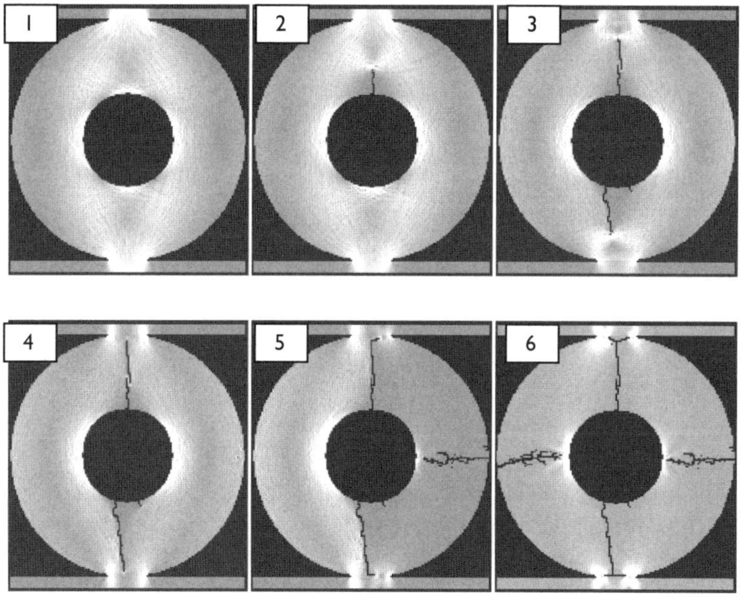

Figure 3.14 Numerically obtained typical failure processes of two types (indicated by (a) and (b)) for ring specimens (specimen size D is 150 mm). In this Figure the gray degree indicates the magnitude of the shear stress. The numbers indicate the sequence of the failure process).

causes the specimen to lose the majority of its loading capacity after the secondary crack at one edge of the ring propagates. Then, the whole specimen cannot support any more load to induce the higher tensile stress at the opposite edge.

The numerical simulations capture well the two typical failure patterns as observed in the experiments of Mellor and Hawkes (1971). For convenience of comparison, the failure patterns observed from these experiments are shown in Figure 3.16. For the same specimen geometry and load conditions, the experimental results are sometimes different (see Figure 3.16b and c). This phenomenon can also be observed in the numerical results—Figure 3.17 illustrates an example for the failure patterns of the ring test with the same specimen geometry and loading conditions. Rock is an inhomogeneous material and the wide scatter of test results concerning the fracture

Rock failure in indirect tension 47

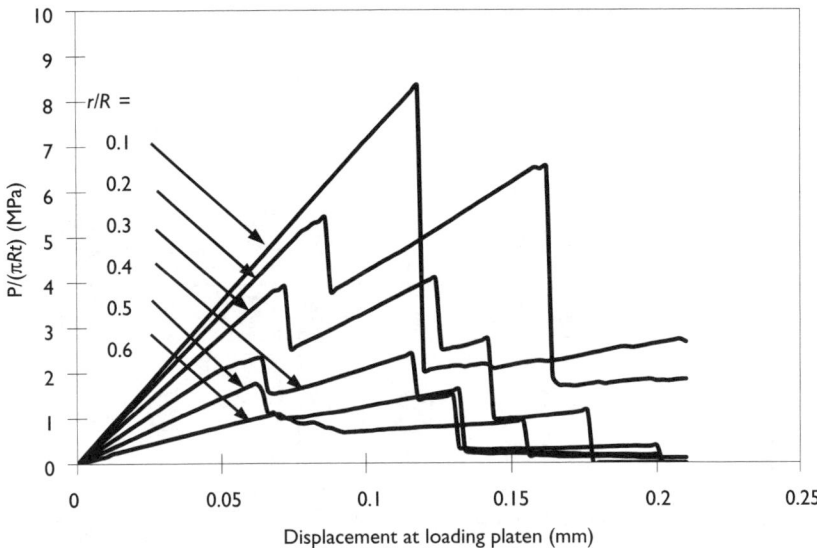

Figure 3.15 Numerically obtained load–displacement curves for ring specimens with different internal radii (the specimen size D is 150 mm and $b/D = 0.16$, with $m = 5.0$, $E_s = 62.2$ GPa and $v_s = 0.28$).

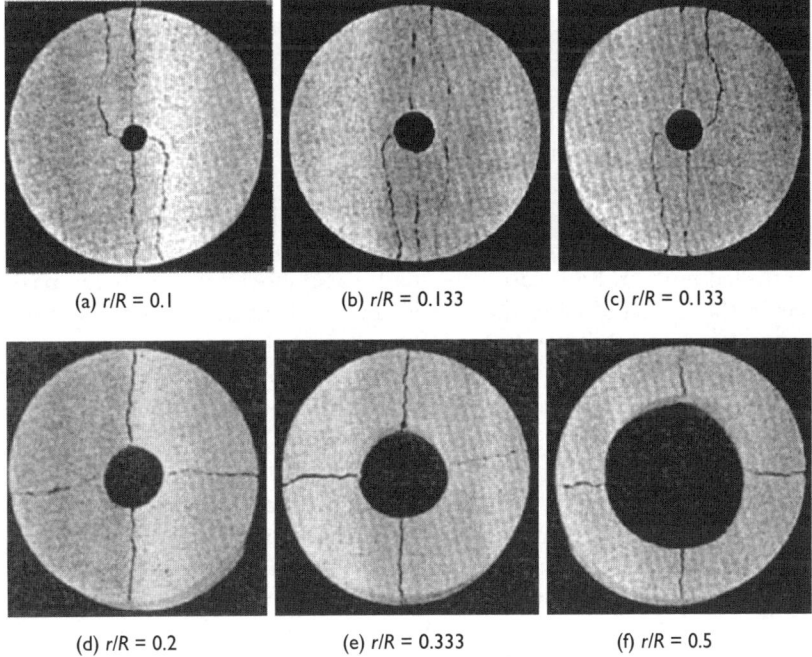

Figure 3.16 The typical failure patterns of ring specimens observed in experiments (from Mellor and Hawkes, 1971).

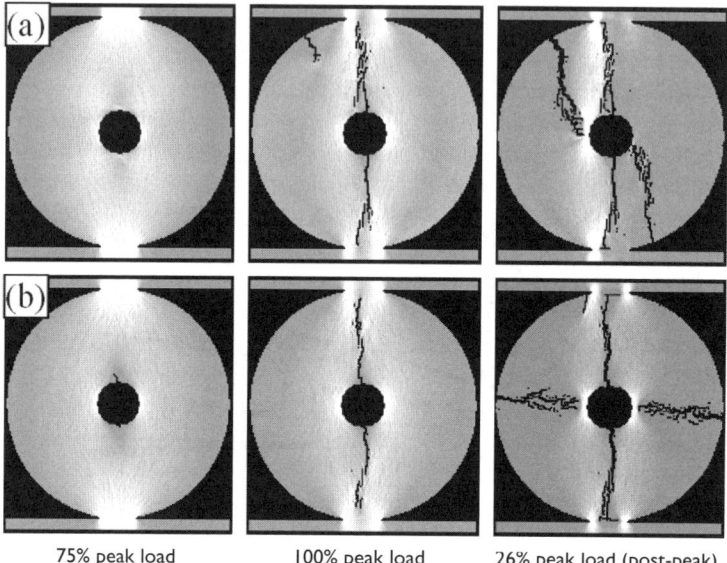

75% peak load   100% peak load   26% peak load (post-peak)

Figure 3.17 The different failure patterns for two ring specimens numerically simulated with the same geometry and loading conditions. (For these two specimens, D is 150 mm, b/D is 0.16 and r/R is 0.2).

behaviour of rock should be considered as a characteristic of its material properties, i.e. the tensile strength is not a single value but a distribution of values, which itself depends on the type of tensile test, the size of the rock specimen and other factors.

For a very small internal hole (for example, $r/R = 0.03$), failure can be initiated at different locations depending on the loading conditions, as has been observed in experiments (see Figure 3.18). In the left-hand case in Figure 3.18 (Specimen 52), the load was applied directly via steel platens; in the right-hand case (Specimen 53), sheets of copper and Teflon were inserted between the platens and the disc. During the servo-controlled tests, in both cases, the load was removed just after the peak load. Note that in the left-hand case, the disc is cracked beneath the loading points and there is no failure around the central hole; whereas, in the right case, there is no damage beneath the loading points, but an axial crack is visible above the central hole.

This phenomenon is also found in numerical simulations. For example, when $r/R = 0.01$ and $R = 150$, the size of the hole is basically commensurate with the grain size, due to the spatially heterogeneous distribution of material properties. The elements with high applied stress or low strength may damage early, which means that the primary crack may not initiate at the surface of the hole, but bypass it, as in the physical plaster specimen in Figure 3.19. The ability to vary the conditions in the numerical simulations thus provides a useful tool for sensitivity studies of the heterogeneity effect—which are much quicker than conducting physical tests.

However, when the specimen's grain size is very fine, even a small hole can affect not only the primary crack but also the secondary cracks, as in the physical plaster tests shown in Figure 3.20.

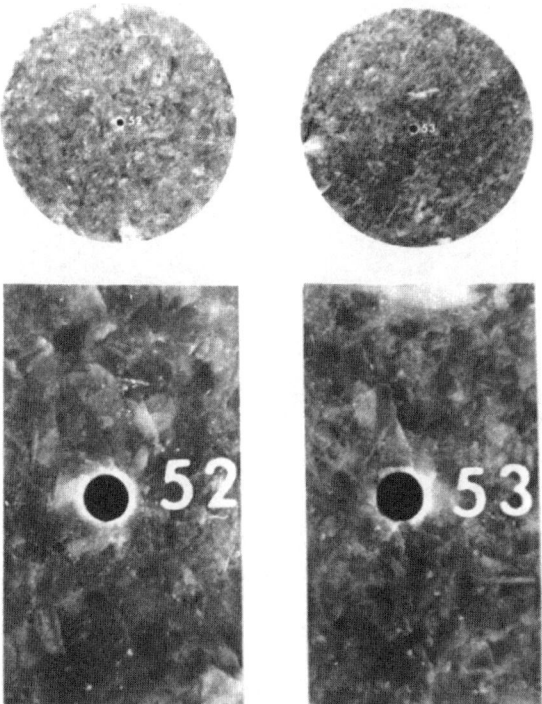

Figure 3.18 Ring tests on Tennessee marble 50 mm diameter discs with a central hole ($r/R = 0.03$) illustrating the effect of direct loading via steel platens (for Specimen 52) or cushioning the loading points (for Specimen 53) on the location of failure (Hudson, personal collection).

3.3.3.3 Effect of hole diameter on the ring test indirect tensile strength

Numerical results on the tensile strength of rings with different hole diameters and loading conditions are summarised in Figure 3.21, where ($P/\pi R t$) is plotted against r/R. For comparison, the experimental results from Mellor and Hawkes (1971) are also given in this Figure, indicating reasonable agreement between experimental and numerical results. With the decrease of the internal radius of the ring, the apparent tensile strength increases, as shown in Figure 3.21. It can be seen that there is no apparent discontinuity between the results for solid discs and results for rings with small values of r/R. However, we might expect that, below a certain hole size commensurate with the grain size, the tensile strength would remain constant. Also, with the increase of internal radius of the ring, the effect of the relative load-bearing strip width becomes negligible.

When the diameter of the hole is small (for example, $r/R = 0.1$), the load–displacement characteristic as shown in Figure 3.15 is similar to that of a Brazilian

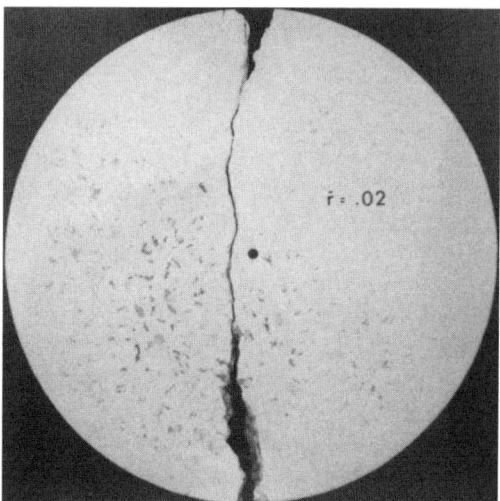

Figure 3.19 Ring test on a plaster specimen, 150 mm in diameter, containing limestone chips to control the heterogeneity, with a central hole with r/R = 0.022, in which the failure path by-passed the hole (Hudson, personal collection).

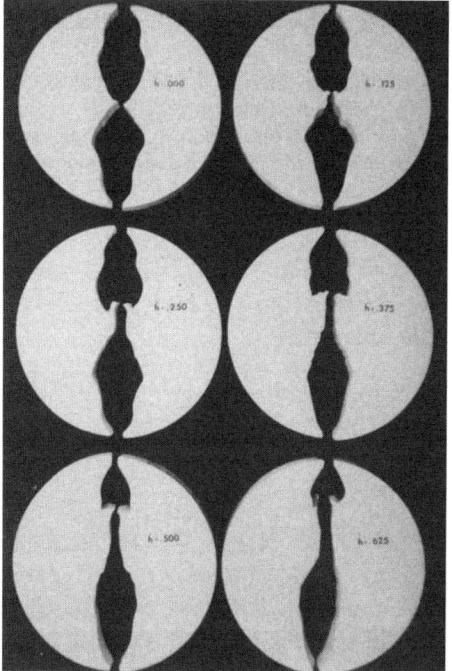

Figure 3.20 Effect of a small hole (r/R = 0.022) located at various points along the 150 mm loaded diameter of a fine-grained plaster specimen on the formation of the secondary fractures (Hudson, personal collection).

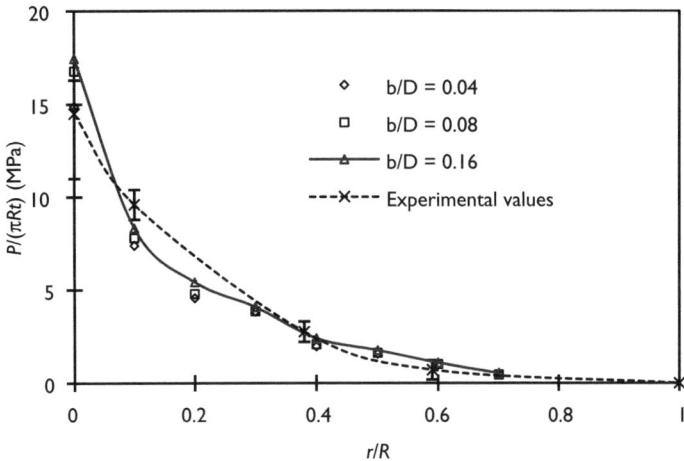

Figure 3.21 Indirect tensile strength of numerical specimens with different internal radii compared with experimental results (after Mellor and Hawkes, 1971).

Table 3.3 The ratio of σ_r/σ_t obtained from the numerical ring tests.

b/D	r/R					
	0.1	0.2	0.3	0.4	0.5	0.6
0.04	3.15	2.29	2.41	1.59	1.61	1.29
0.08	3.32	2.40	2.42	1.67	1.67	1.33
0.16	3.54	2.72	2.56	1.94	1.82	1.43

test on an intact disc specimen, the specimen losing most of its loading capacity suddenly. For larger r/R (> 0.1), there are two or more instances of stress rise and drop in their load–displacement curves, which correspond to the initiations of primary and secondary cracks in the specimen. This numerical result can be explained as follows: in the ring with the larger hole, only a small volume of material is subjected to tensile stresses that approach the peak stress. Conversely, in the disc a considerable volume of material is subjected to the peak tensile stress and there is no steep stress gradient in the critical zone (Figure 3.4). Therefore, under this loading condition, the unstable failure of the disc specimen is caused by failure of many elements at the same loading level.

The tensile strengths of these ring specimens can be obtained from Equation (3.3). In Table 3.3, the values of σ_r/σ_t are listed, from which we find that the ring tensile strength, σ_r, is much higher than the uniaxial tensile strength, σ_t (σ_t = 15.0 MPa) for this rock.

For the calculation of the tensile strength, it is assumed that the crack initiates when the peak load is attained. Therefore, from this peak load, the maximum tensile stress at the surface of the centre of the hole is calculated and considered as the tensile strength of the rock. It is assumed that the rock has attained its peak loading capacity and loses this when the crack initiates at the surface of the hole. But when the crack initiates near the surface of the hole, the stress concentration is released and the stresses in the specimen are redistributed to attain a new equilibrium condition—and it may be possible to then increase the load applied to the specimen. Thus, the initiation of a fracture at the surface of the hole when the local stress reaches the local tensile strength does not mean that the peak load of the specimen has necessarily been fully developed. This is the reason why the tensile strengths, σ_t, established from the experimentally determined peak load and based on Equation (3.3) are higher than the direct tensile strength of this rock. Therefore, it may not be acceptable to calculate the tensile strength according to the theoretically obtained relation between maximum tensile strength on the surface of the hole and the externally applied peak load P.

Nevertheless, the stress when the fracture initiates at the surface of the hole (referred to as the cracking stress and denoted as σ_{rc} hereafter) may be related to the tensile strength of the rock. If this cracking stress can be obtained from experiments, then the tensile strength of rock may be obtained. From our numerical simulations, the cracking stress of each ring specimen can be obtained, and then the cracking load instead of the peak load is substituted into Equation (3.3) to calculate the tensile strength, with the results listed in Table 3.4.

The tensile strengths obtained by this method are much closer to the direct tensile strength of the numerical specimen, especially when the internal hole diameter is relatively large (for example, $r/R = 0.4 - 0.6$). The majority of values of the tensile strength listed in Table 3.4 are also greater than the uniaxial tensile strength of rock, which may be caused by the effect of the load-bearing strip width as discussed earlier. Because of the small critically stressed volume, the cracking load is closely related to the local mechanical properties (usually at the surface of hole in the ring); therefore, the strengths obtained show considerable scatter.

From the above numerical simulations, it is found that σ_d and σ_{rc} are close to the direct uniaxial tensile strength σ_t, and the latter value is slightly greater than the former values. By contrast, the Brazilian test utilising the disc specimen is preferable to the ring test for determination of the tensile strength of rocks. Moreover, it can be concluded that the indirect tensile strength determined with the Brazilian test and ring test is significantly affected by the specimen geometry, specimen size and loading conditions. For this reason, Hudson et al. (1972) pointed out that neither the Brazilian

Table 3.4 The ratio of σ_{rc}/σ_t obtained from numerical ring tests.

b/D	r/R					
	0.1	0.2	0.3	0.4	0.5	0.6
0.04	1.82	1.92	0.92	1.09	1.08	1.19
0.08	1.76	2.06	1.91	1.28	0.99	1.09
0.16	1.69	2.03	1.86	1.24	1.00	1.08

test nor the ring test is recommended as a method for measuring the material property 'tensile strength' of rock.

Indeed, for any testing method for determining the tensile strength of rock (including the uniaxial tensile strength test itself), the obtained values are unavoidably affected by the specimen geometry, specimen size, loading conditions and other factors such as temperature and water content, so that the obtained values should only be used as index properties and are not, strictly speaking, material properties *per se*. However, the values of tensile strength determined in a site investigation using the Brazilian test or the point-load apparatus are often only required as index properties anyway so, providing a constant test procedure is used, the results are useful for characterising the intact rock quality.

Chapter 4

Rock failure in uniaxial compression

4.1 INTRODUCTION

The uniaxial compressive strength of a rock is defined as the value of the peak stress sustained by a rock specimen subjected to uniaxial compression, denoted usually as UCS or σ_c and expressed in MPa. It is the maximum load supported by the specimen during the test divided by the cross-sectional area of the specimen. Thus, the complete stress–strain curve in uniaxial compression and the associated compressive strength are another manifestation of the generic complete force–displacement curve highlighted in the Explanatory Notes at the beginning of the book.

The compressive strength suffers from the same problems as the tensile strength discussed in the previous Chapter in that the value obtained from testing depends on the size, shape and other conditions of the test procedure. Since a material property should not depend on the specimen geometry and loading conditions, we have also to regard the compressive strength as an index or experimental property rather than a material property. For this reason, it is useful to understand why this variability in the compressive strength occurs.

The features of a typical complete stress–strain curve for rock are shown in both Figure 4.1a for an idealised curve and in Figure 4.1b for an actual test on a marble specimen. Initially there is a non-linear region, termed 'bedding down' which is caused by any irregularities in the precise alignment of the loading platens and the closure of cracks in the specimen. Then, there is an essentially linear portion as the rock specimen is compressed elastically. The Young's modulus and Poisson's ratio are determined during this linear portion. Before the peak stress, cracks initiate and propagate in a stable manner and this progressive microstructural breakdown continues throughout the remainder of the complete stress–strain curve. The peak of the curve defines the peak stress and hence the uniaxial compressive strength. At any time in the post-peak region, the specimen can be unloaded and then re-loaded to obtain the local secant Young's modulus of the failed rock. Note that in the pre-peak region, a tangential Young's modulus can be obtained but this does not make sense in the post-peak region where the slope of the curve is negative.

In Figures 4.1a and b, the post-peak curve follows a smooth descent but it is usually somewhat irregular due to the formation of a series of larger cracks. Also, and with reference to Figure 4.1, it is important to note that the strain axis is termed the independent variable and the stress axis is termed the dependent variable. This is

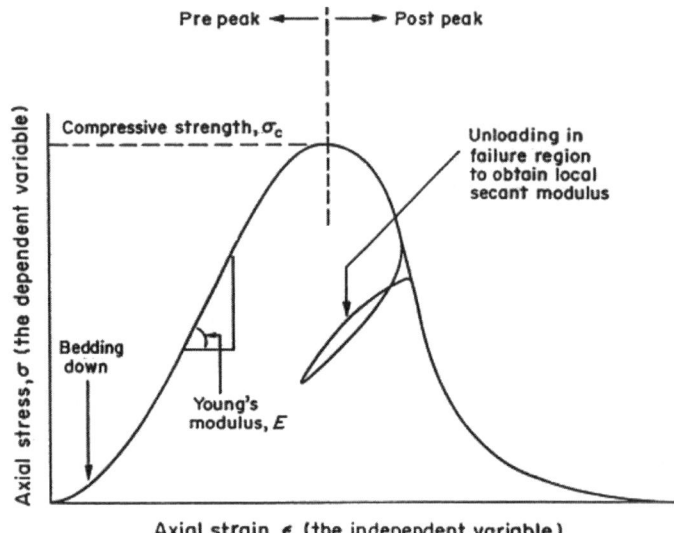

Figure 4.1a Idealised complete stress–strain curve for a rock specimen loaded in uniaxial compression.

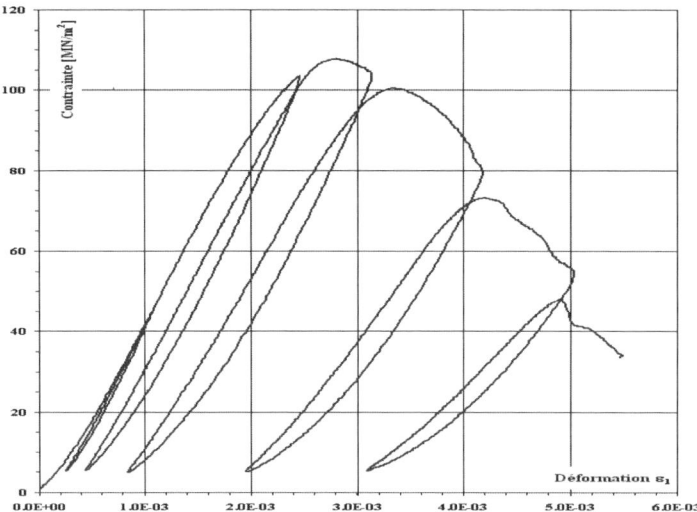

Figure 4.1b Complete stress–strain curve for a marble specimen with unloading and re-loading hysteresis loops (courtesy of EFPL, Lausanne).

because it is the strain that is the controlled variable and the stress that is the result. The complete stress–strain curve cannot be obtained with stress as the independent variable because any increase in the stress after the peak stress, i.e. the compressive strength, would result in violent uncontrolled failure.

Though this mode of failure has been studied in detail for at least five decades, the details of the failure mechanisms, including the micro-fracture initiation, propagation, coalescence, axial splitting, shearing, etc., are still not 100% understood and remain a subject of considerable scientific interest. One of the key questions relates to when the incipient fracture plane can be recognised during the loading stage of an initially intact rock, and at what point does fracture interaction overwhelm the local variations in the stress field or in the local properties and cause the system to fracture either by coalescence of fractures or by extension of one fracture (e.g. Blair and Cook, 1998). The complexity of the failure mechanism can be seen in the section of a Tennessee marble specimen in Figure 4.2.

Although many experimental studies have been conducted to establish the influence of various factors on the compressive strength and the complete stress–strain curve in uniaxial compression, theoretical and numerical studies containing parametrical analyses of the effects of, for example, end constraint of the loading platens and the geometry of the specimens on the strength values and the pre- and post-peak behaviour have been less frequent. As we have seen in the previous two Chapters, numerical models do provide a useful complement to experimental and analytical studies, especially since the failure mechanisms can be studied in detail and sensitivity studies can be carried out rapidly. In this Chapter, numerical studies of uniaxial compressive tests on specimens of inhomogeneous rock material are presented to explain and illustrate the rock failure mechanisms involved.

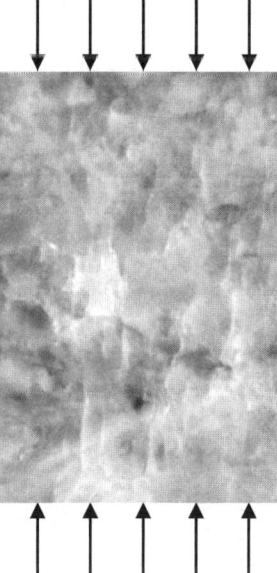

Figure 4.2 Sectional portion of a Tennessee marble specimen unloaded just after the peak of the complete stress–strain curve in uniaxial compression—illustrating the development of axial cracks which are the pre-cursor to the eventual coalescence of micro-fractures and the specimen collapse. (Specimen width 25 mm.)

4.2 NUMERICAL ILLUSTRATIONS OF ROCK FAILURE IN UNIAXIAL COMPRESSION

4.2.1 Model description

Two types of specimen were numerically tested:

1. one specimen was used to investigate the general behaviour of rock failure subjected to uniaxial compression, and
2. five specimens of the same size but with different homogeneity indices were used to study the effect of the rock heterogeneity.

The specimen geometry is 150×100 mm and has been discretised into a 150×100 (15,000 elements) mesh (Figure 4.3). The mechanical properties for all the specimens are as follows: homogeneity index, m, 1.5, 2, 3, 5, 10; mean compressive strength $\sigma_0 = 200$ MPa; mean elastic modulus $E_0 = 60$ GPa; tension cutoff $\lambda = 10\%$; Poisson's ratio $v = 0.25$; frictional angle $\phi = 30°$.

In all cases, the specimens undergo plane strain compression, imposed by a relative motion of the upper and lower loading platens. Although the specimens are heterogeneous on the micro-scale as shown in Figure 4.3, they are statistically homogeneous on the macro-scale because the mechanical property parameters are randomly distributed throughout the whole specimen. In the study, a 'smeared failure' approach is used to predict the occurrence of micro-fracturing when the stress state of an element satisfies a strength criterion. The Coulomb criterion envelope with a tensile cut-off (e.g. Brady and Brown, 1993) is used so that the failure of the elements may be either in shear or in tension.

An external displacement is applied at a constant rate of 0.002 mm/step in the vertical direction and the stress and the deformation in each element are determined. The external displacement is then increased step by step. At each step, the stress states in some elements may satisfy the strength criterion. These elements are damaged and weakened according to the rules specified by the strength criterion. The stress and deformation distributions throughout the specimen are then adjusted instantaneously to reach the equilibrium state. At locations with increased stresses due to stress redistribution, the stress state may reach the critical value and further elemental failures

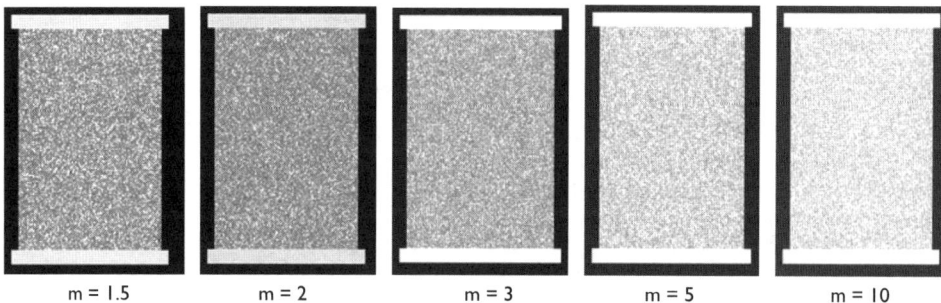

Figure 4.3 Inhomogeneity of the initial five specimens used in the numerical studies.

occur. The process is repeated until there are no longer any failed elements. The external displacement is then increased. In this way the system develops the macroscopic fractures. Energy stored in the elements during the loading process is released as acoustic emissions through the onset of element failures. Owing to the stress redistribution and the long-range deformation induced interactions, a single important element failure may induce a cascade of additional failures in neighbouring elements leading to a chain reaction releasing more energy.

4.2.2 Numerical simulation results

The stress–strain relation for the numerical specimen with $m = 1.5$ is shown in Figure 4.4. The specimen was numerically shortened at a constant displacement rate of 0.002 mm/step, although during the non-linear stages these rates were increased fractionally due to the relaxation of the loading platens. The modelling results in a non-linear complete stress–strain curve similar to the typical curve for a brittle rock shown in Figure 4.1a except that the curve does not have the initial 'bedding down' portion and is visibly more 'jagged' due to the more significant fracture coalescence events. It is also similar to the complete stress–strain curves observed in laboratory tests (e.g. Wawersik and Fairhurst, 1970) when the complete stress–strain curve was initially obtained (in a thermally controlled stiff testing machine, see Section 4.4.4 in this Chapter). It is found that, although the individual elements in the numerical model are essentially brittle (with only 10% residual strength), a substantial non-linearity exists before the maximum stress as illustrated in Figures 4.1a and b and 4.4.

It can be seen from Figures 4.4 and 4.5 that the initial part of the complete stress–strain curve is essentially linear, noting that the intial curvature shown in Figure 4.1 is absent because, as mentioned earlier that is due to 'bedding down' of the specimen which does not occur in the simulation. Approaching the peak stress, the curve deviates from linearity and then, after the peak, becomes irregular as the micro-structure breaks down. The two lower plots, (b) and (c), in Figure 4.5 allow comparison with

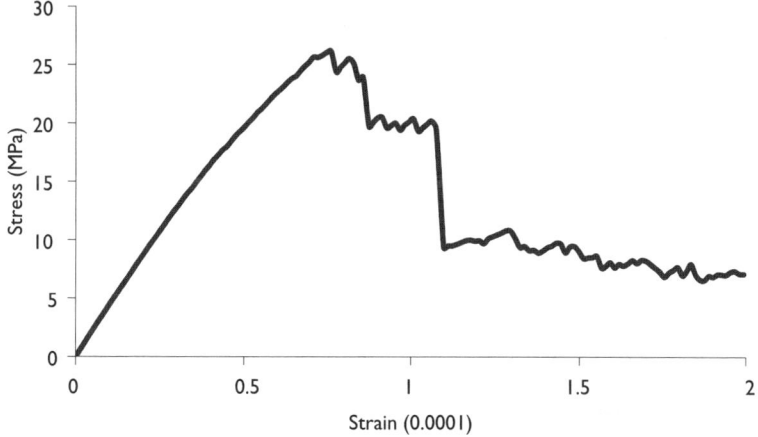

Figure 4.4 Simulated complete stress–strain curve for the $m = 1.5$ specimen subjected to uniaxial compression with an axial displacement rate control of 0.002 mm/step.

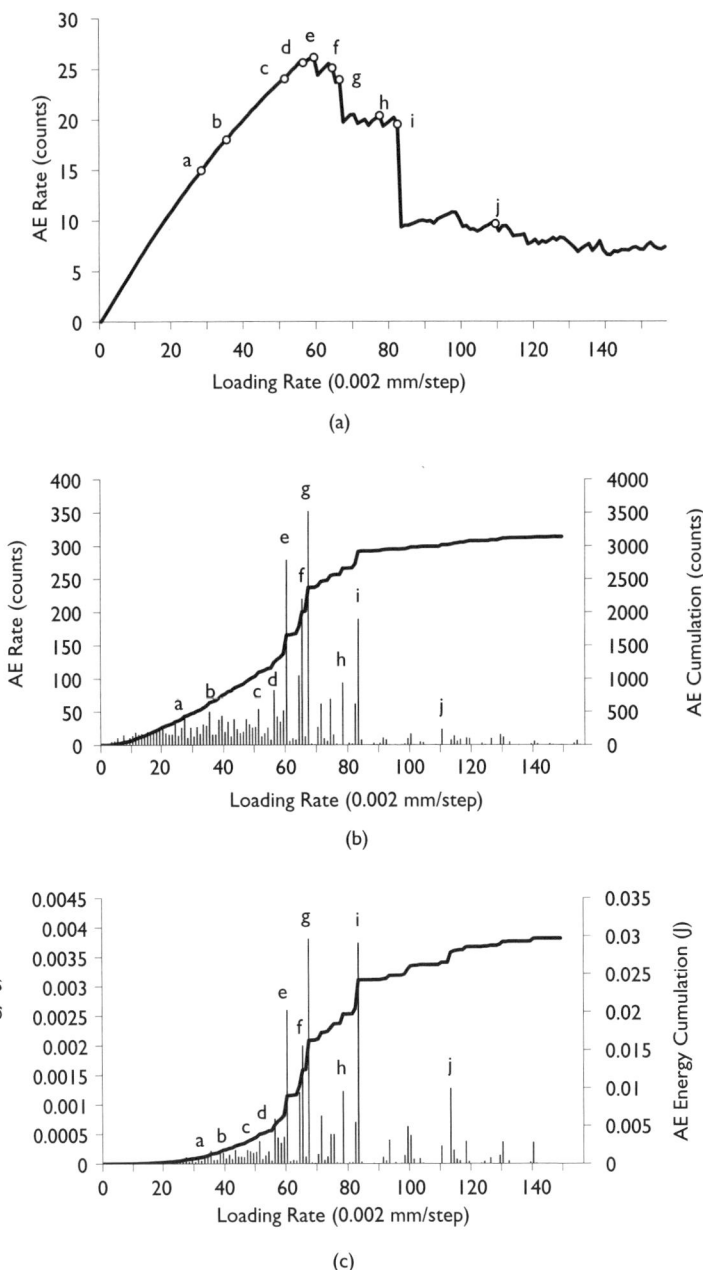

Figure 4.5 Simulated acoustic emission (AE) results plotted vs. loading rate for the $m = 1.5$ specimen shown in Figure 4.3: a) nominal stress (top graph); b) AE event rate and AE event accumulation (middle graph); and c) AE released energy and energy accumulation (lower graph).

Rock failure in uniaxial compression 61

the instantaneous values of two other parameters extracted from the simulation: the acoustic emission (AE) counts (event rate) and the associated energy release.

We note that

1 during the initial deformation or linearly elastic phase (marked with symbol 'a' on the curves), little elastic energy was released, though some AE events occurred;
2 an increasing rate of AE events accompanied the inelastic phase (points 'b' to 'd'). This agrees with the understanding that the AE events are generated by micro-fractures that result in the non-linear deformation behaviour.

It can be seen that a large number of AE events have occurred before reaching the peak load (maximum strength) for this relatively heterogeneous specimen. It is important to note that, as shown in Figure 4.5, although more than 50% of the AE events occurred before reaching the peak load, less than 20% of the acoustic energy is dissipated during this stage. Comparison between the two upper plots in Figure 4.5 show a good relation between the stress–strain curve and the event rate. Note also that every large stress drop (points 'e', 'f', 'g' and 'i') on the stress curve as shown in Figure 4.5(a) corresponds to a high event rate and large event energy release (the two lower plots (b) and (c)). It is also found that the AE released energy seems to have a close relation with stress drop. Although the AE event count at point 'i' is just half of that at point 'f', the AE released energy at point 'i' is nearly the same as that at point 'f'. This implies that more energy is being released per event with the maximum stress drop occurring at point 'i'.

In Figure 4.6, we present the locations of AE events occurring during the stages in Figure 4.5. Each circle represents one AE event and the diameter of the circle represents the relative magnitude of the AE released energy. AE locations for events occurring during loading to 56% peak stress are plotted in Figure 4.6a. Note that events are distributed throughout the specimen, and it is difficult at this stage to predict where the macro-fracture will eventually initiate. As the overall strain is increased, some events are still occurring throughout the specimen, but more events are now clustered near the zone that seems to be the potential nucleation site for the future failure process (stage d). However, it is interesting to find that the next site for more active AE events is not in this predicted potential site but in a zone approximately perpendicular to the fracture zone formed in stage d, as shown in Figure 4.6e. This is believed to be the result of stress redistribution or stress migration from the stress release area to the surrounding micro-seismic inactive area, in which higher stresses are present.

Immediately after that, a large micro-seismic zone develops quickly along the diagonal line from the upper left corner to the bottom right corner (Figure 4.6f), which coincides with the early nucleated micro-fracture zone (stage d). During these stages, the event counts decrease in intensity throughout the rest of the specimen. It is important to discover, and it is undoubtedly of some seismological significance, that the highest AE event counts do not correspond to the maximum stress, but occur in the post-peak region. This phenomenon is attributed to the heterogeneity of the specimen. The AE event plots show that, although most of the AE events that occurred prior to the maximum stress are distributed throughout the specimen, the AE events along the narrow active zone dissipated most of the acoustic energy (Figure 4.6f and g).

62 Rock failure mechanisms

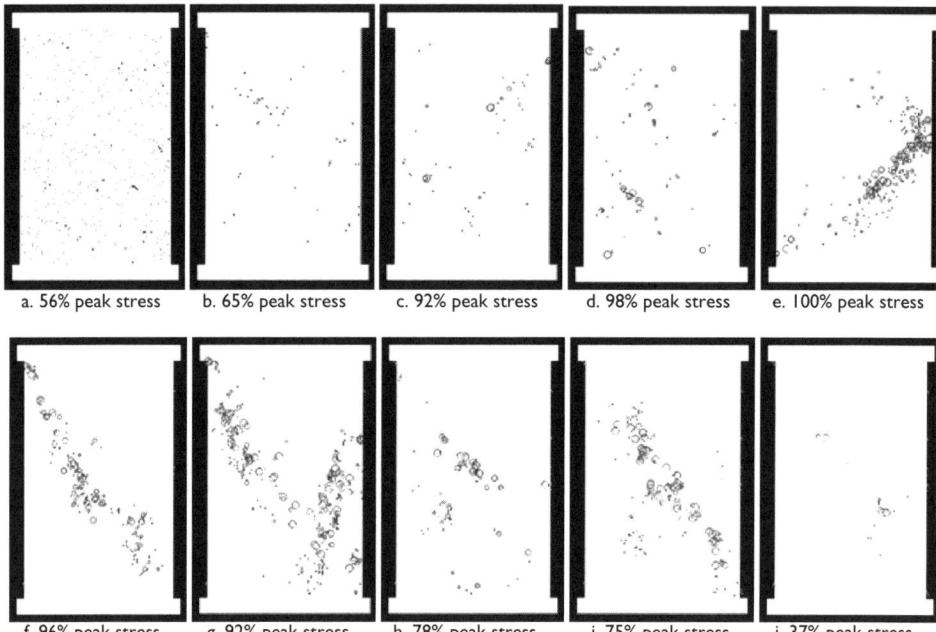

Figure 4.6 Plots of AE locations for the specimen in Figure 4.5. Each circle represents one AE event and its relative magnitude. Stress states for each plot are indicated in Figure 4.5a. Event counts and the corresponding released energy are shown in Figure 4.5b and Figure 4.5c, respectively. The gray colour in plot 'a' shows all the AE locations which occurred before this step. The black colour represents the AE locations at each current step.

Figures 4.7 and 4.8 (with the textures in the two sets of specimen pictures indicating the heterogeneity of the Young's moduli and shear stresses, respectively) illustrate the initiation, propagation and coalescence of the fractures at the different loading steps. It can be seen that the onset of failure is first indicated by the formation of a large number of isolated fractures (Figures 4.7a and 4.8a). Such local fracturing characterises the relief of stress concentration produced by the mechanical inhomogeneities in the specimen. These micro-fractures begin to cluster at stages c and d, and become clearly localised at stage e. The number of micro-fracture sites (AE events) indicates that micro-fracture interaction can cause micro-fractures to propagate and coalesce when as few as 10–15% of the elemental sites are fractured.

This is quickly followed by the development of two macroscopic fracture zones during the post-peak regions (stages f and g). Fracture strips divided by thin columns (bridges) form along the specimen diagonal and, through their propagation, create arrays of fractures. Although most fractures are distributed along these two conjugate zones, the majority of them are tensile fractures orientated parallel to the applied stress. The buckling or shearing of the bridges leads to *en echelon* fracture linkage and the forming of faults. Figure 4.8h shows that there is a highly stressed area between the two macro-fracture zones.

Rock failure in uniaxial compression 63

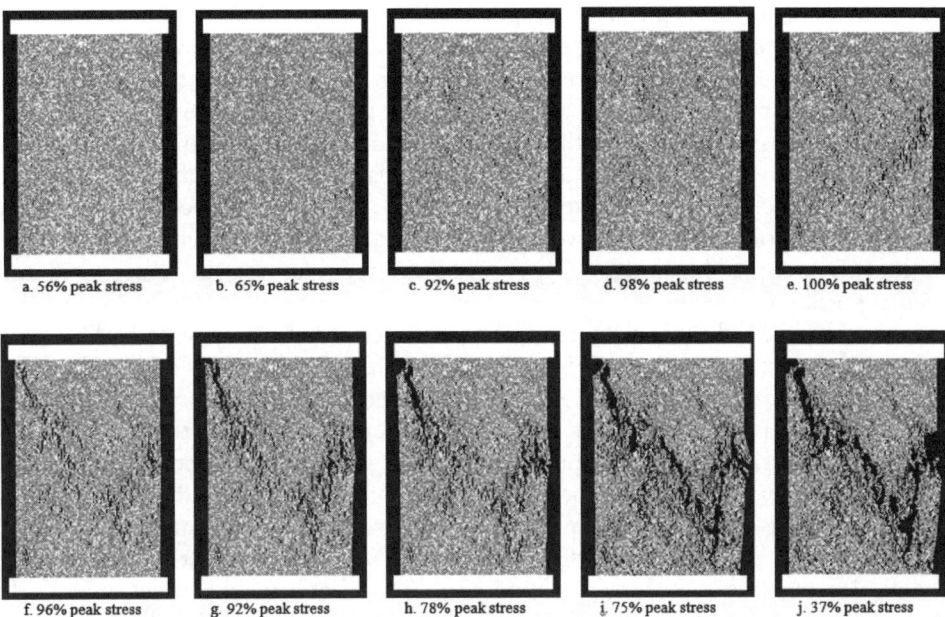

Figure 4.7 Plots of the simulated failure process corresponding to the same loading steps as in Figure 4.5. The gray colour in the images represents the Young's moduli of the elements.

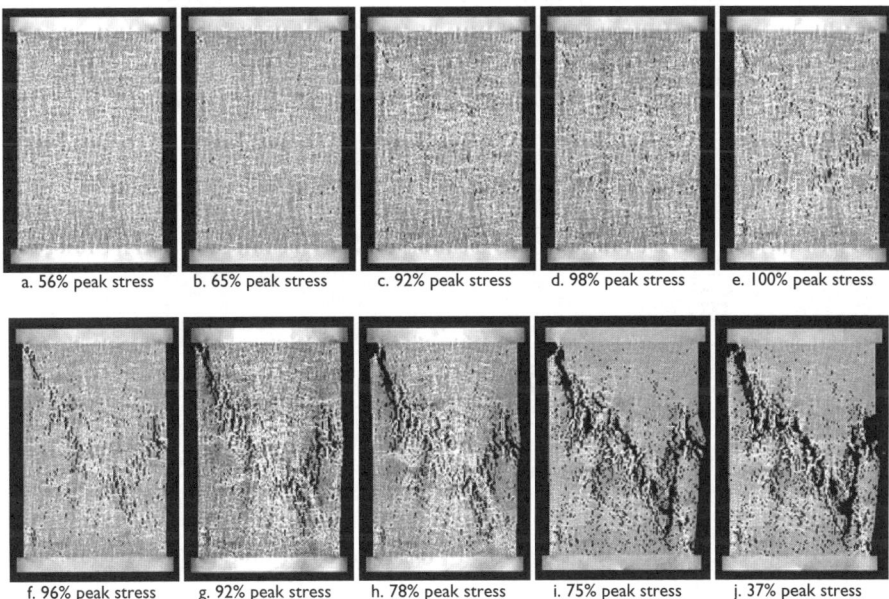

Figure 4.8 Images of the simulated failure process corresponding to the same loading steps as in Figure 4.5. The gray colour in these images represents the maximum shear stress of the elements.

64 Rock failure mechanisms

Finally, the interior macroscopic fracture zones form and become inter-connected to create a V-shaped open fracture (Figures 4.7i and Figure 4.8i). Although the fractures are in a tensile failure mode, most of the fractures are distributed in a highly stressed shear zone. The larger stress drop shown in Figure 4.5a and the higher energy release shown in the lower plot at point 'i' also provide evidence to indicate that most of the tensile fractures at this stage are distributed in a highly stressed shear deformation zone. Since the fractures along this diagonal line become dominant, the often observed cones are not developed at the two ends of the specimen, but rather the specimen ultimately fails in two parts along a diagonal plane, with a block at the bottom right corner breaking away from the specimen (as seen in Figures 4.7j and Figure 4.8j).

As mentioned, there is a strong relation between the sharp increase of AE events and the stress drop, as shown in Figure 4.5. The reason for this stress drop is the accumulation of fracture propagation. After a fracture propagates, stress relaxation occurs and the fracture propagation stops. Following an increase in loading, the concentration of stress at the fracture tip re-attains the local rock strength and fracture propagation is reactivated, and so on. The process continues until interaction with other fractures occurs and more complex failure patterns form, resulting in a large stress drop.

In Figure 4.9, the associated displacement vectors are depicted. It can be seen that the displacement vectors start becoming non-uniform at stage e and become clearly non-uniform at stage g. At stage i, clear faulting is observed. The block at the lower right corner starts to break away from the specimen at this stage, becoming more obvious at stage j.

From the illustrations of this simulation, we can see the mechanisms of failure that occur when a specimen of inhomogeneous rock is uniaxially compressed throughout

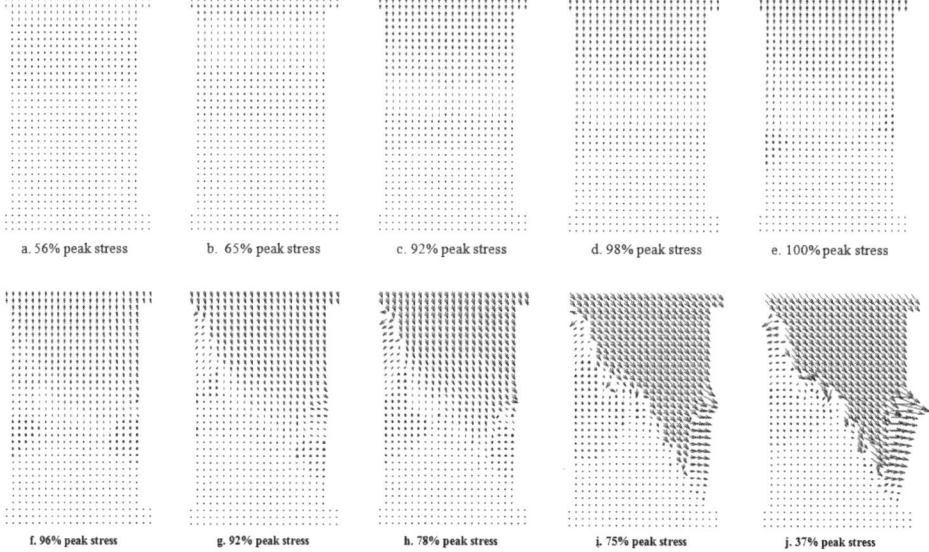

Figure 4.9 Plots of simulated displacement vectors corresponding to the same loading steps as in Figure 4.5. Stress states for each plot are indicated in Figure 4.5.

the complete failure process. It should be remembered that the complete stress–strain curve can only be obtained through controlled loading, i.e. by using a stiff or servo-controlled testing machine in the case of physical specimens, and through the stepwise loading in the simulations. In a real engineering circumstance, the post-peak portion of the complete stress–strain curve can be masked through sudden failure of the rock component—because of 'soft' loading, for example dead weight loading in which the load remains the same after the peak of the curve feeding potential energy into the system and causing 'explosive' failure.

Comparing the stress–strain curve shown in Figures 4.4 and 4.5 and the failure processes shown in Figures 4.6 to 4.9, it is apparent that attaining the maximum stress, i.e. the uniaxial compressive strength, does not necessarily mean abrupt failure of the complete specimen. The stress–strain curve becomes non-linear when micro-fractures begin to occur in the specimen (Figures 4.7a–c and 4.8a–c). At point 'e', the specimen reaches its maximum stress, the peak of the stress–strain curve. Although the load bearing capacity drops significantly from points 'f' to 'h' (Figure 4.5), this does not necessarily indicate complete collapse of the specimen. The specimen does rapidly lose its load bearing capacity, yet the fractures do not propagate drastically until point 'i'. Subsequently, a transition in failure mode occurs and the loading capacity begins to decrease at a much slower rate until it reaches its residual strength, which is about 25% of its maximum strength. This is higher than the residual strength defined for the individual element's constitutive law (10% of the maximum element strength).

4.2.3 Summary of the numerical simulation observations

4.2.3.1 The complete stress–strain curve

As the axial strain is increased, the stress–strain curve is nearly linear in the initial stage. When the load reaches approximately 50% of the peak load, due to the fact that micro-fractures begin to nucleate, the tangential stiffness of the specimen decreases and it reaches zero at the compressive strength. Beyond that point, progressively less and less stress is developed until the macro-fractures occur and the specimen loses most of its strength. Several observations can be made from this curve (Figure 4.4):

- the Young's modulus obtained from the linear portion of the stress–strain curve is 45.9 GPa, which is 24% lower than the mean value of 60 GPa for the elements. In addition, the peak stress is 26 MPa, which is much lower than the mean value of 200 MPa for the elements. The large difference between the specimen strength and the mean value of the constituent elements occurs because the elements with lower strength will fail during loading;
- the complete stress–strain curve for the specimen deformation demonstrates strain-softening behaviour;
- the unstable failure associated with large stress drops is well represented in the simulated curve; and
- the residual strength of the specimen is simulated.
- these results agree with experimental observations [e.g. Wawersik and Fairhurst (1970); Wawersik and Brace (1971)].

The geometry of the stress–strain curves as a function of the heterogeneity of specimens was also evaluated using five specimens with different homogeneity indices. Figure 4.10 depicts the stress–strain curves for the five simulated specimens and Figure 4.11 the related AE events. It is clear that the stress–strain relation and the strength characterisation depend strongly on the heterogeneity of the specimen material. It can be seen that, under uniaxial compression, the shape of the stress–strain curve for a heterogeneous rock has a gentler post-peak behaviour, as was observed in Chapter 2 for the purely tensile case, *cf.* Figure 2.13.

The maximum strength of the specimens is also related to the homogeneity index. The higher the value of the homogeneity index, i.e. the more homogeneous the material, the higher will be the strength of the specimen. The curve becomes more linear and the strength loss is also sharper in the homogeneous case. In the limit, if all the elements had the same modulus and strength, they would all fail simultaneously, *cf.* Figure 2.14.

An interesting feature of the numerical results is that the heterogeneous specimens, although having lower peak strength, as a rock sample are statistically more 'reliable' than homogenous specimens. An explanation for this can be found in the way that defects affect the damage tolerance of a rock. Although such rocks fail at lower peak loads, they typically undergo much greater and more stable damage accumulation before eventual failure because energy is dissipated continuously from an early stage in a controlled process. In contrast, rocks with relatively uniform strength, i.e. with very few defects, are prone to sudden brittle failure. Failure of such a material may often be governed by a single critical defect; once the failure strength is reached

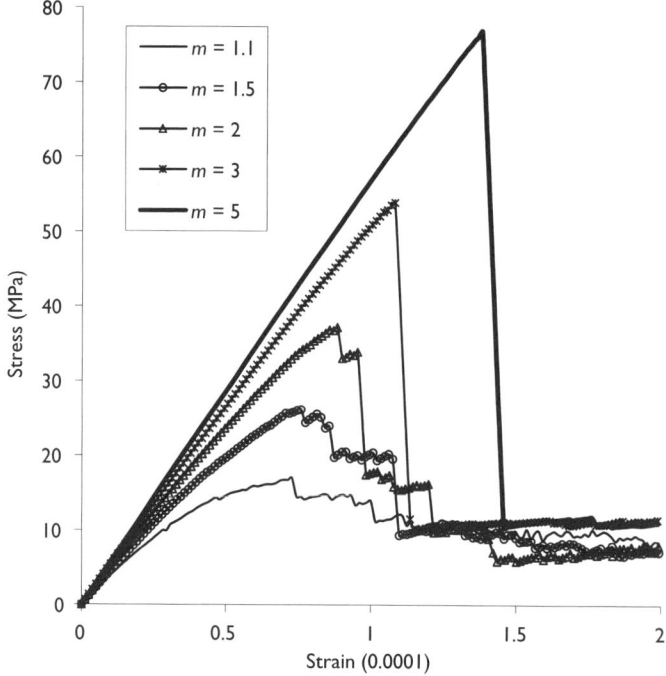

Figure 4.10 Influence of material heterogeneity on the stress–strain curves for five specimens with different homogeneity indices, m.

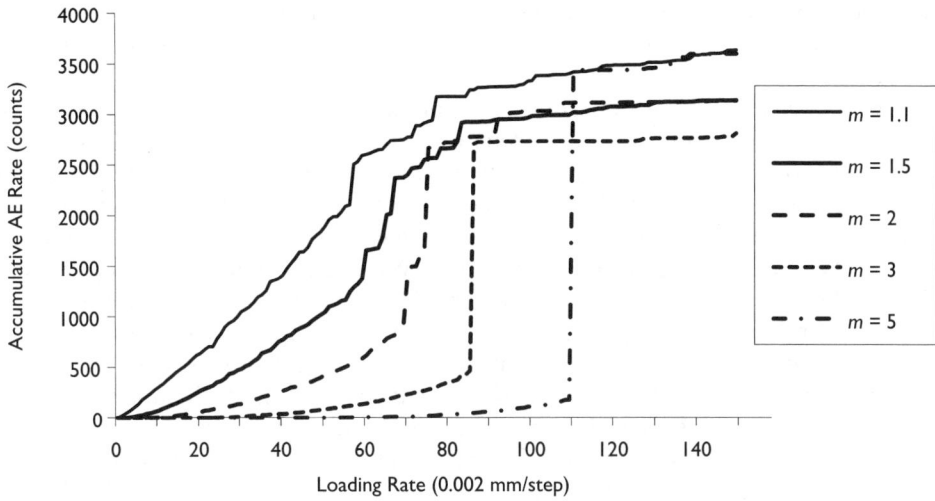

Figure 4.11 Influence of material heterogeneity on the cumulative AE events for five specimens with different homogeneity indices, m.

at the defect, the initiated crack grows unimpeded. The simulations presented in this Chapter will illustrate this point.

4.2.3.2 Acoustic emission (AE) events and their locations

The monitoring of acoustic emission activity produced during deformation is one of the well-known methods for observing damage. A variety of micro-mechanisms, including dislocation activity, twin formation and micro-fracture formation, cause acoustic emissions. In brittle materials such as rocks by far the most important of these is micro-fracture formation. There is a relation between the AE events and micro-fracturing, but the accuracy of this technique for assessing the damage depends on the AE record and the relation assumed for the AE data and fracture geometry (Cox and Meredith, 1993). In the numerical simulation code used to create the illustrations in this Chapter, every elemental failure is considered to be the source of an acoustic event because the failed element must release the elastic energy stored during its deformation. Therefore, by recording the number of failed elements and the associated amount of energy release, the code is capable of simulating the AE activities, including the AE event rate, magnitude and their locations (Tang and Kaiser, 1998).

As we have discussed, the heterogeneity of the rock has an important effect on the shape of the stress–strain curves and strength characterisation; also, the numerical results demonstrate that the AE event patterns are influenced greatly by the degree of heterogeneity of the materials. Figure 4.11 shows that the relatively heterogeneous specimen emits more AE events as precursors to macro-fracture than those of the relatively homogeneous specimen. In Figures 4.5 and 4.6, the simulation results for the relatively heterogeneous specimen (with the low homogeneity Weibull parameter $m = 1.5$) show that the macro-fracture nucleation involves the occurrence of a large number of small magnitude events. In contrast to this specimen, however, and except for some randomly distributed AE events, very few localised events occur during the early stage of loading for the relatively homogeneous specimens with the parameter $m = 5$, Figure 4.11.

These results suggest an interesting new approach for the prediction of failure. There are no obvious nucleation sites for a strongly homogeneous rock, such as the specimen with $m = 5$, prior to the peak load. Once the fracture nucleation occurs, further growth of the fracture can be rapid and uncontrolled and the failure process would be difficult to predict. On the other hand, the results for the specimen with $m = 1.5$, showed that a relatively heterogeneous rock would generate localised zones of intense AE activity well before collapse and hence that unstable failure could then be predicted in advance.

However, in physical experiments, even for a heterogeneous rock specimen, few AE events could be detected at the beginning of the loading stage before the stress state reaches a certain level. The reason for this is that, in most cases, the noise is filtered and a threshold level is set which cuts off emissions with energy less than a prescribed level. In Lockner et al.'s experiments, he found that not until the onset of dilatancy at 50–60% peak stress did significant AE activity occur (Lockner et al., 1992). Taking the specimen ($m = 1.5$) as the example, although the AE *count* accumulation is higher during the loading stage of the pre-peak region, the corresponding AE *energy* accumulation is very low (Figure 4.5). Therefore, these low energy level emissions would be difficult for equipment to detect. As noted before, the reason for lower energy release of AE events in the initial loading stage is simple: most of the elements that failed at this stage have lower strength than those that failed at a later stage.

4.2.3.3 Stress distribution and failure-induced stress redistribution

In a conventional physical experiment, it is difficult, if not impossible, to measure the stress distribution throughout the specimen. One of the important advantages of simulation is that we can obtain detailed information about the internal stress distribution, including failure-induced stress redistribution (Figure 4.8). This provides an opportunity to investigate the effect of fracture initiation, propagation and coalescence on the stress distribution, permitting us to have a better understanding of the formation of macro-fractures.

Although the micro-fractures shown in Figure 4.8 supply the same information regarding the failure process as in Figures 4.6 and 4.7, the stress field illustrated in Figure 4.8 sheds light on the failure mechanism within the specimen. Firstly, these Figures show that, although the stress distribution at the beginning of loading is statistically homogeneous on the macro-scale, it manifests a highly disordered state on the micro-scale—the inhomogeneity of stress being due to the micro-scale heterogeneity of the specimen. Also, the existence of developing micro-structures adds a further dimension to the heterogeneity of the specimen. The stress fields shown in these images represent the first loading stage of the specimen without any micro-fracture having previously been initiated. Therefore, the only reason for the stress inhomogeneities is the local variation of the Young's moduli within the specimen. Figure 4.8 shows how the shear stress distribution evolved as the load acting on the specimen was increased. It can be seen that the macro-fractures result in intensive stress concentrations around the fracture areas.

Two main mechanisms operated as the load was increased: i) the stress carried by the intact rock between fractures progressively increased (Figure 4.8a–h), and ii) new point contacts can transfer some stress, resulting in the establishment of residual strength of the specimen (Figure 4.8i–j). Thus, increasing the heterogeneity of the specimen increases the variation in magnitude of the local stress concentrations, which results in the macro-fractures forming at a lower loading stress level. Secondly, and as shown in Figures 4.5 and 4.8, when the loading stress reaches about 85%–95%, micro-fractures start to interact, coalesce and extend, with the material being weakened as a result. Increasing the heterogeneity of the specimen will cause stress inhomogeneity or concentration and the critical micro-fracture density will be reached at lower levels.

4.3 ROCK FAILURE MODES IN UNIAXIAL COMPRESSION

In laboratory experiments on brittle rock, two failure modes are typically observed. In uniaxial compression conditions, one or more individual micro-fractures expand into one or more large fractures (extension or splitting mode). On the other hand, the confinement conditions in triaxial tests inhibit the expansion of individual micro-fractures and lead to coalescence of a large number of micro-fractures into a shear fracture zone at higher loads. In both cases, the rock specimen is loaded through steel platens in laboratory tests and this causes a local triaxial stress state at the ends of the specimen—because the rock cannot expand laterally and hence faulting (i.e. shear planes forming cones) is induced.

As seen in Figure 4.12b, a relatively heterogeneous specimen ($m = 1.5$) manifests a more random distribution of micro-fractures and a rougher fracture surface than a

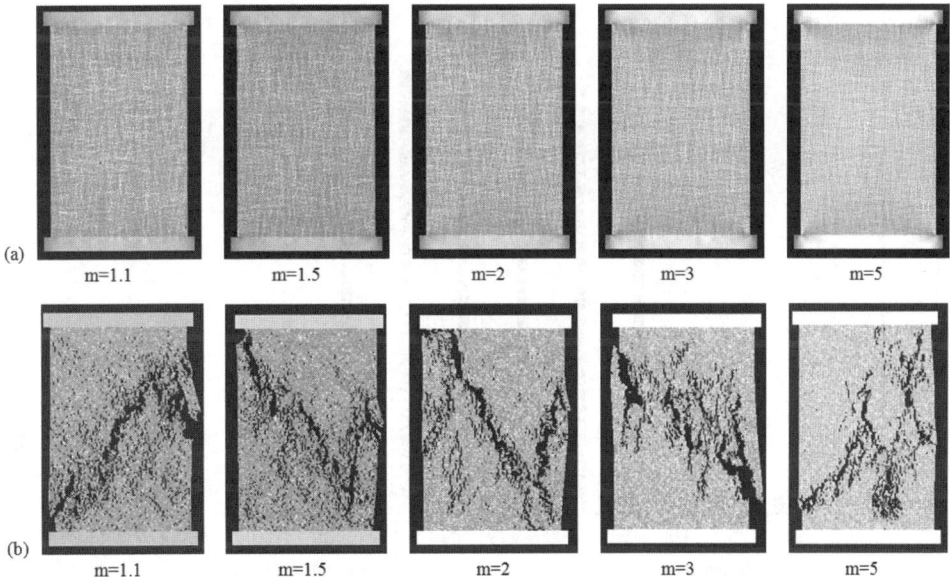

Figure 4.12 Influence of material heterogeneity on the failure modes for five specimens with different homogeneity indices; a) initial stress distributions, and b) final failure modes.

relatively homogeneous specimen ($m = 5$). Thus, the variation in micro-fracture distribution or fracture surface roughness should serve as an indicator of both the degree of the rock brittleness and the heterogeneity of the specimen.

Variation or uncertainty in test results is an important feature in laboratory experiments, and is a problem when a specific value for the uniaxial compressive strength is required during a site investigation for a rock engineering project. The International Society for Rock Mechanics (ISRM) Suggested Method for determination of the uniaxial compressive strength of rock (Ulusay and Hudson, 2007) recommends that, "The number of specimens tested should be determined from practical considerations but at least five are preferred". An illustration of the variation of uniaxial compressive strength variation of a crystalline rock formation in Finland is shown in Figure 4.13.

Because of this variation, a statistical description of the experimental results is often employed. However, a purely statistical approach should not be substituted for an understanding of the rock mechanisms: in other words, to say that the mechanism is not understood and therefore it is a random variable is now unacceptable. Today, we have at our disposal sophisticated numerical codes which allow simulation of the many aspects of rock failure, as illustrated in this book. Hence, we are able to have a much better understanding of the variation of compressive strength as a function of the micro-structural variation in the rock.

Observations from the numerical simulations presented here indicate that the macro-fractures that nucleated in different specimens having the same material parameters were strongly controlled by local variations in the mechanical properties (i.e. the local heterogeneity). Since the computer generates the mechanical parameters of the elements randomly by following the Weibull distribution, the five specimens

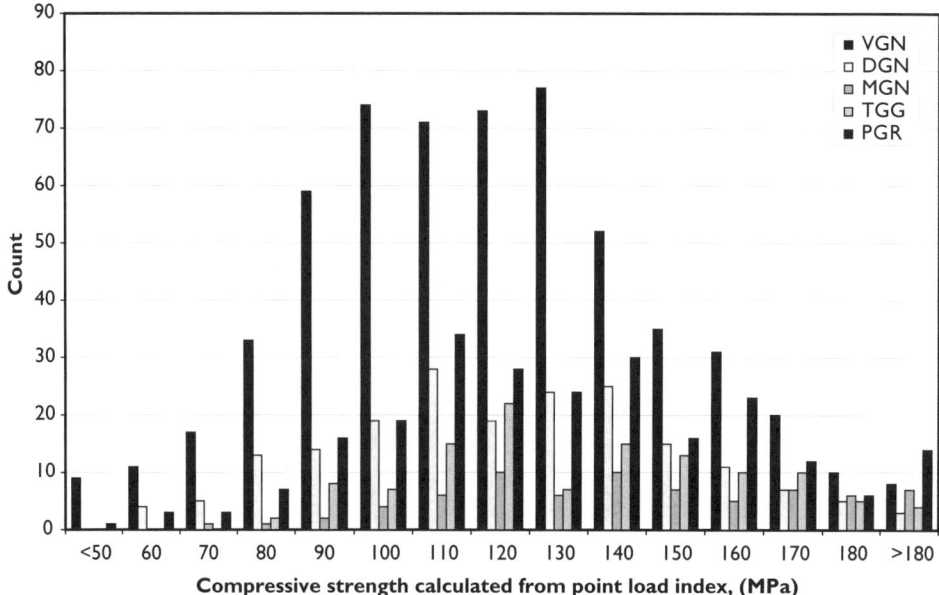

Figure 4.13 Variation of compressive strength for five rock types in the Finnish crystalline basement (from Posiva, 2009).

Rock failure in uniaxial compression 71

with the same macro-properties will have different local characteristics for individual elements. While all the specimens appear to behave similarly during the initial loading stage, considerable differences can be observed in the post-failure region.

The five specimens in Figure 4.14, all having $m = 1.5$, show somewhat similar results for the final failure modes. The results indicate that variation in failure mode is strongly sensitive to the local micro-structural variation in the specimen as soon as the micro-fracture initiates. Although the mechanical properties and the initial stress distribution for the five specimens are statistically the same on the macro-scale (as shown in Figure 4.14a), the localised zones and major fracture positions for the five specimens are different in each case; and consequently, the final failure modes are different (Figure 4.14b). However, as shown in Figure 4.14b, although the failure modes differ considerably, the stress–strain curves for the five specimens have a similar shape.

Figure 4.15 reveals that the stress–strain curves can be divided into three stages. In stage A, from the start of loading to 90% of the sample strength, the stress–strain characteristics for the rocks are approximately the same which indicates that the randomly distributed micro-fractures have little effect on the overall stress field, and that the overall deformation in the specimen is statistically uniform.

In stage B, from 90–100% of the sample strength, slight differences among the slopes of the stress–strain curves can be observed when the load is at about 90% peak load. This implies that the specimen tends to deform non-uniformly (as seen in Figure 4.8c–e) and a large number of predominantly parallel micro-fractures are nucleated. The rocks will then deform non-linearly and the strength of the specimens is affected (as seen in Figure 4.5a). Since the faulting process has not started at this stage, it is concluded that the maximum stress value in the stress–strain curve is not

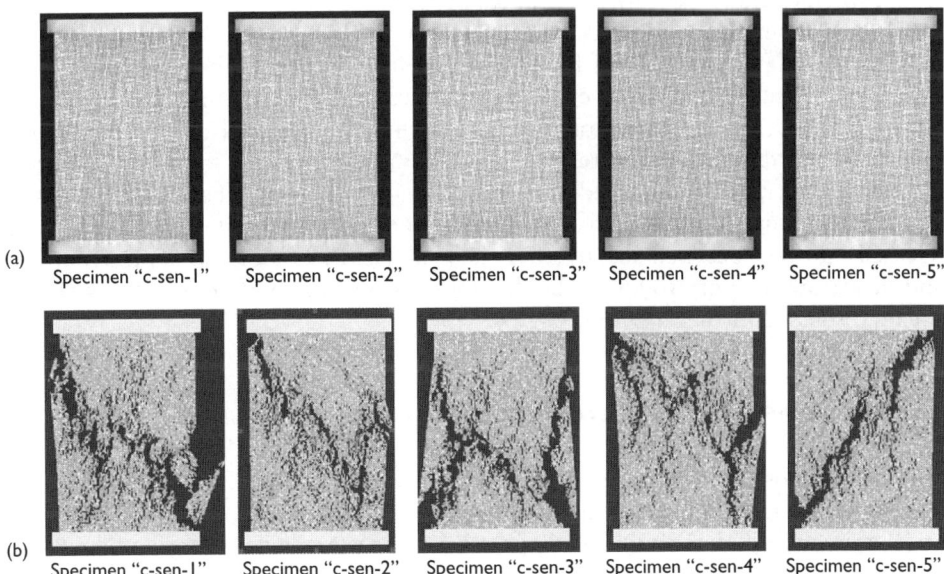

Figure 4.14 Sensitivity of failure modes to local micro-structural variation for five specimens with the same Weibull distribution generating parameter, $m = 1.5$. a) Initial stress distributions, and b) final failure modes.

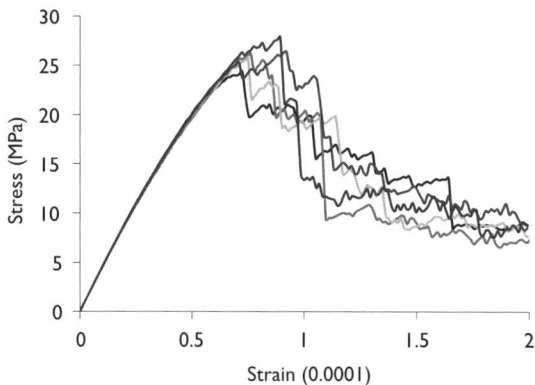

Figure 4.15 Sensitivity of stress–strain curves to the local variation in micro-structure for five specimens with the same overall statistical distribution of local mechanical properties ($m = 1.5$ for all specimens).

attributed to the faulting mechanism. This is found in physical experiments when rock specimens are unloaded at the compressive strength and sectioned to study the crack development—which show a multitude of axial cracks, as are visible in Figure 4.2.

In stage C, from the peak load onwards, significant differences between the curves are observed. The rock is in an unstable condition immediately after reaching the peak load. However, the differences in the curves tends to be smaller in the residual strength region when the failure returns to a stable process.

The experiments conducted by Wawersik and Fairhurst (1970) are in line with the above explanations. It was found from their experiments that the compressive strength of rocks is not governed by the formation of macroscopic faults, as commonly believed, but by local fracturing predominantly parallel to the direction of loading. Their experiments were some of the first in which the failure was controlled using a thermally-controlled stiff testing machine; before that, all rock testing was conducted in soft testing machines where the energy in the testing machine was released into the specimen shortly after the peak stress, resulting in explosive failure and hence the inability to study the development of failure after the rock strength. This is discussed further in Section 4.4.4.

4.4 FACTORS AFFECTING ROCK FAILURE BEHAVIOUR

One of the most important factors affecting rock failure behaviour during sample testing is the height:diameter or slenderness ratio of the specimen. Both loading platens and specimens will deform when compressive stress is applied, but the amount of the lateral expansion in the loading platens is generally different from that of the specimen due to the elastic mismatch. This loading condition causes a multi-axial stress state near the contact zone between the specimen and the loading platen, which can have a considerable effect on the overall specimen behaviour. Changes in the applied frictional end confinement will affect the stress state in the end zone and consequently the whole specimen. For example, the use of brush platens (where the platen is a set of square thin steel rods), rubber, Teflon sheets of different thicknesses, and various combinations of Teflon and metal shims can induce direct tensile stress in the specimen (Wawersik and Brace, 1971). Fracturing can take place in the highly restrained specimen ends at lower height:diameter specimen ratios, while there is a release of elastic strain energy from the unfractured specimen ends to the fractured central zone during post-peak stressing at higher ratios. In this Section, factors affecting the strength characterisation and failure behaviour of rocks, such as specimen geometry (slenderness), size and the end constraint, are numerically examined. The effects of these factors on strength, failure and post-peak characteristics are described and discussed.

4.4.1 Model description

Three types of specimens were numerically tested:

- five specimens of the same size but with loading platens having different elastic parameters were used to examine the end constraint effect;
- five specimens with different values of length to width ratio were modelled to investigate the effect of specimen geometry (the effect of specimen slenderness); and
- five specimens with the same length to width ratio, but of different sizes, were generated to study the effect of specimen size.

In all cases, the specimens undergo plane strain compression, imposed by a relative motion of the upper and lower loading platens. The mechanical properties for all the specimens are the same as in the last Section.

4.4.2 Effect of end constraint in terms of the Young's modulus of the loading platens

In theory, uniaxial compressive testing should involve subjecting the test specimen to a uniaxial stress field. However, this never occurs in practice because of the constraint applied to the specimen ends by the loading platens. There has been considerable controversy about the role of end effects in the occurrence of macroscopic brittle fractures parallel to the direction of maximum compressive loading, the so-called axial splitting or 'axial cleavage' fracture (Blair and Cook, 1998). The simulations for different platen conditions show that this type of fracture is markedly sensitive to the platen/end conditions. Different fracture patterns are obtained, from almost vertical splitting failure in the lower constraint specimens to the faulting mechanism dominating in the higher constraint specimens.

Analysis of the data and observation of the specimens during failure initiation and post-peak states leads to the failure mechanism idealised and schematically shown in Figure 4.16. The value E_p indicates the Young's modulus of the loading platens at the top and bottom ends of the specimen with a higher value of E_p representing relatively stiffer platens. E_p/E_s is the ratio of the Young's modulus of the platen to that of the specimen. Due to this elastic mismatch between the platen and the specimen, the lateral deformations in the platen and specimen are different. As a result, there will be

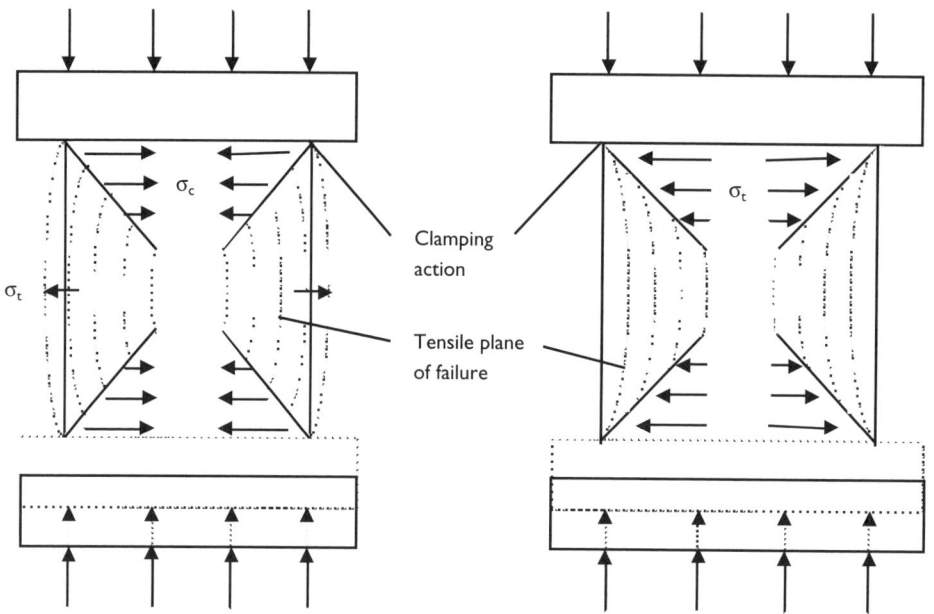

Figure 4.16 Idealised deformation, specimen–platen interaction, stress states in the specimen, and failure modes within the specimen: left, the ratio of platen modulus to specimen modulus $E_p/E_s > 1$; and, right, $E_p/E_s < 1$.

friction between the platens and specimen ends. This friction may cause either a lateral confining compressive stress at the specimen ends if $E_p/E_s > 1$ (a stiffer constraint, as shown on the left side of Figure 4.16), or a lateral tensile stress at the specimen ends if $E_p/E_s < 1$ (a softer constraint, as shown on the right of Figure 4.16).

These effects are a maximum at the ends of the specimen and taper off from the ends toward the specimen centre. So, two cones of compression develop in the specimen when there is the stiffer constraint and consequently fewer fractures are expected in these cone-shaped zones. On the other hand, two cones of tension will develop in the case of the softer constraint and more fractures are expected, resulting in a splitting failure mode in the specimen. Since the axial stress field in the specimen is thus not uniformly distributed, radial tension develops especially in the outer perimeter closer to the middle part of the specimen for stiffer constraint, and a layer in this area may buckle outward resulting in lateral tensile failure. It is predicted that the layer should be thicker in the middle and become thinner at the ends, reflecting the geometry of the two compression cones at the ends. For softer constraint, however, the situation is completely different.

Figure 4.17 provides the results for the stress distributions in five simulated specimens with the same geometrical and mechanical properties, but under different end constraint conditions. It can be seen that, except for the specimen without loading platens, the specimens in uniaxial compression are under a triaxial stress state. Note the change in the gray colour near the platens in Figure 4.17 indicating the changes in inhomogeneous stress field due to the platen effect.

Rock failure in uniaxial compression   75

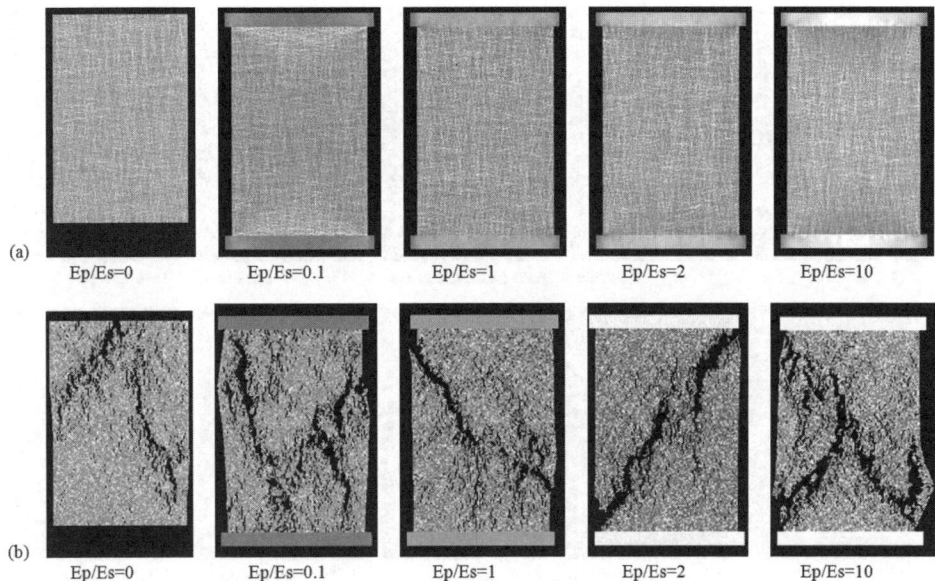

Figure 4.17 Numerical simulation of the effects of end constraint for five specimens with different loading platen Young's moduli, a) initial stress distributions in the specimen and platens; and b) failure modes.

The stress distribution results differ greatly, depending on the contact conditions between the specimen and the platens. Since the platen to specimen ratios of the Young's moduli are not equal for specimens $E_p/E_s = 0.1$, $E_p/E_s = 2$ and $E_p/E_s = 10$, in the interface areas resistance force arises and the radially directed shear stresses produced by the constraint effect are superimposed on the internal stresses within the specimen volume near the contact face. Theoretically, there are no end constraints for the specimen with $E_p/E_s = 1$. For the specimen without platens, however, the ends of the specimen are completely free of constraint (this situation is only achieved in a numerical model and is difficult to approximate in a real experiment).

Note that the specimen surface in contact with the loading platen remains almost uncracked in the case of the stiffer loading platens; whereas, a more distributed failure, with many cracks developing in the top and bottom surface as well, takes place when softer platens are applied. It is clear that the loading platens have an important effect on the failure behaviour, as observed in actual uniaxial compression tests (e.g. Cox and Meredith, 1993).

A detailed comparison of the failure process between the specimen with softer constraint conditions ($E_p/E_s = 0.1$) and the specimen with stiffer constraint conditions ($E_p/E_s = 10$) shows more clearly the effect of end constraint (Figure 4.18). For the stiffer loading platen, the lateral deformation of the specimen is restricted at the specimen ends. In this simulation, full frictional restraint is assumed, i.e., no sliding can occur at the interface between the loading platen and specimen. Thus, shear stresses develop between the specimen and loading platen, causing a two-dimensional state of stress in the specimen ends as shown in Figure 4.18a. Consequently, cracks appear to initiate in

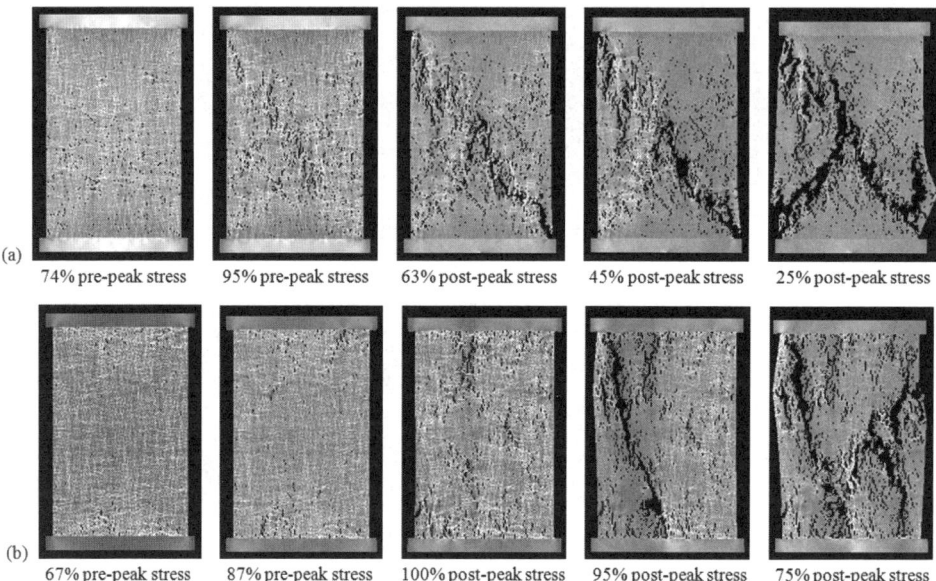

Figure 4.18 Numerical simulation of failure modes for specimens with different loading platens in terms of the relative Young's modulus of platen and specimen (the gray colour represents the shear stress): a) stiffer constraint condition, $E_p/E_s = 10$; and b) softer constraint condition, $E_p/E_s = 0.1$.

the central portion, outside the cone-shaped areas. As the loading increases, some splitting fractures occur along the shear zone, and there is a distinct zone of failure. More longitudinal cracks extend from the zone of failure. The post-peak evaluation of the specimen shows a single faulting plane and a cone-type failure that simply intersects the faulting plane. The cone formed then acts as a wedge splitting the rest of the specimen.

A completely different situation emerges for the softer loading platens ($E_p/E_s = 0.1$). In this case, the platens give rise to large lateral deformations, resembling outward-directed shear forces at the interface, as shown in Figure 4.18b. As the frictional forces are now directed outward, a biaxial tensile/compressive state of stress develops in the specimen ends and a splitting type of failure occurs.

In Figure 4.19, the effect of constraint in the contact areas on the axial stress–strain curve for uniaxial compression is shown. The pre-peak behaviour is found to be essentially independent of the properties of the loading platens; the peak strength and the corresponding strain are not affected noticeably by different contact conditions. However, the post-peak curves are much more influenced by the loading platens. Wawersik and Fairhurst (1970) and Wawersik and Brace (1971) drew similar conclusions from their experiments. In these experiments, although the use of rubber, Teflon sheets of different thickness, and various combinations of Teflon and metal shims induced secondary tensile, rather than compressive, fracture, the observed compressive strengths were essentially the same, regardless of whether the specimens were in direct contact with the loading platens or whether suitable rock end-caps were employed. Although small, the variation of the peak stress is strictly related to the

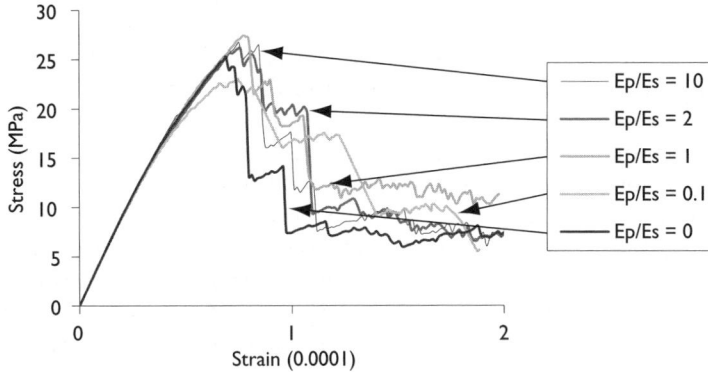

Figure 4.19 Simulated stress–strain curves for five specimens with loading platens having different relative Young's moduli.

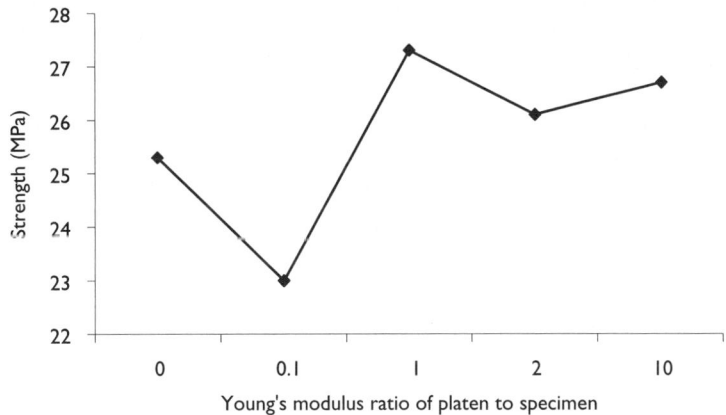

Figure 4.20 Simulated strength reduction with end constraint for five specimens with different loading platens in terms of Young's modulus.

end constraint conditions. For example, as shown in Figure 4.19, the specimen with end constraint condition of $E_p/E_s = 1$ has the highest strength, 27.3 MPa, whereas the specimen with $E_p/E_s = 0.1$ (inducing tensile cracks in the two ends of the specimen) has the lowest strength, 23 MPa.

The numerical results presented here also indicate that platens made of the same material as the test specimen are more suitable for such tests. In fact, this is a step which can be taken in practice: higher strength specimens of the same rock can be selected for the platens, thereby reducing the end-effects caused by the mismatch of the loading and tested materials. Note that the specimen without any end constraint ($E_p/E_s = 0$) has the strength 25.3 MPa (Figure 4.20), which is lower than the value of the specimens with $E_p/E_s \geq 1$.

4.4.3 Effect of height to width ratio (slenderness) of the specimen

The geometry of the specimen can also influence strength characterisation in several ways (e.g. Blair and Cook, 1998). In practice, most of the research on the role of specimen geometry has been concerned with the ratio of height to diameter in compression, especially in relation to the end effect because, the longer the specimen, the greater will be the specimen volume not affected by the end constraints. It has been established that micro-fracturing starts in the middle part of the specimen (i.e. outside the triaxial cones at the specimen ends) if the ratio of height to width is larger than 2 and even more so if larger than 3 (Lockner et al., 1991). With the increase in height, however, the specimen surface instability increases and the danger of buckling arises. As a result, the experimentally obtained strength decreases with increasing height to width ratio. This effect has been incorporated in the term 'scale effect', which is unfortunately an ambiguous term since it incorporates both the specimen slenderness and volume effects.

In Figure 4.21a, the initial loading stresses for the five specimens with different ratios of height to width are shown. The simulations are conducted for specimens with slenderness ranging from $H/W = 3, 1.5, 1, 0.67$ to 0.5. The effect of slenderness is clearly reflected in the changes in the fracture patterns, as shown in Figure 4.21b and as observed in many physical tests on rock. Although the observed vertical splitting failure is combined with inclined fractures, because of the differences in the internal stress fields for the specimens with different heights, the fracture patterns differ from one another. The longer specimens show more dominance of the splitting failure mode (as seen in Figure 4.21b for the $H/W = 3$ specimen). The shorter specimens show more complicated failure modes because, although surface splitting occurs first, more internally inclined faulting fractures dominate the overall failure process (as seen in Figure 4.21b for the $H/W = 0.5$ specimen).

When the specimen slenderness increases, (i.e., H/W increases) the specimen surface instability is enhanced (reminiscent of the surface spalling and slabbing around

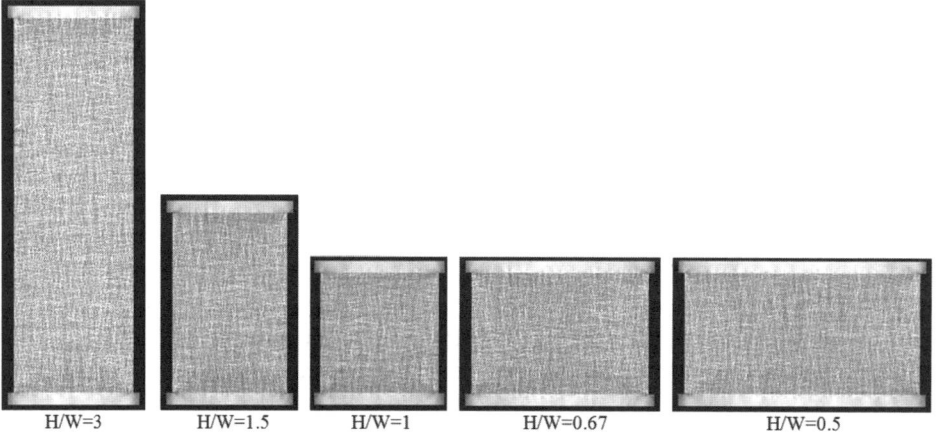

Figure 4.21a Specimens used in the numerical modelling to study the effect of specimen height: width ratio, illustrating the initial stress distributions in the specimens and platens.

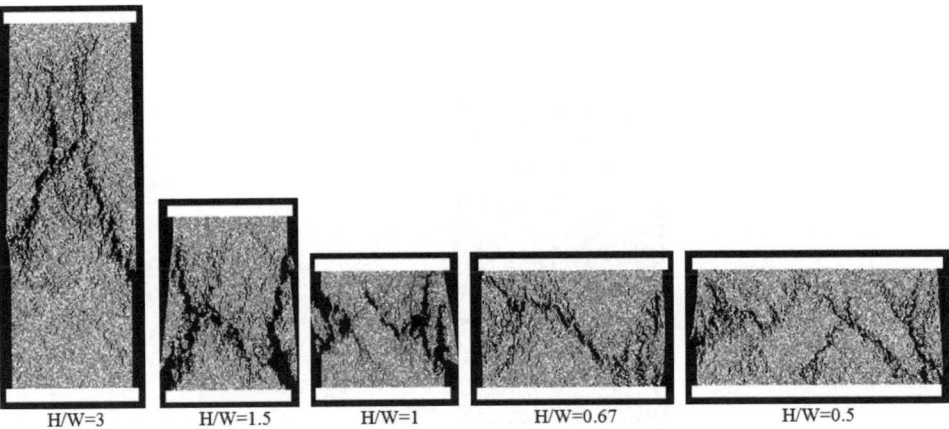

Figure 4.21b Failure modes for the five specimens with different height to width ratios shown in Figure 4.21a.

excavations first described by Fairhurst and Cook (1966)), the stress state becomes more homogeneous in the middle of the specimen, and the failure mode changes: axial splitting occurs instead of faulting (as seen in Figure 4.21b).

As a comparison, the failure processes of specimens with ratios of height to width $H/W = 0.5$ and 3 are shown in Figure 4.22. Both specimens fail in combined failure patterns, both splitting and faulting. However, the height to width ratios of the specimen have a great effect on the overall failure patterns. As can be seen in Figure 4.22a, when the specimen is short and wide ($H/W = 0.5$), slabs closer to the surface of the specimen form first and buckling failure occurs. Within the specimen's central portion, a large zone of triaxial compression is present and the failure patterns change to localised faulting. Almost an exact matching failure pattern was observed from the experiments using a plaster sample with $H/W = 0.5$ (Lockner *et al.*, 1991).

Increasing the H/W ratio, as seen in Figure 4.22b for $H/W = 3$, induces a different, more brittle, failure mode. In Figure 4.22a for short specimens, a large number of macro-fractures occur well before the peak stress is reached and a more ductile failure pattern is simulated. For these longer specimens (Figure 4.22b), however, very few macro-fractures occur before the peak stress (only one crack nucleates immediately before the peak stress is reached). The macro-failure occurs immediately beyond the peak stress, resulting in a brittle failure process. It is found that, for the specimens with $H/W = 3$, the macro-cracking which starts in the middle of the specimen does not depend on the contact conditions. This is due to the diminished influence of the end effect and, more significantly, to the increasing surface instability. The failure pattern clearly indicates that splitting failure dominates (Figure 4.22b).

In Figure 4.23, the complete stress–strain curves for the five specimens with $H/W = 0.5, 0.67, 1, 1.5$ and 3 are plotted to indicate the effect of specimen geometry. Figure 4.24 plots the peak strength as a function of height to width of the specimen. The numerical simulations show that peak stress, i.e. the uniaxial compressive strength as determined using each specimen, depends strongly on the specimen height, with the shortest specimen giving the highest strength. As well as the peak stress

80 Rock failure mechanisms

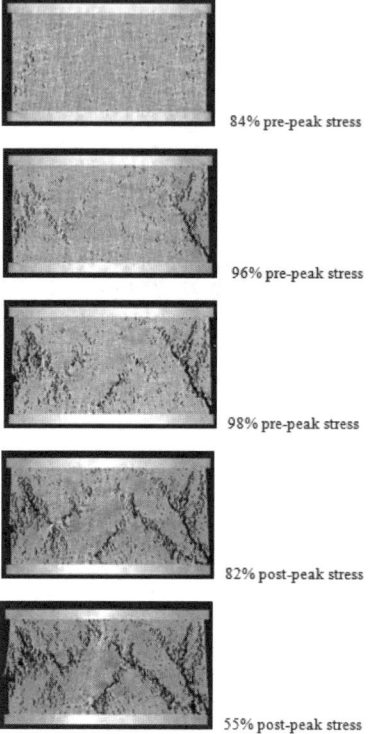

Figure 4.22a Numerical simulation of failure modes for specimens with a height to width ratio of 0.5.

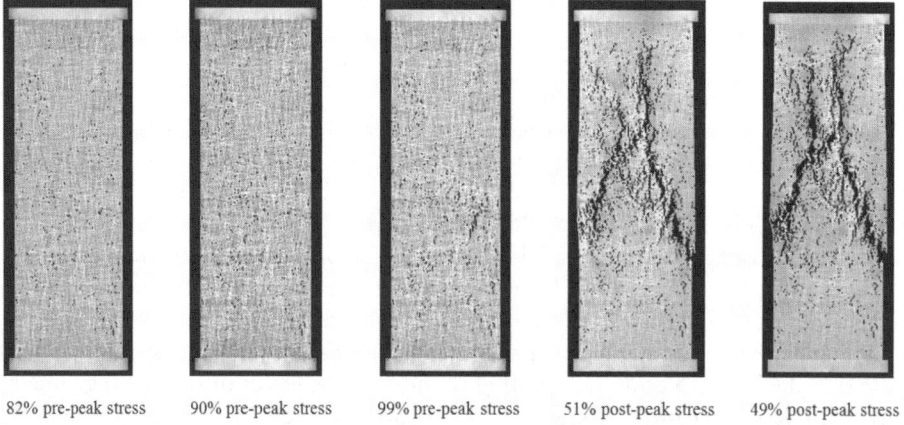

Figure 4.22b Numerical simulation of failure modes for specimens with a height to width ratio of 3.

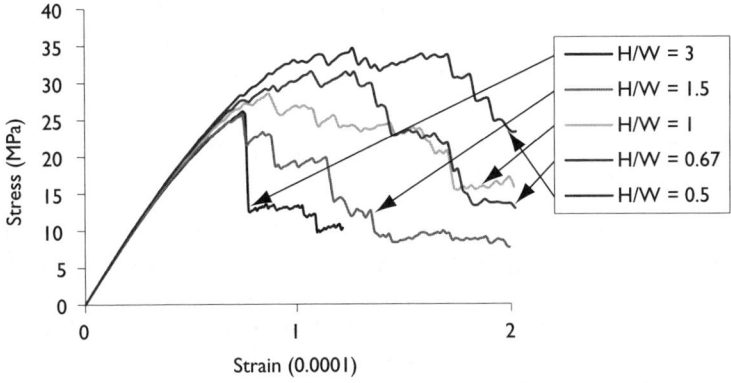

Figure 4.23 Stress–strain curves for five simulated specimens with different shapes in terms of the height to width ratio (*cf.* Figure 4.21).

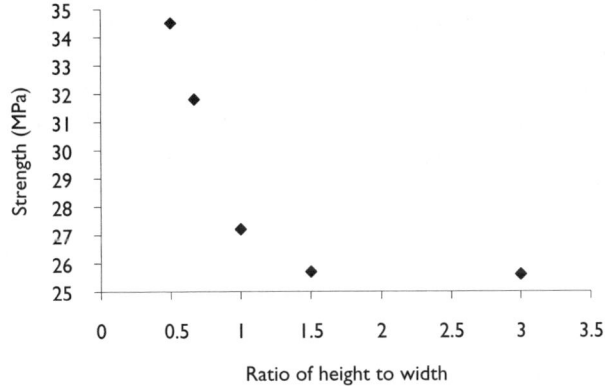

Figure 4.24 Strength reduction with specimen size for five simulated specimens (*cf.* Figure 4.21).

effect, the specimen height to width ratio also affects the strain softening behaviour of the rock: increasing the height to width ratio results in a steeper slope of the post-peak stress–strain portion, even leading to severe stress drop. These numerical results are in harmony with the observations obtained from experiments carried out by Van Mier (1997) and Cox and Meredith (1993): namely, that the descending branch becomes steeper with increasing specimen height. Again, it is found from numerical results that the pre-peak behaviour of the specimens is independent of the specimen slenderness, as shown in Figure 4.23.

Failure of rock in uniaxial compression is a localised phenomenon. Due to this localisation, fractured areas are more deformed, while unfractured portions recover their deformations. Specimens with a higher height to width ratio have more undamaged portions for the potential deformation reversal during failure localisation (see

Figure 4.22b). For these specimens, the combination of elastic recovery and localised damage results in a steeper drop after the peak in their stress–strain curves (Figure 4.23). In fact, it is possible for the complete stress–strain curve not to increase monotonically in strain—which is the subject of the next section.

4.4.4 Class I and Class II curves in uniaxial compression

The fact that elastic energy is stored in the specimen (as strain energy) and can be released via unloading to supply growing fractures with the necessary surface energy (Cook, 1965) causes two types of stress–strain curve to occur: Class I and Class II, first termed by Wawersik and Fairhurst (1970). As indicated in Figure 4.25, a Class I curve monotonically increases in axial strain; whereas, a Class II curve does not. The reason why a Class II curve can characterise the specimen failure is because, at the peak stress, when the uniaxial compressive strength is reached, there can be more elastic energy in the specimen than is required to fail it.

If a rock specimen is loaded from point O in Figure 4.26 to point A, then the energy in the specimen is represented by the area OAC. If micro-fracturing in the specimen then causes the specimen to reach point B, the energy contained in the specimen is represented by the area OBD. Thus, at Point A, there is more energy contained in the specimen itself than is needed to continue failure. Area OAB represents the energy required to change the specimen's state from A to B and area ABDC represents released kinetic energy. So, without any further control on the experiment, the Class II behaviour will result in uncontrolled specimen failure, however stiff the applied strain loading.

The thermally controlled testing machine shown in Figure 4.27 was used for the first systematic series of tests on a suite of rocks to obtain the complete stress–strain curves (Wawersik and Fairhurst, 1970). This type of machine was developed because strain rather than stress is the independent variable when obtaining the Class I curves. The whole testing machine was heated and the specimen then locked between platens

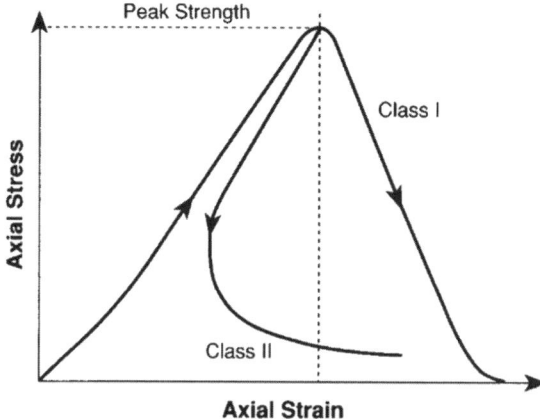

Figure 4.25 Class I and Class II complete stress–strain curves—a Class II curve does not monotonically increase in strain.

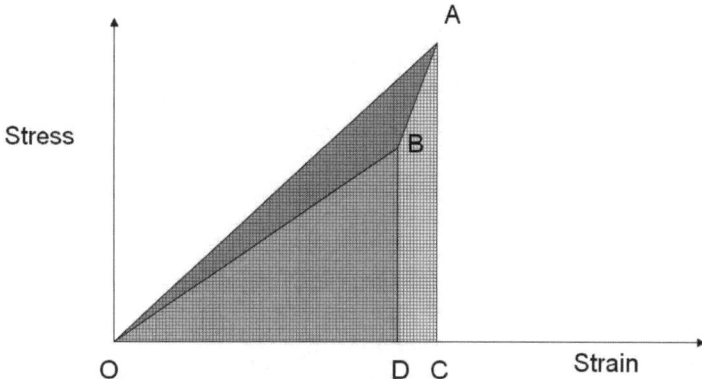

Figure 4.26 Energy changes during micro-fracturing leading from stress–strain state A to stress–strain state B.

Figure 4.27 Thermally controlled stiff testing machine at the University of Minnesota in 1970.

inside the three main columns. As the machine cooled down, strain was applied to the specimen and the complete stress–strain curves obtained, see Figure 4.28. In addition to the Class I curves, it was also possible to obtain Class II curves by hydraulically pushing back on the platens when failure began to be uncontrolled (this is possible because the fracturing within the rock micro-structure starts with a low velocity).

Since that time, servo-controlled machines have been used because there is much more flexibility in their operation. These testing machines operate by the continuous comparison of a pre-programmed variable's value and the actual experimental value or feedback of that variable's value—as determined by the experimental condition. This means that there are many experimental values that can be used as feedback, including

84 Rock failure mechanisms

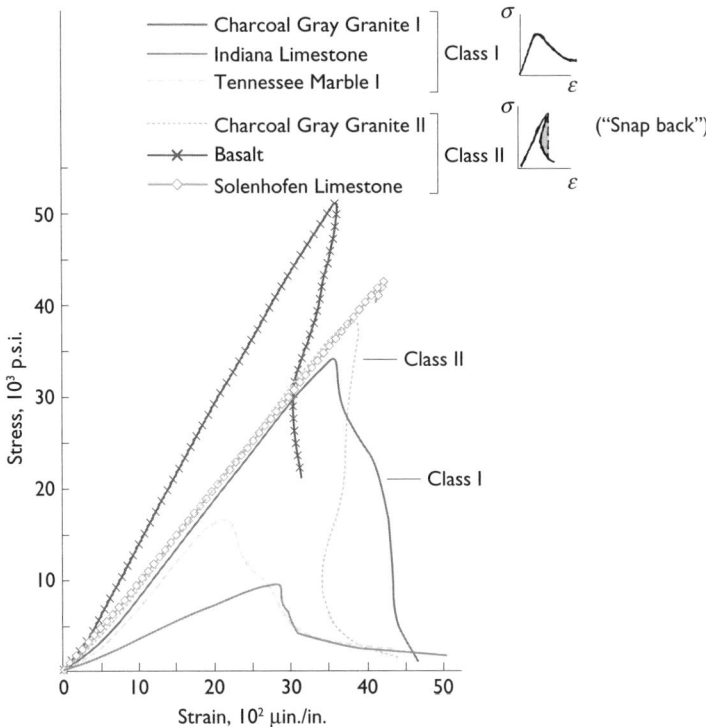

Figure 4.28 Complete stress–strain curves obtained in a thermally controlled stiff testing machine (Wawersik and Fairhurst, 1970).

lateral strain, volumetric strain, energy absorption, etc. (Hudson *et al.*, 1972). The commonest way of obtaining a Class II curve in a servo-controlled testing machine is to use the circumferential displacement as the feedback signal. In this way, the machine applies hydraulic pressure to the loading platens such that the circumferential displacement linearly increases with time while the axial displacement is independently monitored, allowing Class I or Class II curves to be obtained. In other words, in order to observe a stable complete stress–strain curve for a Class II circumstance, energy has to be incrementally removed from the specimen, *cf.* Figure 4.26. An example of a crystalline rock sample that has been tested in this way is shown in Figure 4.29. Note the circumferential chain displacement transducer around the rock specimen.

Complete Class II stress–strain curves have been obtained using numerical simulation with a variety of control techniques analogous to the feedback in the physical tests. Curves obtained by Hazzard *et al.* (2000) using the Particle Flow Code (PFC) are shown in Figure 4.30.

Another numerical technique has been used to obtain Class II curves in numerical simulations, as seen in Figure 4.31 from Pan *et al.* (2006). For these simulations, an elastic-plastic cellular automaton (EPCA) was used. The m values shown in Figure 4.31 represent the same Weibull distribution m values as used in the simulations already presented in this book. The higher the m value, the more homogeneous

Figure 4.29 The measurement of circumferential displacement as feedback in a closed-loop servo-controlled testing machine to obtain a Class II curve (photo from the SP Laboratory, Borås, Sweden).

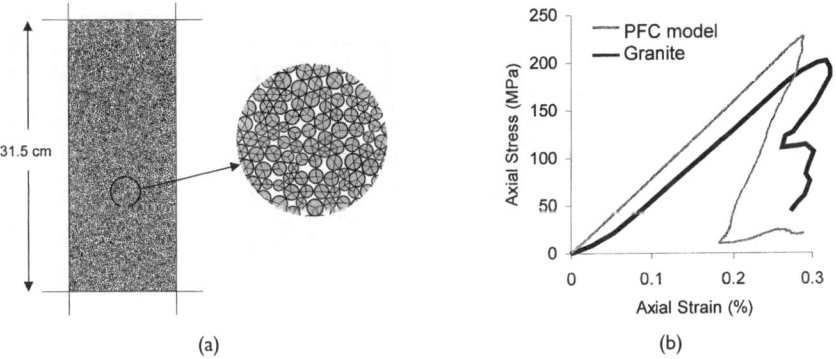

Figure 4.30 Use of the Particle Flow Code (PFC) to obtain the complete stress–strain curve for Lac du Bonnet granite. (Hazzard et al., 2000).

the simulated rock will be. Thus, the degree of inhomogeneity is directly reflected in the shapes of the simulated complete stress strain curves in Figure 4.31: within the suite of curves simulated, only for $m = 2.0$ is the curve of Class I type; all the others are of Class II type.

4.4.5 The size effect

The dependence of rock strength on specimen size is generally attributed to the presence of a greater sample of fractures and cracks in larger specimens than in smaller specimens. Generally, this means that the largest crack in a large specimen will be bigger than the largest crack in a small specimen—and so failure will be initiated earlier in a larger specimen. However, the exact nature of this dependence in practice is poorly known. The literature presents a number of contradictory

86 Rock failure mechanisms

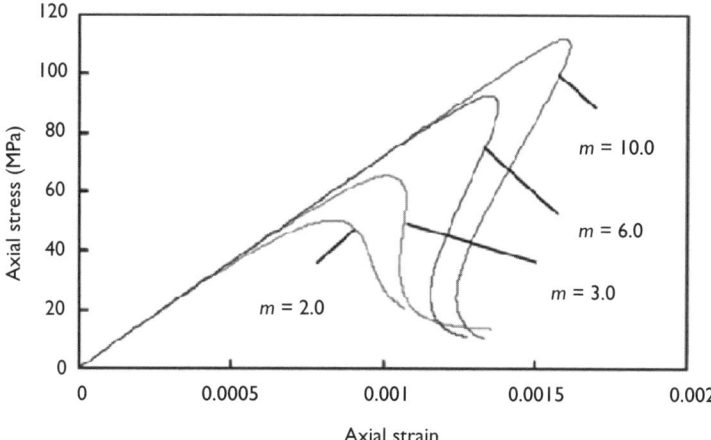

Figure 4.31 Complete Class II stress–strain curves numerically simulated with an elastic-plastic cellular automation (Pan et al., 2006).

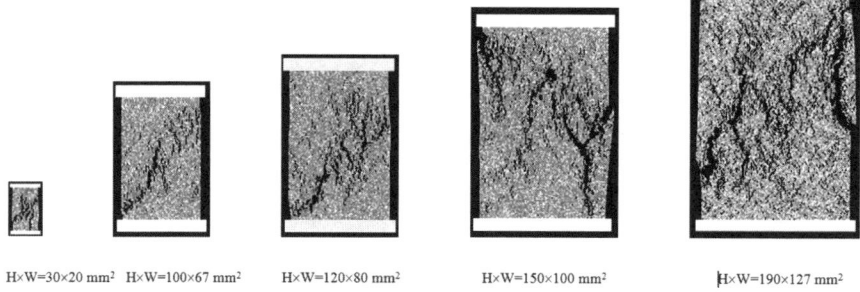

Figure 4.32 Five specimens used for numerical simulation of the size effect with the same height to width ratio but different sizes.

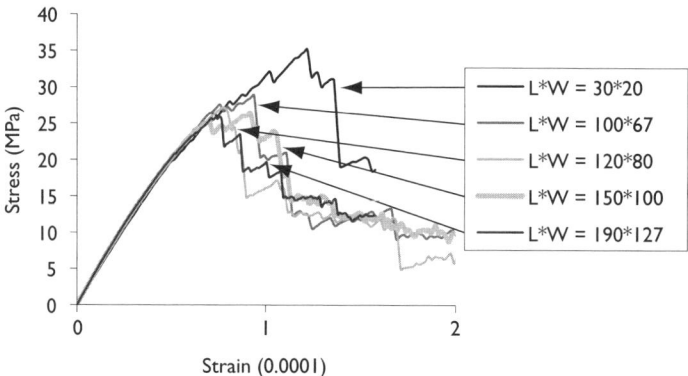

Figure 4.33 Stress–strain curves for five simulated specimens with different sizes but with the same height to width ratio tested in uniaxial compression.

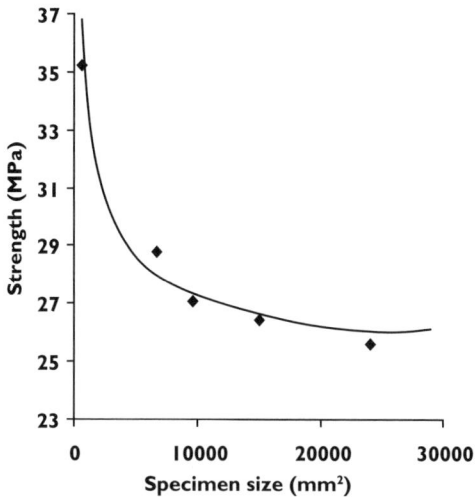

Figure 4.34 Simulation of strength reduction with specimen size.

opinions and observations: while some authors have reported no influence of the specimen size, others have reported a reduction in strength with an increase in specimen size.

The numerically simulated results for the five specimens with the different sizes, $H = 30, 100, 120, 150$ and 190 mm, are presented in Figure 4.32. The effect of specimen size on the stress–strain curves is presented in Figures 4.33 and 4.34. It is evident from this latter Figure that the dependence of peak strength on specimen size follows an inverse power-law relation. Although the effect of size on strength characterisation is clear, as shown in Figures 4.33 and 4.34, the change in specimen size does not obviously change the failure patterns, as can be seen in Figure 4.32: the specimens have similar failure patterns. This subject is discussed further in Pan *et al.* (2006, 2009).

Chapter 5

Confinement and shear

5.1 THE EFFECT OF CONFINEMENT

In the previous Chapters, we have explored the nature of rock failure in uniaxial tension and uniaxial compression. However, we also have to consider the mechanisms of rock failure when a sample is subjected to more complex stress states. Experimental evidence indicates that rocks are significantly strengthened by confinement and that there is a brittle-ductile transition zone as the confining pressure is increased, i.e., beyond this zone the complete stress–strain curve continues to ascend, rather than descending as we have discussed for the uniaxial loading case in the previous Chapter. The behaviour of rock under the confinement condition is of fundamental and practical significance in both the structural geology and engineering fields. However, despite many years of theoretical and practical research, we still do not have a universally accepted failure criterion for the general case of a stressed rock sample, see Yu *et al.* (2009) where the main developments leading to a unified strength theory for geomaterials are described. Hence, the exploration of rock failure via numerical models continues to be helpful, as we will describe in this Chapter.

Figure 5.1 shows the numerical model set-up for confinement pressure loading and the mesh for the plane strain numerical sample consisting of 200 × 100 elements with equivalent overall dimensions of 100 mm × 50 mm. The corresponding numerical complete axial stress versus axial strain curves for rock at constant confining stresses up to 16 MPa are presented in Figure 5.2. It can be seen from these stress–strain curves that the rock deforms linearly elastically at axial stresses below the yield strength which is dependent on the confining pressure. Further compression leads to inelastic deformation up to the peak strength. At low confining pressures, the curves show a defined peak strength and a strength decrease in the post-peak region. At higher confining pressures, the Young's modulus of the rock is higher than for the lower confining stresses and the rock exhibits work-hardening. The transition from brittle to ductile deformation in rock with an increase in confining pressure is seen to occur for this simulation between 8 and 10 MPa confining pressure.

The numerical results further indicate that the confining stress also influences the brittleness of the specimens with the brittleness decreasing as the confining stress increases, see Figure 5.2. Under uniaxial compression when the confining stress is zero, the microscopic processes develop quickly, so that the specimen collapses within a small strain range and the behaviour of the material is brittle.

90 Rock failure mechanisms

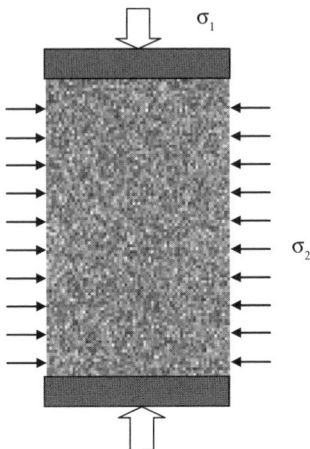

Figure 5.1 The numerical model for studying failure of rock in biaxial compression.

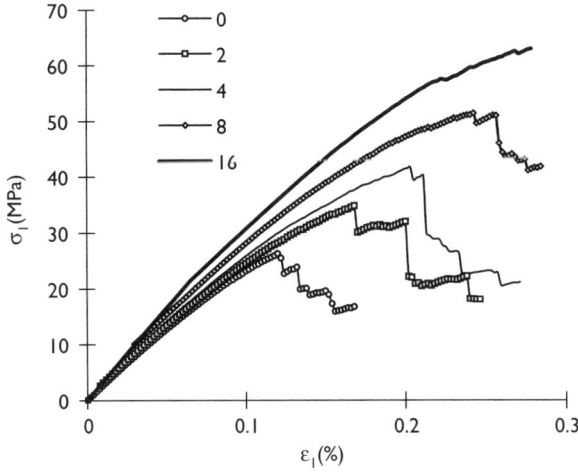

Figure 5.2 Complete stress–strain curves for numerical rock specimens with varying confining stress.

In general strength theory, failure is assumed to take place at the maximum stress ordinate of the complete stress–strain curve, thus defining the uniaxial strength of the rock. This has been generalised into the statement that, if any particular component of stress is increased under specified conditions until failure takes place, the magnitude of that stress at failure is known as the strength of the material under those conditions. Thus, we may speak of the uniaxial tensile or compressive strength, the triaxial compressive strength at a certain confining stress, and so on. However, when polyaxial stresses are being considered, a more general approach is necessary, and it will be assumed that failure takes place when a definite relation between the stresses characteristic of the material is satisfied. Such a relation is called a criterion of failure, and its geometrical representation in principal stress space is the failure surface,

Yu et al. (2009). Figure 5.3 gives the numerically obtained relation between the peak strength of the simulated rock specimens and the confining pressure at failure and Figure 5.4 shows the numerically obtained failure envelopes of the rock specimens for the 2-D simulation.

As can be seen from Figures 5.3 and 5.4, the ultimate compressive failure strength, i.e., peak strength of the numerical rock specimens, gradually increases with confining pressure. Even though a linear Mohr–Coulomb failure criterion with tension cut-off is adopted for the individual elements in the finite element simulation model,

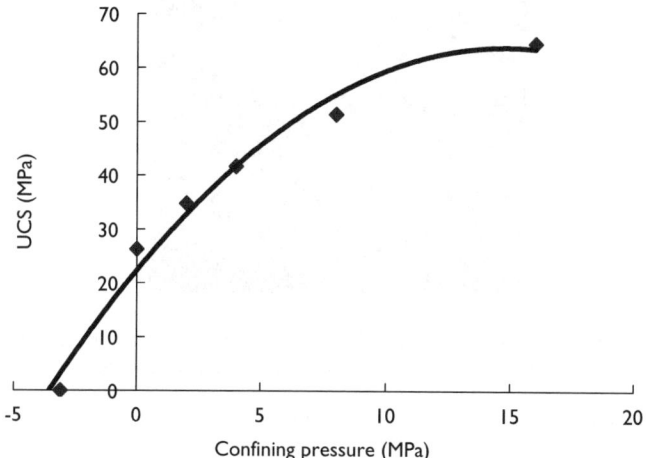

Figure 5.3 Relation between the compressive strength and confining pressure for the simulated rock specimens.

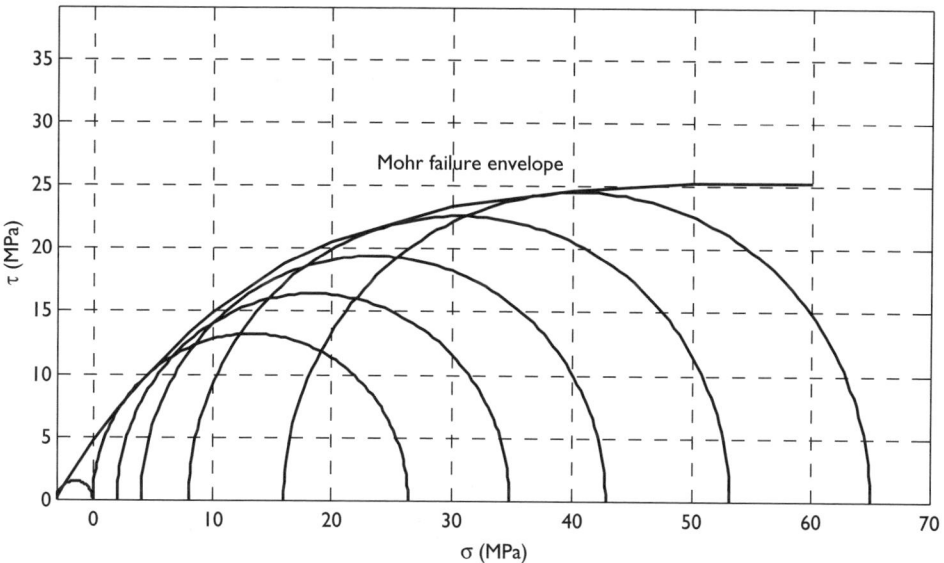

Figure 5.4 Failure envelope in normal stress–shear stress space for the simulated rock specimens.

92 Rock failure mechanisms

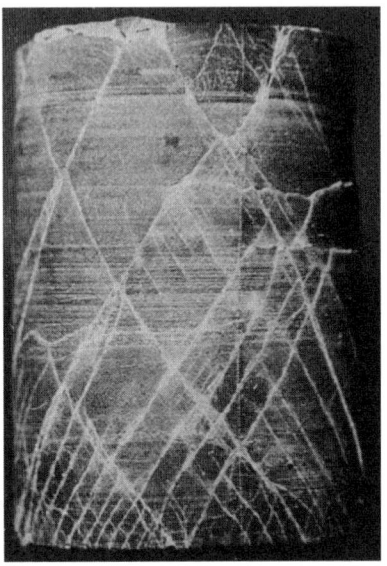

Figure 5.5 Shear planes developed during a physical test on rock at a high confining pressure, from Stavrogin and Tarasov (2001).

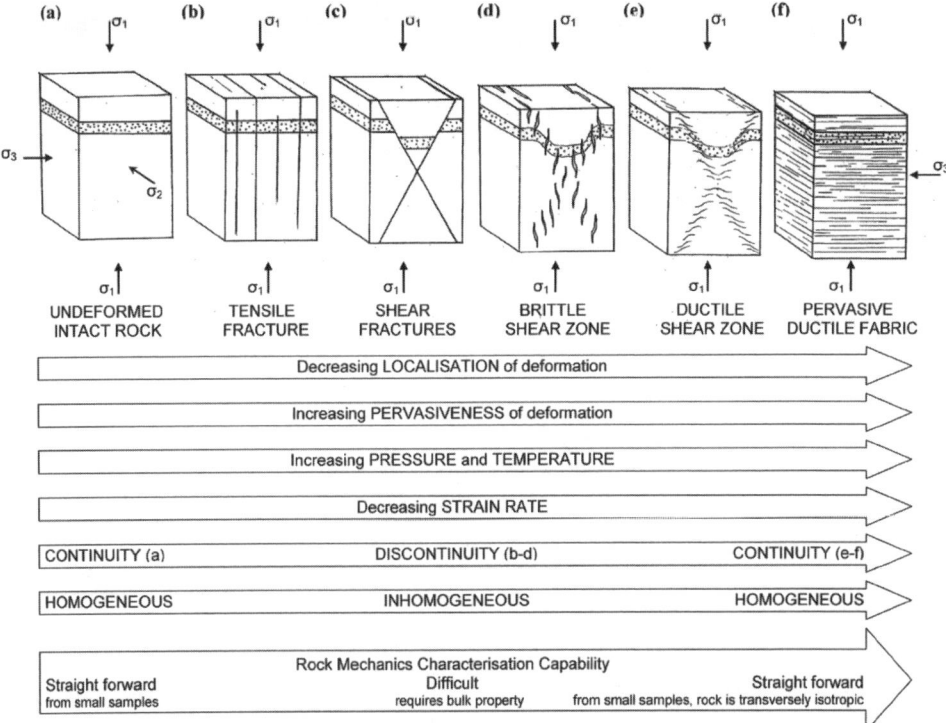

Figure 5.6 The formation of fractures and shear zones in rock over geological time (diagram developed in association with Prof. J.W. Cosgrove of Imperial College, London, UK).

the macroscopic failure envelope is concave towards the normal stress axis. These numerical results indicate that macroscopic non-linear phenomena, such as rock failure in nature, can be described and revealed through simple linear rules at the microscopic level. In addition, it is noticed that the residual strength of the rock, also dependent on the confining pressure, increases with confining pressure.

In Chapter 1, we described the strong link between engineering rock mechanics and structural geology. Shear slip planes, such as those evident in the photograph of a rock specimen tested at high confining pressure in Figure 5.5 and the brittle-ductile transition are of significant interest in geology and geophysics in connection with the behaviour of materials in the lower crust where higher confining stresses exist. In Figure 5.6, the types of brittle and ductile features that are developed naturally during geological processes are illustrated. The horizontal bars across the diagram

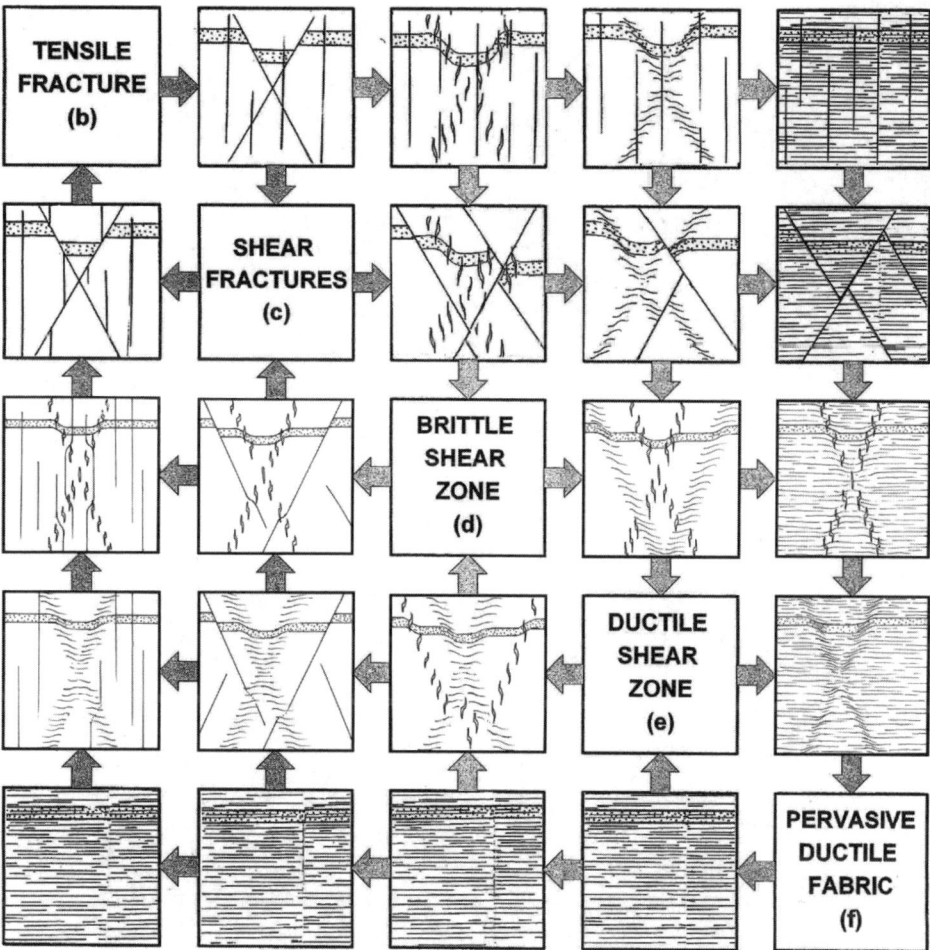

Figure 5.7 Interaction matrix illustrating the effect of a second phase of fracturing or shearing on a rock that has already experienced an initial fracturing or shearing episode (developed in association with Prof. J.W. Cosgrove, Imperial College, London, UK).

indicate the conditions under which the brittle and ductile features are developed. Note especially the decreasing of localisation of deformation and increasing of pressure from left to right across the diagram. The effect of confinement is thus directly illustrated in these sketches, although it must be remembered that there are many variables involved, including temperature and strain rate, so the final deformational mode of a rock mass is a coupled function of interacting variables.

Moreover, both in structural geology and rock engineering a rock mass can experience a succession of fracturing/shearing events, which will be superimposed. The second event will be influenced by the rock mass structure left by the first event, and there could then be further events following, so that interpretation of the consequential rock mass configuration can often be difficult. In Figure 5.7, the result of a second fracturing/deformation event on the first is illustrated for the b) to f) samples in Figure 5.6 through an interaction matrix presentation.

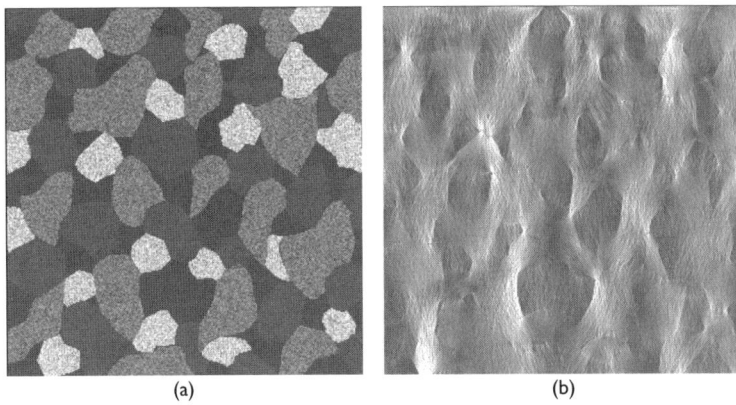

Figure 5.8 a) Rock material numerically modelled; b) Shear stresses in the simulated rock when subjected to a vertical to horizontal stress ratio of 3.

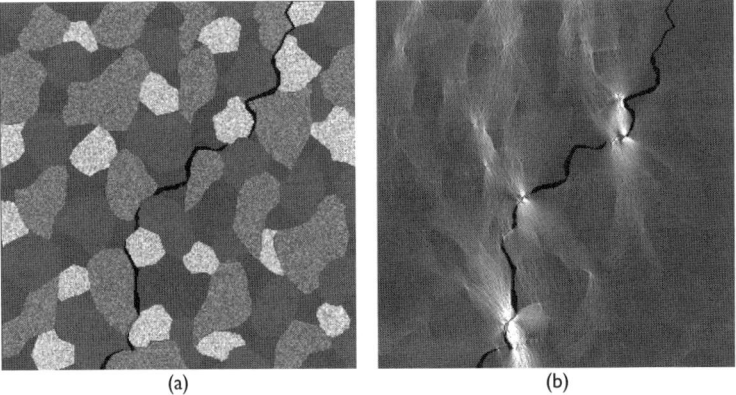

Figure 5.9 a) Fracture in the rock; b) Shear stresses around the fracture.

The matrix is read clockwise: for example, the effect of a subsequent tensile fracturing episode on a previous shearing episode is shown in the second box along the first row; and the effect of a subsequent shearing episode on a previous tensile fracturing episode is shown in the first box of the second row. This matrix presentation could be extended to account for the variety of factors involved but it is sufficient here to emphasise how complexities in the natural rock mass fracturing pattern can quickly arise through concatenated deformational events.

As one example of the factors involved, Goodman (1980) has given an explanation of the effect of confining pressure in fissured rock which can be imagined as a mosaic of matching pieces. Sliding along the fissures is possible if the rock is free to displace normal to the mean surface of a rupture. But, under confinement, the normal displacement required to move along such a jagged rupture path requires additional energy input, as illustrated by the numerical simulations in Figures 5.8 and 5.9. Thus, and as pointed out by Goodman, it is not uncommon for a fissured rock to achieve an increase in strength by many times through the application of a small increment in the mean stress. This is one reason why a small amount of confining pressure is so effective in preventing spalling in highly stressed rock surfaces (Andersson, 2007) and why rockbolts are similarly effective in strengthening tunnels in weathered rocks.

5.2 ACOUSTIC EMISSION DURING SHEARING

The monitoring of acoustic emission (AE) or seismic events has proved to be one of the powerful tools available for analysing damage or brittle fracture during rock deformation. There is generally a good correlation between the AE rate and the damage events—so that the AE rate can be used as a measure of the damage accumulation occurring in the rock sample. Cox and Meredith (1993) have analysed catalogues of AE events recorded during compression tests on rock in terms of the information they provide. By combining the measured damage state with a model for the weakening behaviour of cracked solids, predictions for the mechanical behaviour are possible. Based on this concept, and by recording the counts of failed elements, the AE associated with progressive failure can be simulated (Tang, 1997; Tang and Kaiser, 1998).

Figure 5.10 shows the complete stress–strain curves and corresponding AE characteristic curves of model specimens at different constant confining pressures. Figure 5.11 shows the AE and corresponding normalised cumulative AE energy (AEE) curves for model specimens under different confining stresses. Comparison of the curves in Figure 5.10 shows a good relation between the simulated stress–strain curves and the modelled curves of event rate. It can be seen from Figures 5.10 and 5.11, that in general a sharp increase of the AE event rate in the AE characteristic curves corresponds to an abrupt stress drop in the complete stress–strain curve, with the maximum rate of AE events appearing in the post-peak range. This indicates the initiation and propagation of mesoscopic fracturing preceding the final stage macroscopic fracture development in the rock.

The results also show that the maximum AE event or main shock emitting from rock can be regarded as the precursor of macro-fracturing in the rock, which is essential for the location of earthquake sources, and the search for earthquake precursors

96 Rock failure mechanisms

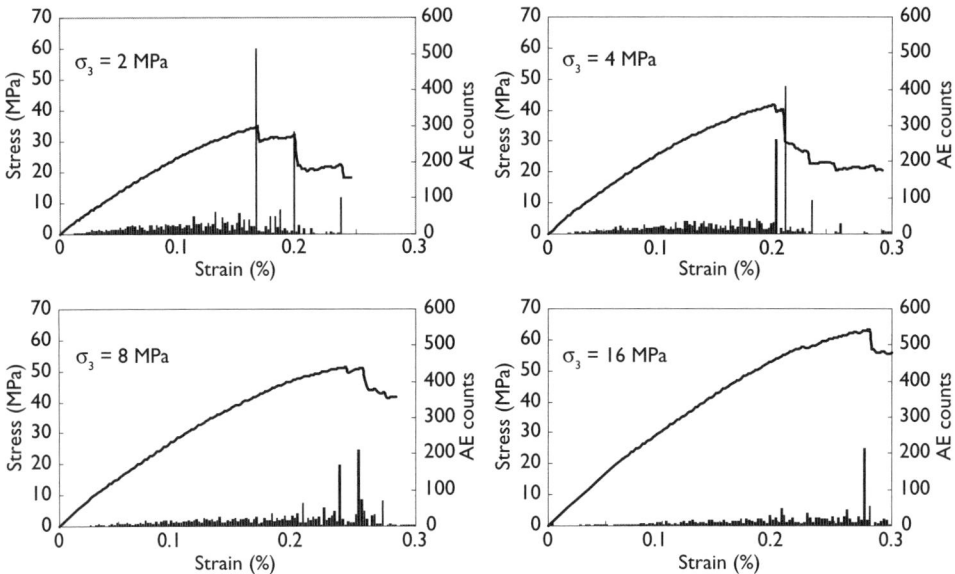

Figure 5.10 Complete stress–strain curves and AE characteristic curves for the simulated specimens subjected to different confining stresses.

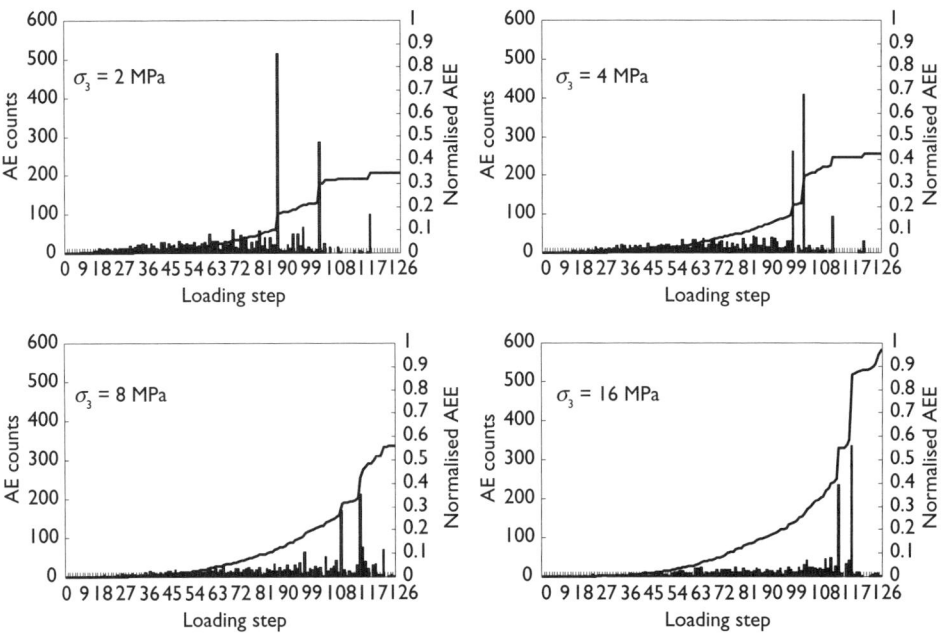

Figure 5.11 AE curves and normalised cumulative AE energy (AEE) curves for model specimens under different confining stress.

and earthquake prediction. In addition, the main fracturing will occur later for rock specimens that are subjected to an increased confining stress. The confining stress increases the ultimate compressive strength and defers the occurrence of the main fracturing at failure—which is the confining stress effect on rock failure.

5.3 BIAXIAL LOADING

In the first section of this Chapter, the confinement was implicitly considered in the context of being applied via a fluid, as is the case in triaxial rock testing (although 2-D tests were modelled here). Now, we explore the general biaxial stress situation where general stresses σ_x and σ_y are applied to a specimen. The numerical specimen, which has been previously subjected to uniaxial loading is now subjected to biaxial loading. The two principal stresses, σ_x and σ_y, are applied in the vertical and horizontal directions. Monotonically increasing, proportional, biaxial loading is gradually applied with load control for the (compression positive) cases of σ_x/σ_y = 0.5/1, 0.2/1, 1/1, −0.1/1, −0.2/1 and −1/−1. These six stress ratios cover the range of compression–compression, compression–tension, and tension–tension loading conditions. In combination with the numerical results of specimens subjected to uniaxial tension and compression, as described in previous Chapters, the biaxial strength envelopes for the numerical rock specimens are obtained, as shown in Figure 5.12 with experimental results.

With the numerical results being normalised by the uniaxial compressive strength, the strength envelopes from numerical simulation compare favourably with those of experimental results for rock specimens when a 'brush platen' was used to transfer the testing machine load to the specimen (Brown, 1976). A 'brush platen' is a platen composed of many steel rods, so that the triaxial stress state associated with solid platens, as discussed in the previous Chapter, is reduced to only small triaxial stress states beneath each rod. Note that the descending branch of the complete stress–strain curve as well as the complete failure process could not be obtained in the experimental tests because of the load-controlled condition.

By means of the displacement-controlled loading scheme in the numerical model, the descending portions of the stress–strain curves, the crack propagation process, the localisation of deformation, as well as the cracking patterns during the complete fracture process for the numerical specimen could be modelled in further simulations. The displacements applied in the two directions are designated by u_x and u_y. Here b represents the ratio of u_x/u_y, and b = 0.5/1, 0.2/1, 1/1, −0.1/1, −0.2/1, and −1/−1.

Representative examples of the typical failure patterns for this specimen under the different stress combinations for the biaxial loading conditions including uniaxial compression and tension (see the sketches in Figure 5.13a), are shown in Figure 5.13b. Of course, sometimes the stresses in the two directions attain their peak values at different loading levels due to the specimen heterogeneity; a quite different stress path is applied to the numerical specimen in these simulations because of the displacement control. We find that the strength values obtained from the displacement-controlled simulation are also in agreement with the experimental results. It is concluded that the loading path has a minor effect on the strength

98 Rock failure mechanisms

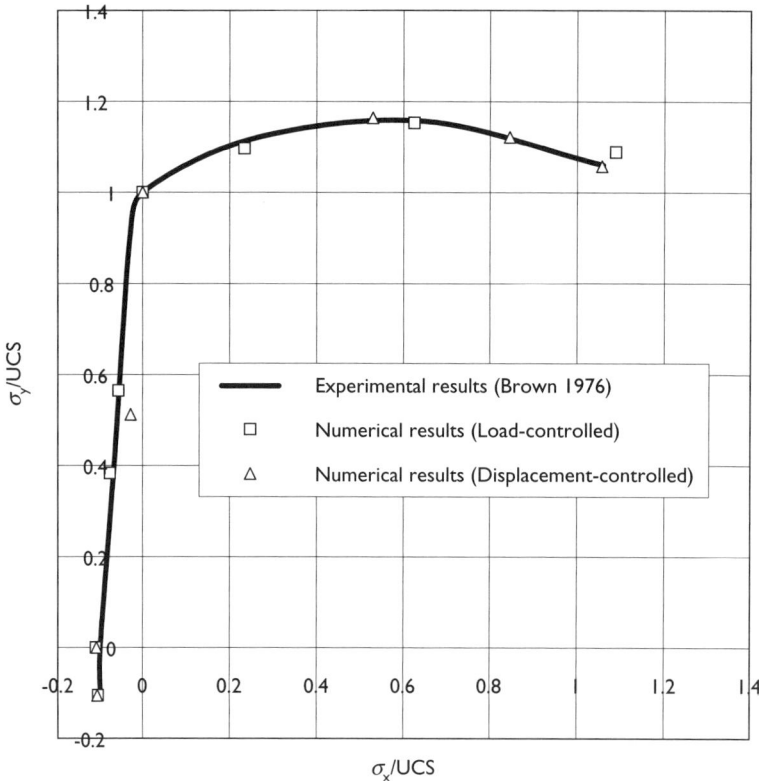

Figure 5.12 Biaxial strength envelopes for rock (experimental and numerical results).

envelope of rock subjected to biaxial loading. This conclusion was also made for concrete, as reported by Kupfer and Gerstle (1973).

The transverse tensile strains and tensile stress concentrations induced in numerical specimens under uniaxial compression are greatly decreased when a compressive stress is additionally applied in the lateral direction. Consequently, application of a lateral stress improves the load-carrying capacity of rock specimens. On the other hand, such a lateral stress on the specimen will cause some additional tensile strain in the third direction. This can explain why the enhancement of the biaxial compressive strength of rock is somewhat limited when compared to its uniaxial compressive strength. Under biaxial compressive loading, there are generally one or two inclined fault planes as shown in Figure 5.13b. Because the numerical simulation is carried out as a plane stress problem, we cannot simulate the failure phenomena orthogonal to the loading plane.

Under combined compression and tension (Case e in Figures 5.13a and 5.13b), cracks parallel to the applied compressive stress form more easily compared to the uniaxial compression case because the lateral tensile stress promotes the initiation and propagation of such splitting cracks. One or more continuous cracks normal to the

Confinement and shear 99

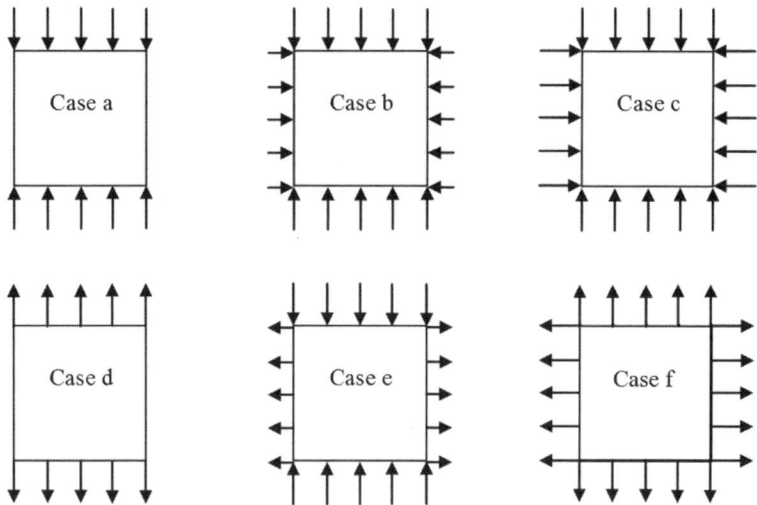

Figure 5.13a Sketches of the loading cases for the specimens shown in Figure 5.13b.

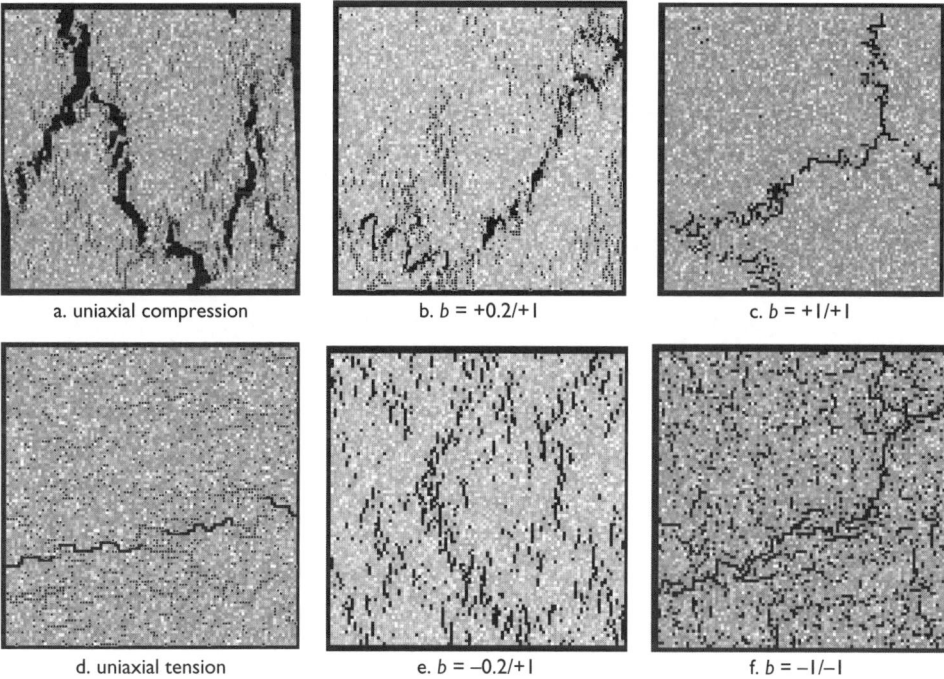

Figure 5.13b Failure patterns of simulated rock specimens under uniaxial and biaxial loading (b = the ratio of the applied displacement rates, u_x / u_y).

principal tensile stress are usually formed, and show similar cracking characteristics to uniaxial compression tests. Under biaxial tension, failure occurs by the formation of a single crack perpendicular to the direction of the maximum tensile stress. For equal biaxial tension, case f in Figures 5.13a and 5.13b, there is no preferred direction for the fracture surface; a zigzag crack crosses the whole specimen.

For further descriptions of shear failure in rocks, we refer the reader to structural geology orientated books: e.g., Hobbs *et al.* (1975); Price and Cosgrove (1990); Davis and Reynolds (1996); Mandl (2000); and Paterson and Wong (2005).

Chapter 6

Effect of heterogeneity on rock failure

6.1 INTRODUCTION

We emphasised earlier on in the book that the nature of rock as a natural material causes it to be discontinuous, inhomogeneous, anisotropic and inelastic. Also, in earlier Chapters, we have demonstrated how the inhomogeneity (having different properties in different locations) affects the rock failure mechanisms through the numerical models with the characterisation of the properties of the elements via the Weibull probability distribution and the homogeneity index, m. In this Chapter, we provide a more extended discussion of the effect of rock inhomogeneity because of the importance of this factor.

The conventional concept of brittle failure is based on considering defects (inhomogeneities, pores, cracks, etc.) as stress concentrators (Cook, 1965; Dyskin, 1999). For example, pre-existing cracks as strong stress concentrators can, under certain circumstances propagate, resulting in ultimate failure. There is, however, another role that the defects and the micro-structural characteristics can play in fracture processes (Dyskin, 1999): they induce stress fluctuations which are essentially random unless the micro-structure has a definite pattern. This perturbed stress field can include high local tensile and compressive stresses, and initiate local fracture—even if the magnitude of the external loading is less than the overall rock strength. Dyskin (1999) has emphasised that the presence of high local tensile stresses does not necessarily have to be associated with a particular geometrical stress concentrator (e.g., pre-existing crack), but is rather an integral effect of heterogeneity present in the rock material. Failure will initiate where the local stress reaches the local rock strength—which may or may not be at a pre-existing crack location.

The conclusion, therefore, is that fracture initiation should be viewed as the combined effect of the local stress fluctuations and the local stress concentrators (e.g., pre-existing micro-cracks). Rock failure mechanisms will be investigated in this Chapter by considering fracture initiation and growth under spatially random stress fields due to the existing heterogeneity in the rocks. We will highlight the random stress fluctuations produced by the defects and other heterogeneities in uniaxially loaded specimens, the crack propagation path and the heterogeneity-associated fracturing event patterns.

6.2 HETEROGENEITY-INDUCED STRESS FLUCTUATIONS

The determination of the statistical distribution of the stress fluctuations generated by heterogeneities in the rock material is a complicated task because of the necessity to account for the interactions between the many inhomogeneities present. Although this has been studied, the problem of how to measure and analyse stress distributions in heterogeneous solids, such as rock, concrete or other composite materials, remains largely unresolved. No available technique can effectively measure such stress fields in real solids, particularly in rocks. The widely used photo-elastic method (Frocht, 1941) is only valid for observing stress fields in homogeneous and transparent non-crystalline solids. Stress analyses using continuum mechanics theory are usually performed by either assuming the materials are homogeneous or averaging the heterogeneous features so that the materials can be treated as homogeneous. Such simplifications do not allow modelling of the stress fluctuations induced by the inhomogeneities within the solids.

Thus, we continue to investigate the stress distribution in rocks using the numerical approach. Two cases will be highlighted: the first case extends the studies of the stress distribution in a diametrically loaded disc; and the second case demonstrates that complex shear stress fields can exist in a heterogeneous matrix even when it is subjected to hydrostatic stress, i.e. when the axial stress, σ_y, and horizontal stress, σ_x, are equal. The hydrostatic loading condition is selected for the second case for the following reason: according to continuum mechanics theory, if the material is homogeneous or the heterogeneity is averaged out, the material will be free of shear stress when a hydrostatic stress state is applied. Using the numerical approach, we wish to demonstrate clearly that, if the influence of the rock heterogeneity is taken into account, the calculated shear stress fluctuation may be high enough to cause failure in the rock, even under hydrostatic loading conditions.

6.2.1 Discs subjected to diametral loading

In order to demonstrate the difference in stress fields between models with and without the material heterogeneity, we have analysed a homogeneous and a heterogeneous disc subjected to diametral loading (Figure 6.1), as was first shown in Chapter 3. The diameter of each disc is 180 mm. As before, the mechanical properties of the elements in the numerical code are defined by two independent parameters: Young's modulus, E, and Poisson's ratio, v. The mean Young's modulus of the elements is 60 MPa and the mean Poisson's ratio is 0.25. These mechanical parameters are the same for all the elements in the homogeneous disc. For the elements in the heterogeneous disc, randomly distributed mechanical parameters following the Weibull probability density function are assigned. A compressive load is then applied through the two contact areas between the loading platens and the disc.

Figure 6.1c and d illustrates the computed fringe patterns of shear stress in the homogeneous and heterogeneous discs, respectively. These two images clearly show that the material heterogeneity has a strong influence on the stress fields in the discs.

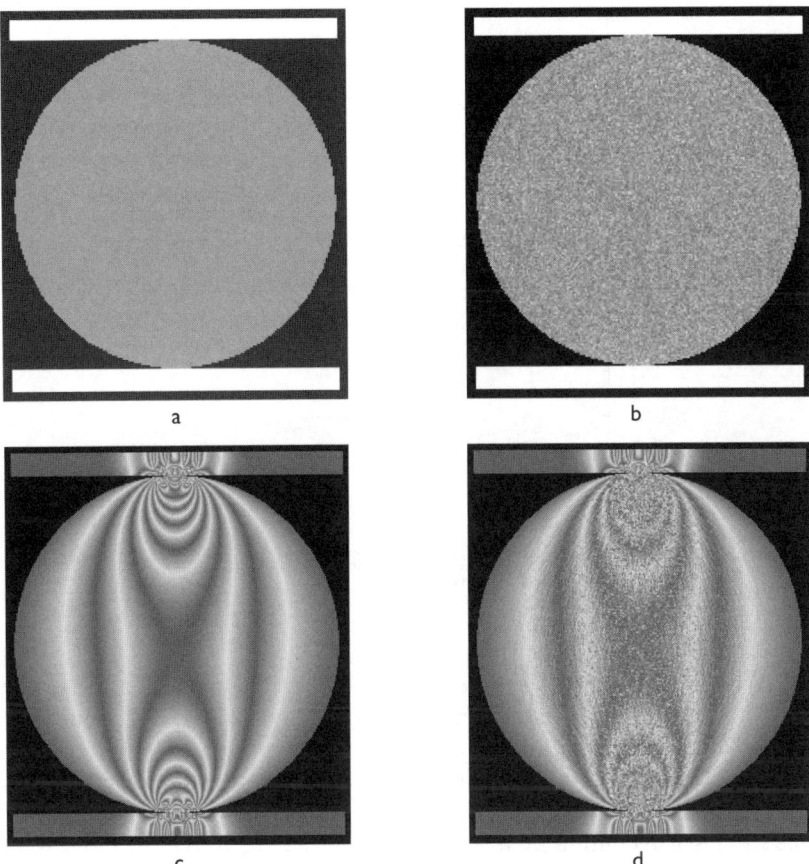

Figure 6.1 Simulated shear stress fields (equivalent to photoelastic fringe patterns) in discs subjected to vertical loading. Images a and b: finite element models with 40,000 elements with vertical loading conditions for homogeneous and heterogeneous discs respectively. The gray colour represents the variation of the elastic moduli of the individual elements; Images c and d: shear stress fields in the homogeneous and heterogeneous discs respectively.

Since it is not possible to obtain the fringe patterns in heterogeneous materials, the numerical results assist in our understanding of such stress fields. For a quantitative evaluation of the heterogeneity effect, Figure 6.2 highlights the normalised stress distributions, σ_x and σ_y, along the loading axis in the homogeneous and heterogeneous discs. Again the influences of material heterogeneity on the stress distribution in the discs are clearly illustrated.

6.2.2 Rock blocks under hydrostatic stress

For a heterogeneous rock block (Figure 6.3) with the same mechanical properties as the heterogeneous disc, the distributions of the shear stress $\tau = (\sigma_1 - \sigma_3)/2$ and the

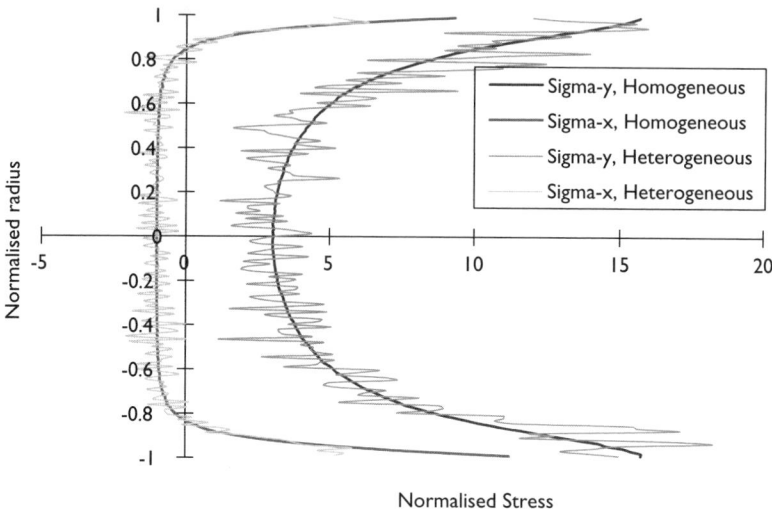

Figure 6.2 Simulated shear stress changes along the line between the loading points of the discs shown in Figure 6.1.

principal stress σ_1 are shown in Figure 6.3b and c. The stresses are calculated by assuming that the external stresses in the vertical and horizontal directions (σ_y and σ_x) are both 1.0 MPa. The magnitudes of the stresses are represented by the gray colour in the images. Most interesting results are found relating to the unexpected stress fields obtained when the heterogeneity in the solids is taken into account. The shear stress computed by classical continuum mechanics theory will be zero if the block is assumed to be homogeneous; however, the results indicate that the shear stress fluctuates and is not necessarily zero, the values depending on the stiffnesses of the individual elements comprising the simulated rock matrix (Figure 6.3b). Furthermore, the local principal stresses are not equal to the applied load (Figure 6.3c); they depend on the stiffness of the elements and vary in both magnitude and orientation.

If we use the relations $\tau = \tau_o + \tau'$ and $\sigma_1 = \sigma_o + \sigma'_1$, where τ_o and σ_o are the mean shear and principal stresses and τ' and σ'_1 are the fluctuations of the associated stresses, we find that the fluctuations of τ' and σ'_1 are up to 70% of the mean stress for the material assumptions made in the simulations. Figure 6.4 also shows the shear stress fluctuation along the cross-section A-A' of the block in Figure 6.3a.

These numerical results illustrate clearly that fractures may develop in a rock even under hydrostatic loading conditions, which, based on classical continuum mechanics, would be considered to be completely safe according to a strength theory based on a shear failure mechanism. Ignoring these stress fluctuations could therefore result in an erroneous conclusion regarding the failure potential. In rocks and rock masses, with a hierarchy of such internal heterogeneities, including the presence of pre-existing fractures on all scales from intra-crystalline flaws to brittle deformation zones several kilometres long, we would therefore expect a hierarchy of such stress perturbations.

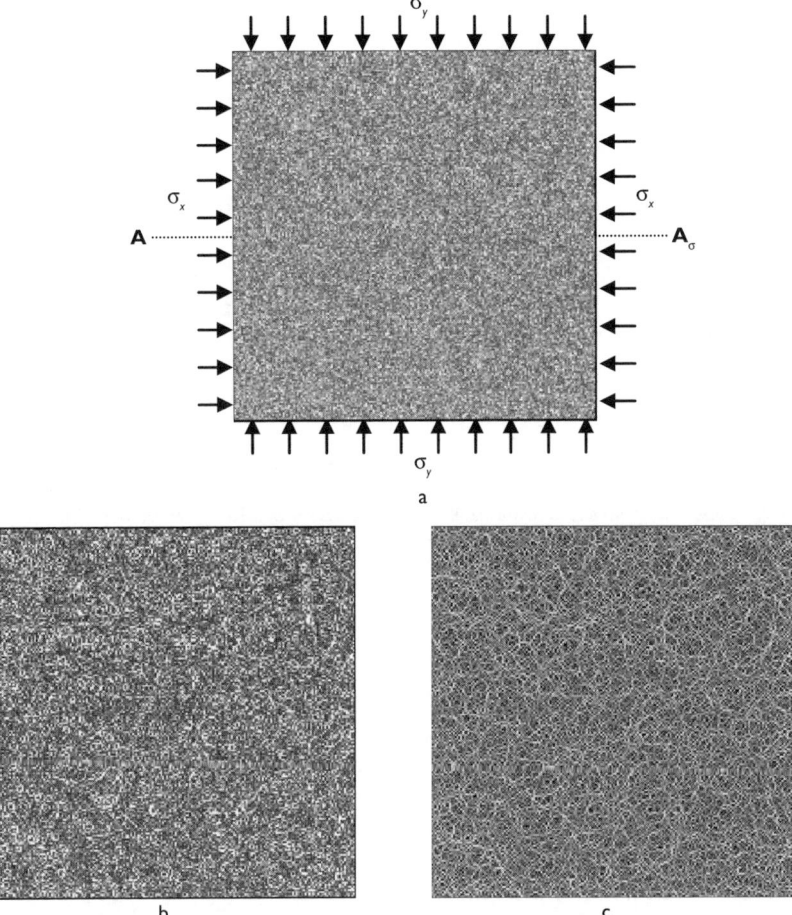

Figure 6.3 Simulated stress fields in a heterogeneous rock block. a) Finite element model with 40,000 elements with boundary conditions, $\sigma_x = \sigma_y = 1$ MPa, and with the gray colour representing the variation of the elastic moduli of individual elements. b) Image of the resulting shear stress τ, with the gray colour representing the value of τ of the individual elements. The value from dark to light is 0–0.7 MPa. c) Image of the simulated principal stress σ_1. The gray colour represents the value of σ_1 of the individual elements. The value from dark to light is 0–1.7 MPa.

This suggests that extra care must be taken in interpreting the results of stress analyses of materials if a highly heterogeneous material is simply assumed to be homogeneous and is loaded close to its failure load. In earthquake research, it has been suggested that monitoring the build-up of stress should be a third component of the strategy for predicting earthquakes, i.e. in addition to detecting precursors and detailed modelling of the earthquake physics (Nature debate, 1999). For such an approach, one has to take into account the influence of heterogeneity on the larger

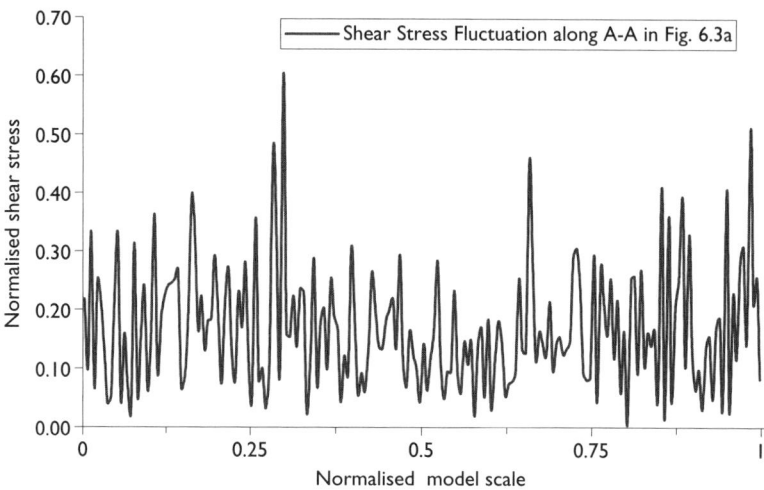

Figure 6.4 Shear stress fluctuations along the cross section A-A of the simulated rock block shown in Figure 6.3a.

scale, i.e. with reference to the stress fields in the Earth's crust. Assuming a highly heterogeneous solid to be homogeneous can result in a misunderstanding of the real failure mechanisms, and this indeed happened in the 1960s when physicists, based on their experiments on glass, claimed that earthquakes were difficult to predict. But the properties of rock are different from those of glass (Eberhardt, 2003), and consequently the failure mechanisms of rock will be different. It is now considered that there will be detectable precursors before large-scale failure in a complex heterogeneous medium such as a rock mass (Nature debate, 1999).

6.3 HETEROGENEITY-RELATED SEISMIC PATTERNS

Five numerical specimens with different homogeneity indices (m = 1.1, 2, 3, 5, and 10) are subjected to uniaxial compressive loading to investigate the micro-activity (acoustic emission), as well as the failure evolution within the stress fields developed. The specimen with m = 1.1 has the most heterogeneous mechanical properties; while the specimens with $m > 5$ are much more homogeneous. The specimen with m = 10 can be considered as an essentially homogenous rock specimen. The stress is imposed by a constant displacement loading rate of 0.002 mm in each step.

Following on from the presentations in Chapter 4 on uniaxial compression, five complete curves of stress and displacement were obtained as shown in Figure 6.5. As the homogeneity index increases, the peak strength of the specimens increases accordingly and it has taken more displacement steps before the peak of the curves is reached. As the displacement and hence the axial strain increases, the stress-displacement curves are close to linear in the initial stage. When the load reaches the

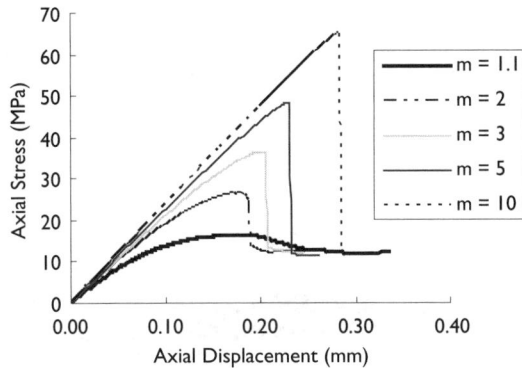

Figure 6.5 Complete stress–displacement curves with the more homogenous specimens showing higher peak strength and more linearity during the failure process.

peak stress, the specimens begin to exhibit crack nucleation into clear micro-fractures. In the cases where the homogeneity index increases, the curves descend more sharply from the peak strength stress, manifesting a more brittle failure mode.

The homogeneity index, *m*, not only influences the failure modes of the rock specimens but also has a strong influence on the acoustic emission patterns. Figures 6.6–6.8 show the sequences of acoustic emission of the specimens with $m = 1.1, 3$, and 10 during the displacement loading. The results show that relatively heterogeneous specimens emit more acoustic emission as precursors of macro-fractures than from relatively homogenous specimens during the first loading stage. According to the degree of inhomogeneity, three basic patterns of seismic activity can be identified: swarm shocks, pre-main shocks, and the main shock.

- Swarm shocks can be found in significantly heterogeneous rocks; micro-fractures and AE activity are scattered here and there throughout the specimen and the acoustic emission could be detected at the initial stage.
- Pre-main shocks can be found in more homogenous rocks and acoustic emissions can be detected both before and after the main macro-fractures are formed.
- Main shock patterns can be found in most homogenous rocks although, in the more homogenous rocks, it is hard to predict the pre-cursors because they manifest only a small number of acoustic emissions. Unlike fracture development in the more heterogeneous specimens, the main cracks resulting in collapse are smoother and more regular. A greater number of acoustic emission events was recorded in the more homogenous numerical rock specimens than in the heterogeneous ones at the peak stress.

The numerical simulations show good agreement with the trends observed in laboratory experimental results (Figure 6.9) and the three basic earthquake patterns observed by many researchers (Mogi, 1985).

108 Rock failure mechanisms

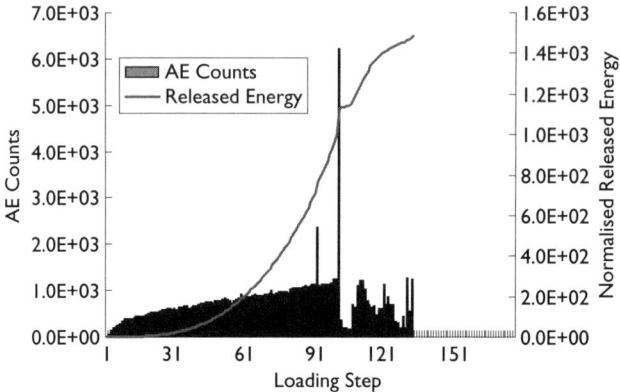

Figure 6.6 Acoustic emission (AE) vs. displacement of the specimen for $m = 1.1$ (significantly inhomogeneous).

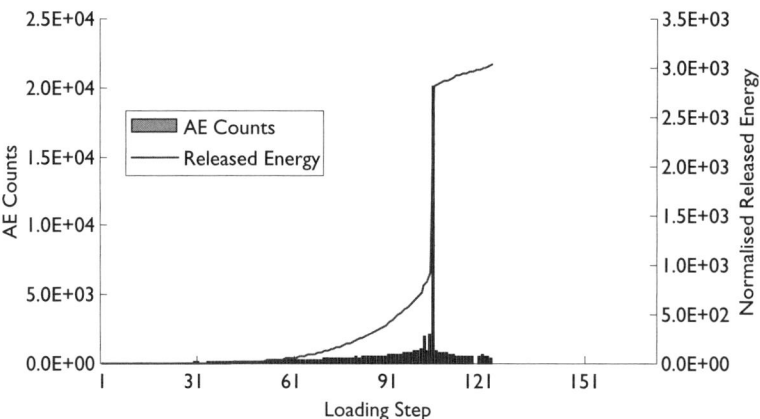

Figure 6.7 Acoustic emission vs. displacement of the specimen with $m = 3$ (intermediate heterogeneity).

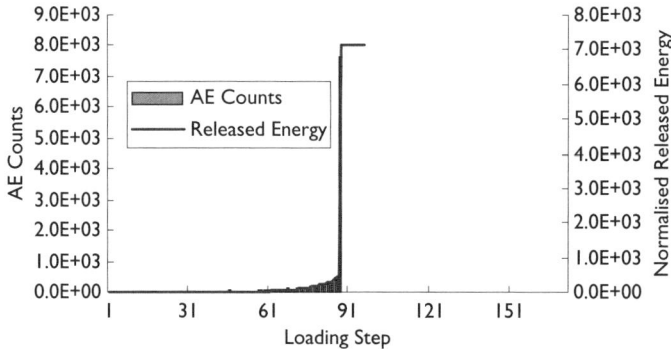

Figure 6.8 Acoustic emission vs. loading step/displacement for the specimen with $m = 10$ (strongly homogeneous).

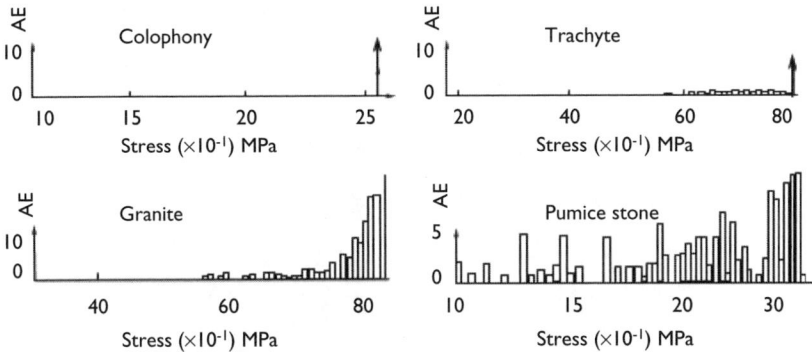

Figure 6.9 Acoustic emissions of resin and three rock types with different variations in their mechanical properties subjected to stress loading. Colophony (pine resin) is the most homogenous, and pumice stone is the most heterogeneous among these four materials. They manifest different patterns of seismic activities: the main shock, pre-main shock, and swarm shock (Mogi, 1985).

6.4 INFLUENCE OF HETEROGENEITY ON CRACK PROPAGATION MODES

The development of fracture mechanics and the application of the concepts, theories and methods of fracture mechanics undoubtedly play an important role in helping the understanding of fracture mechanisms. However, although fracture mechanics does provide this fundamental basis for understanding crack behaviour, the application of fracture mechanics in describing the crack initiation and propagation in heterogeneous materials is onerous. From the viewpoint of studying the progressive failure of rock, as we have seen, the heterogeneity of rock is one of the most important factors, not only having influence on the seismic behaviour induced by the micro-fractures, but also on the paths of crack propagation. Because of the complex nature of the problem, few theoretical approaches have been found in the literature regarding crack propagation in heterogeneous materials. Accordingly, we continue with the numerical simulation approach and study the mode of crack propagation in rocks.

6.4.1 Numerical specimen

In order to study the influence of heterogeneity on crack propagation in rock, two specimens with different homogeneity indices and containing a single pre-existing flaw were numerically loaded in uniaxial compression (as shown in Figure 6.10). The pre-existing flaw was located in the centre of the specimen, with a flaw angle of 45°. The specimen geometry is 200 × 100 mm and has been meshed into 200 × 100 = 20,000 elements. The specimen undergoes plane strain compression, imposed by the relative motion of the upper and lower loading platens at a displacement rate of 0.002 mm per step. Altogether, 200 steps are used to load the specimen until the specimen undergoes complete collapse.

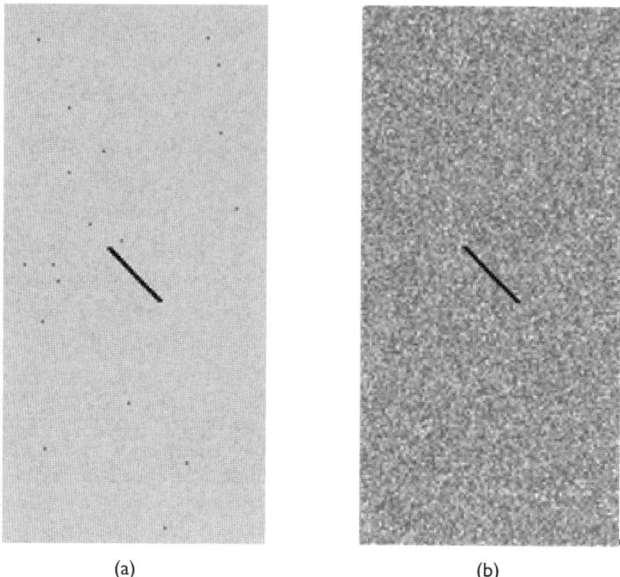

Figure 6.10 Numerical models for homogeneous and heterogeneous rock specimens containing a pre-existing flaw. a) relatively homogeneous model, m = 20 b) relatively heterogeneous model, m = 2.

The mechanical properties for the two specimens are as follows: homogeneity index (m), 2, 20; mean compressive strength (σ_0), 200 (MPa); mean elastic modulus (E_0), 60 GPa; tension cutoff (λ), 20%; frictional angle (ϕ), 45°.

6.4.2 Numerical results and discussion

For comparison, the numerical result for the relatively homogeneous specimen ($m = 20$) is shown in Figure 6.11. The wing-crack propagation in the homogeneous material is in good agreement with the Griffith theory of crack propagation, the typical result calculated from fracture mechanics and as observed in experiments. In Figure 6.12, the numerical results for the relatively heterogeneous specimen are shown. Only six steps from the 200 steps are selected.

Figure 6.12 shows that, although the stress distribution at the beginning of loading is statistically homogeneous on the macro-scale, it is disordered at the micro-scale—due to the micro-scale heterogeneity. As the micro-crack grows and turns into a meso-crack, the path deviates to the regions with higher local tensile stresses. As a result, the crack grows further under the action of the concentrated stresses. Thus, depending on the scale of interest, micro or macro, the rock's micro-structure has a significant effect on the stress and strain fields.

It is interesting to find that the stress field is irregular and is similar to water waves, with the irregular extent depending on the heterogeneity of the specimen. The failure-induced stress redistribution (and particularly the migration of high stress concentration locations due to the crack initiation and propagation) is displayed

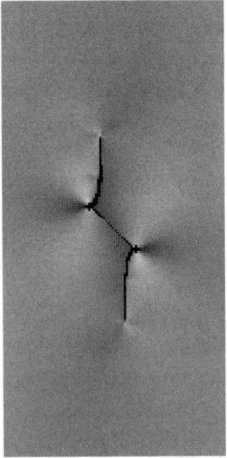

Figure 6.11 Numerical simulation of the crack propagation path in a relatively homogeneous rock specimen ($m = 20$) containing a pre-existing flaw (The gray colour represents the value of the minimum principal stress).

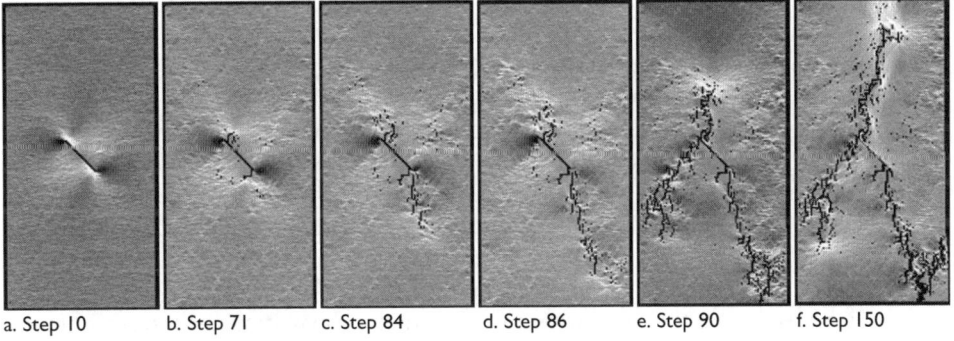

a. Step 10   b. Step 71   c. Step 84   d. Step 86   e. Step 90   f. Step 150

Figure 6.12 Numerical simulation of the crack propagation path in a heterogeneous rock specimen ($m = 2$) containing a pre-existing flaw (The gray colour represents the value of the minimum principal stress).

clearly in these Figures. Figure 6.12 shows how the stress distribution evolves as the load acting on the specimen is increased. It can be seen that the macro-fractures result in intensive stress concentrations around the tips of the fractures.

The load–deformation curve in Figure 6.13a is similar to the typical curve for brittle rock without a pre-existing flaw as discussed in Chapter 4. It is found that, although the constitutive law for the individual elements in the numerical model represents brittleness (with only 10% residual strength), a substantial non-linearity exists before the peak load, and the curve has a clear post-peak region (strain softening) as we have presented before for the un-flawed rock material.

The fracture event rate as a function of deformation is given in Figure 6.13b. The event rate shows the expected features: during the initial deformation or linearly elastic

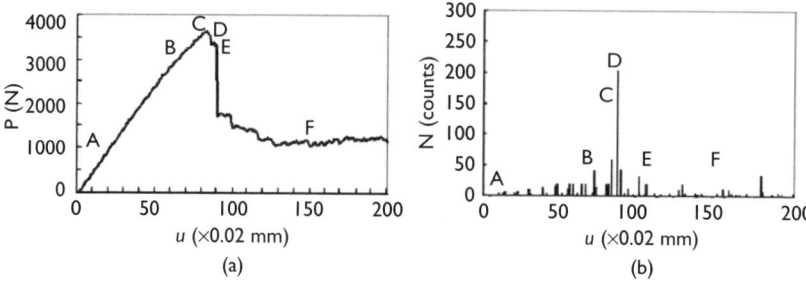

Figure 6.13 a) Load–deformation and b) AE event rate–deformation curves obtained from the simulation of the heterogeneous rock specimen.

phase (marked with the symbol A in the curves), few AE events occur and then an increasing rate of fracture events accompanies the inelastic phase (points B, C and D), consistent with the understanding that the fracture events are generated by the micro-fractures. It can be clearly seen that, for this relatively heterogeneous specimen, many fracture events occurred before reaching the peak load. Comparison between Figures 6.13a and b indicates a good relation between the load–displacement curve and the event rate. Note that the largest stress drop (point E) in the stress curve corresponds to the largest event rate.

In Figure 6.14, we present the locations of the fracture events that occurred around the peak load (i.e. for steps 84, 86 and 90). Each circle represents one fracture event with the diameter of the circle representing the relative magnitude of the released energy. The individual plots of the simulation shown in Figure 6.14 are indicated on the stress–step curve by the marks C, D and E, as shown in Figure 6.13. Note that events are not just distributed near the flaw tips—which is because of the heterogeneity of the specimen material. It is important, and undoubtedly of seismological significance, that the highest fracture event counts do not correspond to the maximum stress, but occur in the post-peak region (point E).

It can be seen from the simulation that the crack propagation path for this heterogeneous specimen is more complex than that of the homogeneous specimen shown in Figure 6.11. In fact, from Figures 6.12 to 6.14, four stages of crack propagation can be identified as follows.

1 In the initial loading stage, the stress at the flaw tip is not high enough to initiate crack propagation. However, from Figure 6.12a for step 10, it is seen that a concentration of high tensile and compressive stress fields is established at the tip in a four-quadrant distribution. The load–deformation curve is linear in this stage.
2 Due to the lower ratio of tensile to compressive strengths of the rock ($\sigma_t/\sigma_c = 1/5$), it is predictable from the stress distribution shown in Figure 6.12a that the cracks will initiate from the two quadrant areas with higher tensile stress around the tips of the flaw (as shown in Figure 6.12b for step 71). Unlike the situation for the homogeneous specimen (as shown in Figure 6.11), micro-fractures also occur in the lower stress area away from the flaw tip. The reason is that in this area there exist some lower strength sites due to the heterogeneity and hence local variations of the material. It is seen from Figure 6.13b that, although the fracture events

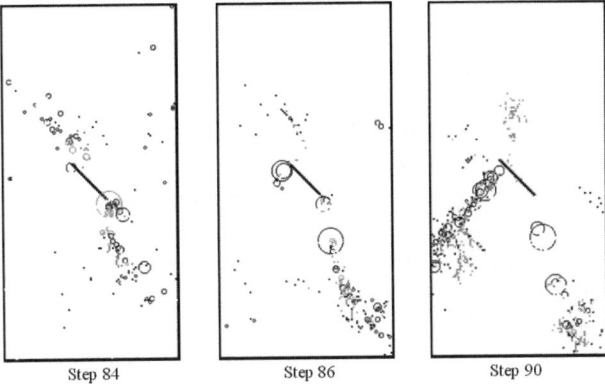

Figure 6.14 AE event locations in the numerical simulation of the heterogeneous rock specimen ($m = 2$) containing a pre-existing flaw.

increase steadily (mainly due to the crack propagation), the deformation and the process of crack propagation are stable before point B (approximately 80% of the peak load), and the load–deformation curve is approximately linear, but a slight non-linearity can be detected in the curve.

3   The crack propagation speed accelerates abruptly when the load reaches point C (step 84), which results in the occurrence of a large number of micro-fractures, five times as large as that in step 71 (point B). At point C, the load reaches its peak. After that, the crack propagation speed continues to increase and the failure process becomes unstable. Two large events with an associated large number of micro-fractures related to the high speed propagation of cracks occur, which results in the related large stress drop in the load–deformation curve (as indicated by point D and E shown in Figures 6.12 to 6.14 for steps 86 and 90). At loading step 90, shear failure dominates and a macro-fracture zone in a different direction is formed, as shown in Figure 6.14.

4   Afterwards, a residual strength, approximately 30% of the peak strength, is gradually reached. The crack propagation speed returns to the normal level and the failure process of the specimen becomes stable again. Only small fluctuations of the crack propagation speed are found in steps 150 and 179.

There is a good relation between the sharp increase of the fracture events and the stress drop because of the 'jump' phenomenon of crack propagation. After a fracture propagates, stress relaxation occurs and the fracture propagation stops; on increasing the displacement, the concentration of stress at the fracture tip may attain the elemental strength again, and the fracture propagation starts anew, etc. The process continues until the macro-fractures occur, which induce a large stress drop (as seen in Figure 6.12e and Figure 6.13 at point E).

Another aspect is that the propagating crack becomes impersistent, forming a rock bridge. This is one of the important phenomena occurring in rock materials and it is believed that the *en echelon* cracks occurring in the Earth's crust are related to this mechanism. Because the material is heterogeneous, crack propagation may cease at sites where the local strength is higher. When further loaded, some weaker sites away from but in the vicinity of the crack tip may reach their failure strength and isolated fractures may initiate—resulting in an impersistent pattern of the crack propagation path.

We have demonstrated here that the natural heterogeneity of rock should be considered in the study of rock failure mechanisms because crack propagation patterns are strongly influenced by the heterogeneity of the rock. To ignore this influence will result in the concealment of many important features of rock deformation and the failure process which include the patterns of the crack propagation path (persistent or non-persistent), fracture event patterns, and rock bridge phenomena. The inclusion of heterogeneity in the studies opens up an extra path for approaching the fracture mechanics of rock and even the faulting mechanisms in seismology.

6.5 THE INFLUENCE OF HETEROGENEITY ON THE MESO-SCALE

In real geomaterials, isolated pre-existing Griffith-type cracks are seldom found to be the major source of micro-cracking, and many other sources have been identified. Grain boundaries, stiff or soft inclusions, low-aspect ratio cavities, as well as suitably oriented interfaces of two different minerals, will initiate micro-cracks when loaded. Therefore, it is essential that the numerical model simulating the failure process of rocks should be able to consider these types of inhomogeneities. Some numerical analyses of the micromechanical behaviour of granular materials have been made but few methods have been developed to model details of the interactions between grains, inclusions and filling materials in multiphase geomaterials.

Figure 6.15 is a photograph illustrating the texture of Lemunda sandstone (Kou *et al.*, 1999) which is quarried on the NE shore of Lake Vättern in Sweden. Around 95% of the rock composition is quartz; the rest is rock fragments (quartz, granite), weathered feldspar grains and minor amounts of small mica flakes, heavy minerals and clay minerals. The mean grain size is 0.35 mm (minimum 0.05, maximum 1.09 mm). It seems likely that the stochastic nature of the arrangement of the strong grains and weak inclusions in the sandstone will influence the formation and interaction of micro-cracks: the strong grains can act as crack barriers, whereas the weak grains or grain boundaries may enhance crack growth. A numerical approach to such characteristics is important for the understanding of the mechanisms of crack initiation, propagation, and interaction in such a rock.

Using digital image processing technology, a new numerical method based on digital image processing has been developed (Chen *et al.*, 2004). The mineral grains in the rock are coloured, so the image from the cross-section of a rock sample can directly reflect the micro-structure of rock through the colours, or through the gray degree. In this way, the actual inhomogeneity of a rock in 2-D can be established through digital image processing technology and a mapping relation established with a numerical method for mechanical simulation. Yue *et al.* (2003) combined digital image processing technology with the finite element method (FEM) and employed digital image processing based FEM to study the effect of geomaterial inhomogene-

Effect of heterogeneity on rock failure 115

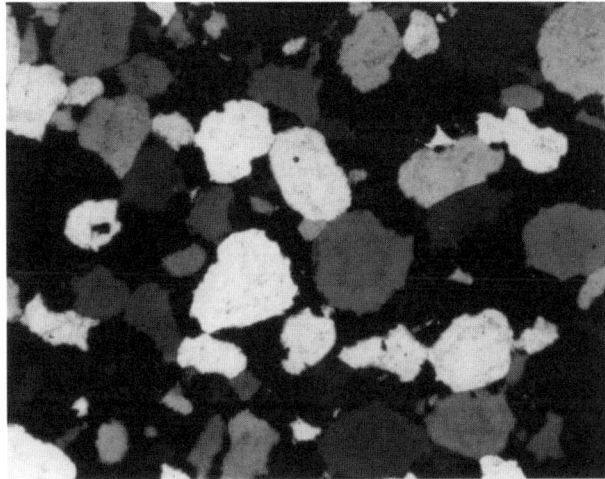

Figure 6.15 Lemunda sandstone (mean grain size is 0.35 mm; minimum 0.05 mm, maximum 1.09 mm).

ity on the tensile stress distribution along the loading axis of the Brazilian indirect tensile test.

Here, we incorporate digital image processing into the RFPA code used to generate some of the diagrams in this book and develop a new meso-mechanical analysis method for rock failure mechanism studies. The method can be used to study the mechanical behaviour of rock failure at the meso-scale efficiently and accurately. To demonstrate this, we conduct a numerical test under uniaxial compression using a granitic rock as an example and use this method to study the fracture phenomena by considering the interface strength between mineral grains, which has a significant influence on the strength of a sample and its failure modes.

6.5.1 Digital image based modelling method

Figure 6.16 shows the process of obtaining the image of granitic rock. If necessary, the sample can be locally magnified for photographing. Its size is 264×163 (height × width) pixels and the actual size is 7.75 mm × 4.79 mm. There are three main minerals in the sample: feldspar, mica and quartz.

The image contains many pixels and in fact is a pixel matrix. When the image has been scanned, the minerals and their actual geometrical distribution in specific rock samples can be individually identified. Figure 6.17 shows the results after the image analysis. In the analysed image, the dark gray/black indicates the interface between mineral grains, which is obtained by digital image processing software.

6.5.2 Numerical model based on the digital image

In order to carry out the numerical analysis, the analysed image micro-structure must be transformed into the mesh data that can be accepted in the finite element method

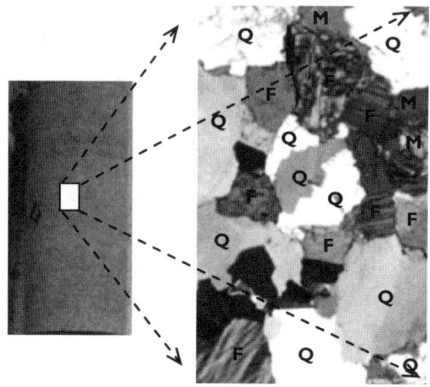

Figure 6.16 Process of obtaining the rock image (Q = Quartz, F = Feldspar, M = Mica).

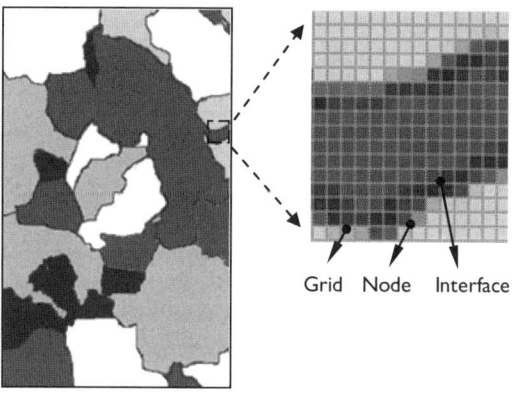

Figure 6.17 Analysed image and an example of the transformation from pixel to grid.

(FEM). When the image is magnified several times, it can be seen that every pixel is a square (Figure 6.17), so every pixel in the analysed image can be regarded as a square element in the FEM and the image can be divided into 264 × 163 elements for the FEM calculations. By using the algorithm adopted in the code, the micro-structure of the sample has been identified. Each mineral type and the interface can be assigned parameter values by an assignment function in the code. These mechanical parameters are shown in Table 6.1. In an extension of the study, Section 6.5.4, the elastic modulus of the interface is half the weak mica mineral's value, and its strength is taken as 10%, 20%, 30%, 40%, and 50% of the mica's strength. Five numerical models under uniaxial compression were carried out to study the influence of the interface strength.

Figure 6.18 is a picture of the RFPA model with interface strength of 0.12 GPa after transformation, in which the elastic properties of the elements are expressed in the gray shading system. In the model, the stronger the brightness of an element, the higher are the parameter values. In the numerical simulations, the plane stress problem is considered. In

Effect of heterogeneity on rock failure 117

Table 6.1 Mechanical properties of minerals and interface for the reference granite.

Mineral type	Elastic modulus [GPa]	Compressive strength [GPa]	Poisson's ratio
Feldspar (F)	69.7	2.4	0.301
Mica (M)	88.1	1.2	0.248
Quartz (Q)	95.6	2.8	0.079
Interface (I)	44.0	0.12, 0.24, 0.36, 0.48, 0.60	0.25

Figure 6.18 Numerical model used for the simulation.

order to simulate the progressive failure process of a model under loading, the exterior load is applied step by step in a displacement controlled manner at a rate of 0.0003 mm/step. It takes approximately 100 steps to load the sample to its load capacity.

6.5.3 Simulation results for uniaxial compression

The load–displacement curve and associated AE events obtained from the simulation are shown in Figure 6.19. Its characteristics are similar to those that have been reported by many researchers using either an experimental or numerical method (e.g., Wawersik and Fairhurst, 1970; Tang *et al.*, 2000). For convenience of description and discussion, the curve is annotated with the letters A to E in Figure 6.19. The curve can be divided into four stages, a linearly elastic deformation stage (AB), a non-linear damage deformation stage (BD), a post-peak stage (DE), and a shearing and slipping stage (after E), with the C point being an additional point for later discussion. Although in the simulation we regard a single mineral grain as an homogeneous material, the curve presents typical heterogeneous characteristics which have the substantial non-linear deformation stage (BD) and the post-peak stage (strain softening: DE+). The reason is that each kind of mineral grain will express a different mechanical response, even when under an identical loading level.

From Figure 6.19, it is seen that there is a good relation between the force–displacement curve and the AE event rate: every large stress drop in the curve has

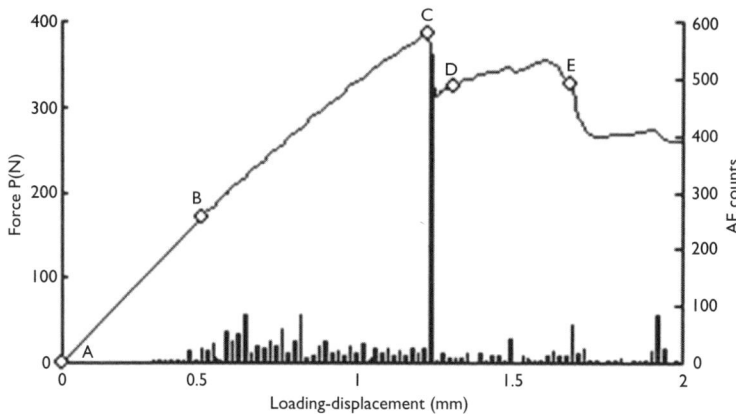

Figure 6.19 Force–displacement curve and associated AE events.

a high AE event rate associated with it. During the initial deformation, the linearly elastic stage (AB), there are scattered AE events, caused by weak minerals (e.g., mica) and grain boundaries. An increasing rate of AE events accompanied the non-linear damage deformation stage (BD). The highest AE event counts occurred in the post-peak phase, corresponding to a large stress drop.

Figure 6.20 shows the shear stress distribution during the loading process, with the associated fracturing shown in Figure 6.21. The individual images in Figure 6.20 and Figure 6.21 are indicated on the force-loading displacement curve by the letters A-E. When the displacement is applied, the shear stress in the model is not uniformly distributed but concentrated at the boundaries of the quartz-quartz grains and quartz-feldspar grains (Figure 6.20A). Both Figures 6.20 and 6.21 show the simulated initiation, propagation and coalescence of the fractures at different loading levels. Stress-induced cracks are mainly located in the direction of loading. The isolated fractures occur firstly at the location of the mica and feldspar grains boundaries (Figures 6.20 and 6.21B) and begin to cluster at stage C. Cracks propagate along the grain boundaries so most of them are inter-granular cracks. If a crack encounters a strong grain, it may stop propagating.

In Figure 6.21, the ellipses indicate the locations of the cracks at different loading levels. Because some mineral grains and the background colour of the picture are both black, the cracks are somewhat hidden by their colour. But, by contrasting Figures 6.20 and 6.21, it can be seen that most of cracks occur along the grain boundaries.

6.5.4 Influence of interface strength

The stress–strain curves resulting from the simulations with the mica interface strength being taken as 10%, 20%, 30%, 40%, and 50% of the mica value are shown in Figure 6.22. These curves can also be divided into four stages: a linearly elastic deformation stage, a non-linear damage deformation stage, a post-peak stage, and a shearing and slipping stage. Although in the simulation we regard a single mineral grain as homogeneous material, the curves present typical heterogeneous material

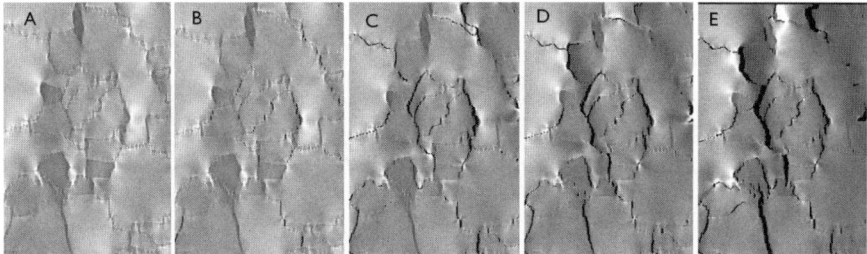

Figure 6.20 Distribution and redistribution of shear stress in the simulation (the lighter the gray colour, the higher the shear stress). See Figure 6.19 for the location of these images relative to the complete load–displacement curve.

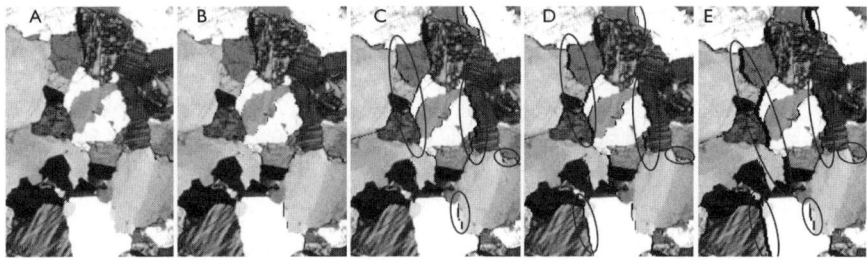

Figure 6.21 Progressive fracture process in the simulation.

characteristics—because each type of mineral grain will exhibit different mechanical responses, even for identical loading levels. With the interface strength increasing, the curves present an increasingly brittle trend.

Figure 6.23 indicates the influence of local interface strength on the overall strength of the numerical samples. When the interface strength is increased from 10% to 50% of the mica's strength, the strength of the samples increased from 51 MPa to 373 MPa, i.e. an increase of more than six times.

Figure 6.24 shows representative failure modes for each numerical test under uniaxial loading. The individual images in Figure 6.24 are indicated on the stress–strain curves by the letters A-E in Figure 6.22. In the numerical tests, we find that, when the load is applied, the stress in the model is not uniformly distributed but concentrated at the boundaries of the quartz-quartz grains and quartz-feldspar grains and the initial damage occurs at the boundaries. As failure progresses, the micro-cracks preferentially propagate along the boundaries in the direction of loading, so intergranular fractures are mainly observed.

However, when the strength of the interface reaches 30% of the mica's strength, trans-granular cracks appear, and the cracks can propagate through the feldspar grains. With the interface strength increasing further, some cracks also propagate through quartz grains. Hopefully, we have demonstrated here that this digital microstructural replication coupled with the numerical FEM simulation is a powerful method of studying the influence of the detailed micro-structural parameters on the overall rock failure mechanisms.

120 Rock failure mechanisms

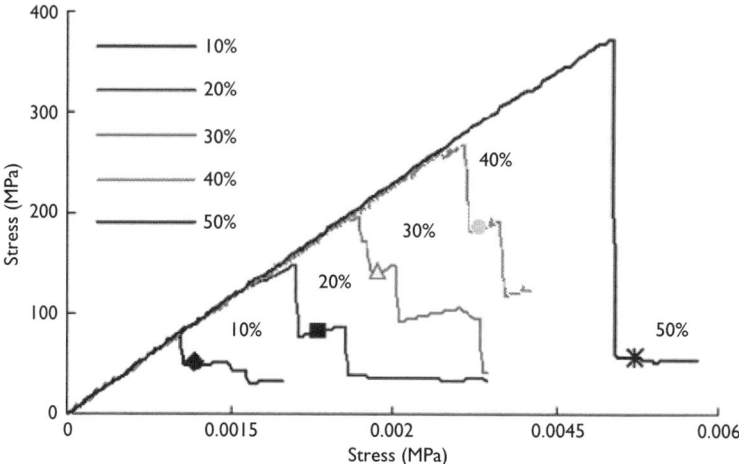

Figure 6.22 Numerically obtained stress–strain curves for the rock with the interface strength being taken as 10%, 20%, 30%, 40%, and 50% of the mica strength.

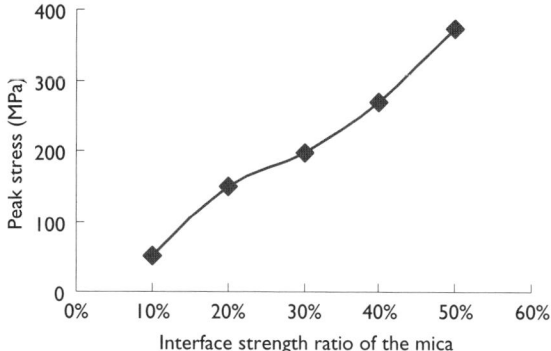

Figure 6.23 Influence of interface strength on the overall strength of the samples.

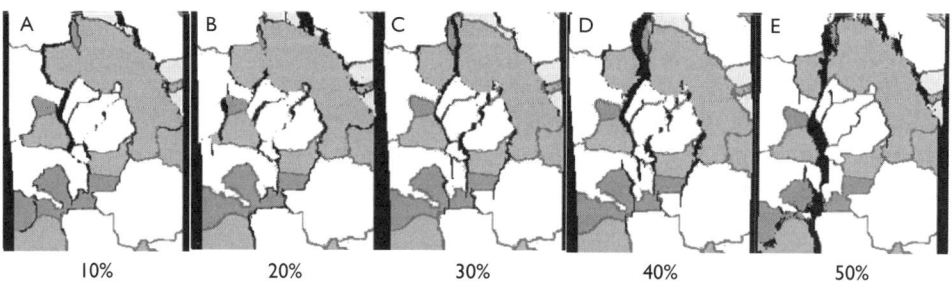

Figure 6.24 Influence of interface strength on the sample failure modes (percentages indicate the interface strength vs. the mica strength).

Chapter 7

The effect of rock anisotropy on rock failure

7.1 INTRODUCTION

Anisotropy (having different properties in different directions) is a well-known feature of rock and is of interest in the field of rock mechanics and engineering. In the context of elasticity theory, the numbers of mechanical properties required to characterise a rock are as follows: isotropic (2—Young's modulus and Poisson's ratio); transversely isotropic (5—two Young's moduli, two Poisson's ratios and a shear modulus); orthotropic (9—three Young's moduli, three Poisson's ratios and three shear moduli); and general anisotropy (twenty one constants). These cases are illustrated in Figure 7.1 and the associated properties are explained further in Hudson and Harrison (1997) via the elastic compliance matrix.

Most elastic analyses assume that the rock is isotropic. In some cases, transverse anisotropy can be accommodated, as for example when using hollow inclusion overcoring devices for stress measurement. Orthotropy is rarely considered and, to the authors' knowledge, no one has ever attempted to measure all 21 elastic constants for a generally anisotropic rock.

Many types of rock, such as sedimentary rocks and metamorphic rocks consisting of a fabric with preferentially parallel arrangements of flat or long minerals may be considered to be transversely isotropic having different properties parallel and perpendicular to the dominant fabric, Figure 7.1b. Shale, siltstone and claystone exhibit strong inherent anisotropy, manifesting itself in a directional dependence of the deformational characteristics. Also, isotropic rock material which is cut by regularly orientated discontinuities (joints and faults) may be considered to have transversely isotropic mass properties, e.g., granite and basalt. We expect the overall anisotropic mechanical behaviour to be partly a reflection of the anisotropic micro-structure. The study of the mechanical behaviour of sedimentary rocks, especially shale and mudstone, is of particular interest to the oil exploration industry, as well as to civil and mining engineering. These materials are often unavoidable in the foundations of a broad range of civil structures, in tunnelling and other underground excavations.

Experimental studies on transversely isotropic rocks indicate that the maximum axial compressive strength is associated with specimens in which the bedding planes are either parallel or perpendicular to the loading direction. The minimum strength is typically associated with failure along the weakness planes, which corresponds to specimen orientations within the range 30°–60°. Failure criteria have been developed for transversely

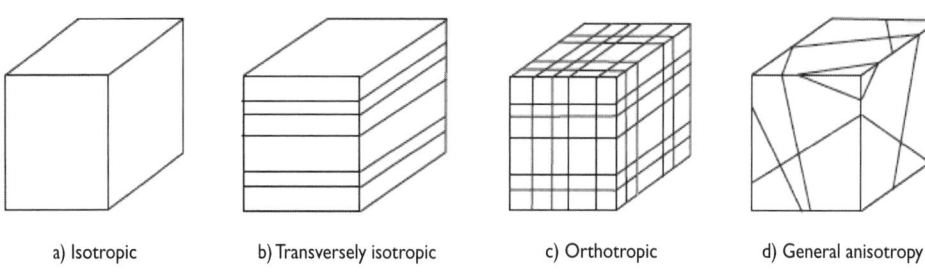

Figure 7.1 Types of elastic anisotropy.

isotropic rocks in order to characterise the variation of rock mechanics properties with the orientation angles and when subject to various confining pressures. Jaeger (1960) introduced the basic analysis for rocks containing a well-defined discontinuity (a single plane of weakness) and there have been many modifications to the basic idea.

Since, by definition, the mechanical properties of anisotropic rocks must be established by testing in different directions, the number of specimens required for a full characterisation is much greater than that required for isotropic rocks. Moreover, for laboratory testing it is often difficult to obtain an adequate number of specimens with uniform properties and at different angles to the main anisotropy direction because of the effects of heterogeneity, weathering, mineral composition, texture, fracture or joint characteristics, and core drilling directions. For this reason, many experimental studies have focused on the preparation and mechanical properties of artificial transversely isotropic rocks. Different researchers have obtained different results with different components introduced into isotropic rock-like materials. Unlike natural anisotropic rocks, the artificial transversely isotropic rock-like specimens are more uniform and easy to produce to meet the requirements of the researchers.

In this Chapter, we will use numerical simulations to illustrate the failure process and strength behaviour of transversely isotropic rocks subjected to uniaxial loading. Seven rock specimens are simulated with varied dip angles of the layers.

7.2 NUMERICAL MODELS

The mechanical properties of transversely isotropic rocks depend on the mechanical properties of the individual constituent materials. So for this demonstration, we consider seven simulated rock specimens composed of two different materials and with dip angles of 0°, 15°, 30°, 45°, 60°, 75°, and 90° with respect to the major principal stress. The specimen geometry is 50 mm × 100 mm which has been discretised into $100 \times 200 = 20,000$ elements. The strength and the elastic modulus of material A are about double those for material B. The mechanical properties of the specimens are as follows.

Material A: homogeneity index (m), 4; mean compressive strength (σ_0), 200 MPa; mean elastic modulus (E_0), 50 GPa; Poisson's ratio (v), 0.25; tension cutoff (λ), 10%; frictional angle (ϕ), 30°.

Material B: homogeneity index (*m*), 3; mean compressive strength (σ_0), 100 MPa; mean elastic modulus (E_0), 25 GPa; Poisson's ratio (*v*), 0.25; tension cutoff (λ), 10%; frictional angle (ϕ), 30°.

The light gray stiffer and stronger rock layers and the dark gray softer and weaker rock layers in Figure 7.2 are composed of materials A and B respectively.

In all cases, the specimens undergo plane strain compression, imposed by a relative motion of the upper and lower loading platens and by applying an external displacement at a constant rate of 0.002 mm/step in the axial direction, analogous to a servo-controlled laboratory test. We also assume that each element has a Mohr-Coulomb failure criterion envelope with a tensile cut-off on the mesoscale, and the failure of elements may be either in shear or in tensile mode. No anisotropic behaviour is considered for each of the individual simulation elements in the FEM program. At each step, the stress states in some elements may satisfy the strength criterion. Such elements are then damaged and weakened according to the rules specified by the strength criterion.

The complete stress–strain relations for the simulated rock specimens are illustrated in Figure 7.3. In Figures 7.4 to 7.6, the following relations are shown: peak strength and rock layer dip angle with respect to loading direction; peak strain and dip angle; elastic modulus and dip angle. Figure 7.4 shows that the uniaxial compressive strength for the specimen with dip angle $\beta = 0°$ is close to that for the specimen with dip angle $\beta = 15°$, but the strength then decreases as the dip angle increases. The peak strength reaches the minimum when the dip angle increases to 60°. Note that the single plane of weakness theory predicts the weakest strength when the dip angle in degrees is (45 + (ϕ/2)) (Hudson and Harrison, 1997) which is equal to 60° for these simulations where ϕ has been assumed to be 30°. When the value of the dip angle continues to increase to 75° or 90°, the strengths increase. This is because at these angles slip on the interface or in the weaker material cannot occur.

It is also found that the axial strain corresponding to the peak strength for specimens follows the same trend as the peak strength (Figure 7.5). With the dip angle increasing, the peak strain decreases at first and increases with higher dip angles. When the dip angle is 0°, the strain is controlled by the weak rock layers but, when the dip angle increases to 45°–60°, slip occurs more easily. When the value of the dip

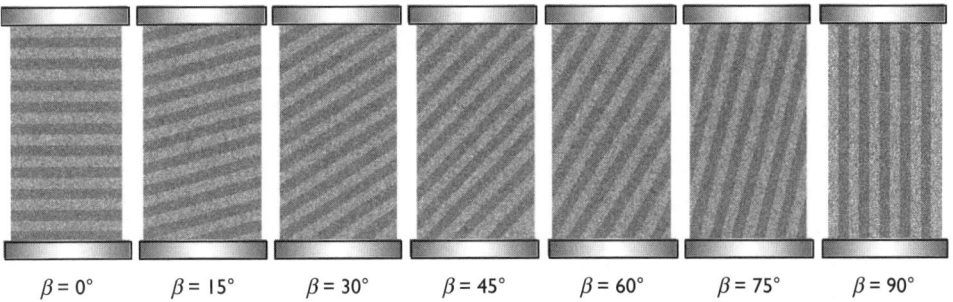

Figure 7.2 Seven transversely isotropic rock specimens with layers having different dip angles and composed of two materials, the lighter gray material being stiffer and stronger; β is the angle of the layers with respect to the axial direction of loading.

124 Rock failure mechanisms

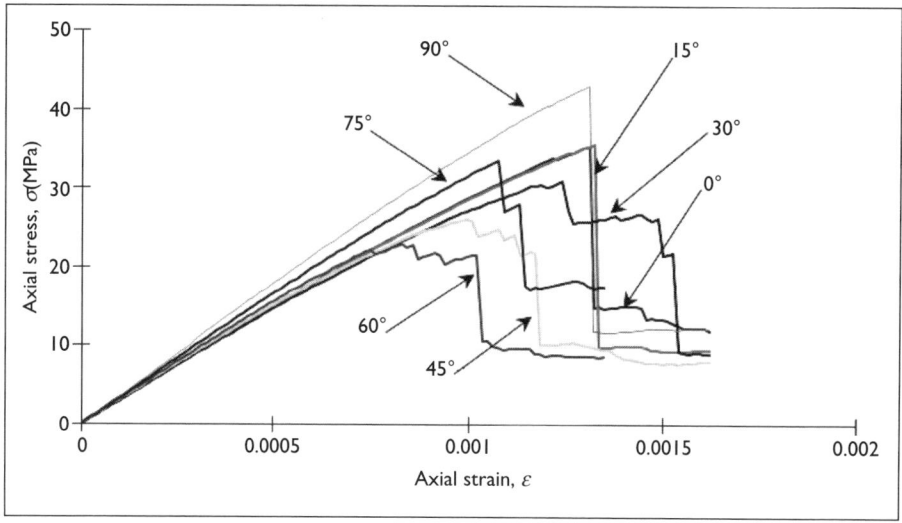

Figure 7.3 Complete stress–strain curves for seven rock specimens with different dip angles relative to the transverse anisotropy plane (cf. Figure 7.2), showing that the dip angles influence the shape of the curves during the failure process.

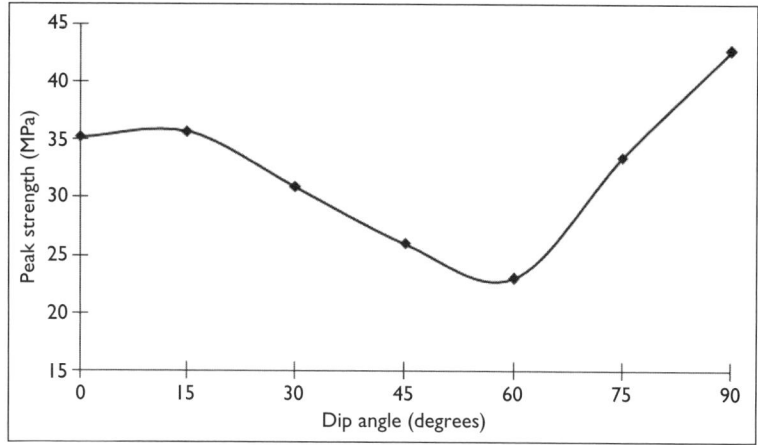

Figure 7.4 Plots of peak strength vs. dip angles for the seven rock specimens. The peak strength of the transversely isotropic rock decreases for dip angles 15° to 60°, and increases for dip angles 60° to 90°.

angle increases to more than 75°, the rock specimens have a higher stiffness and the strengths are controlled by the stiffer material because it carries more stress. Higher dip specimens have a greater stiffness than lower dip specimens.

As pointed out by Tien et al. (2006), the observation of failure processes and failure modes can provide the feedback necessary to verify a new failure criterion. Therefore, it is helpful to numerically model the failure processes and failure modes of

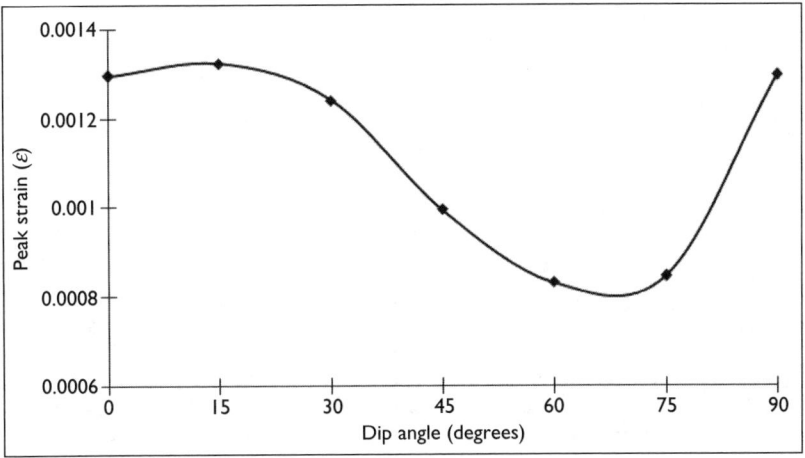

Figure 7.5 Plots of maximal axial strain vs. dip angles for the seven simulated rock specimens.

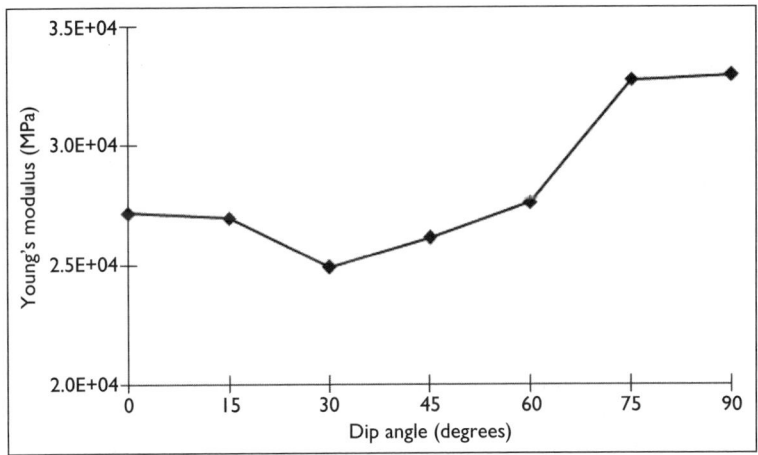

Figure 7.6 Plots of Young's modulus vs. dip angles for the seven simulated rock specimens.

anisotropic rocks at different stress levels. We have modelled seven specimens under uniaxial loading by reproducing the failure process associated with the increase of loading steps. It is not practical to illustrate pictures for all the loading steps but in Figure 7.7 we present four typical images to show the minor principal stress fields and the fracturing during the failure process of the rock specimen with the layer dip angle at $\beta = 45°$. The gray degree indicates the value of the minor principal stress (a lighter shade represents a higher stress value), with the lightest parts appearing in the weak layers composed of material B. At the 70th loading step, many cracks appear. At the 75th loading step, unconnected cracks nucleate and propagate to coalesce one by one

126 Rock failure mechanisms

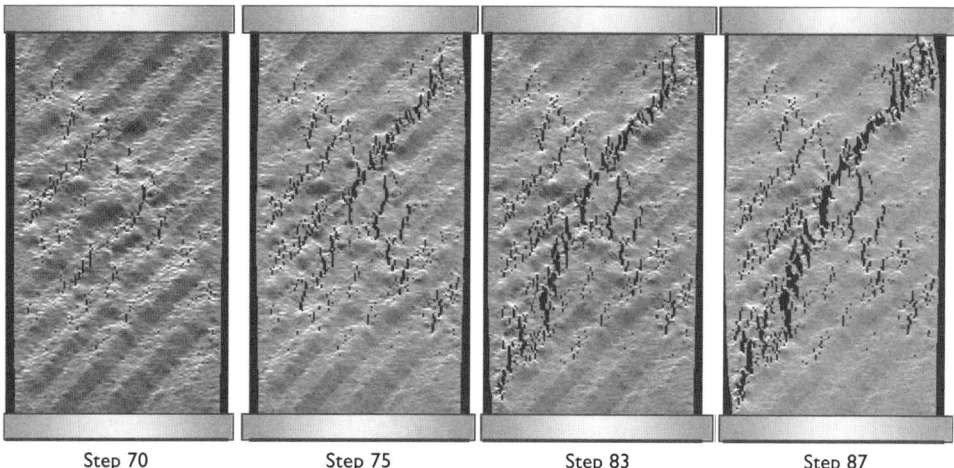

Figure 7.7 Minor principal stress and fracturing during the failure process of the simulated rock specimen with a dip angle $\beta = 45°$.

along the line from the upper right point to the centre of the specimen. The macro-crack forms and then propagates down through the stiffer layers. The macro-crack from the lower left corner to the centre forms and propagates upwards at the 83rd loading step. These two cracks gradually coalesce and lead finally at the 87th loading step to sudden failure along the fracture. However, due to the existence of the residual strength, even at this point we cannot consider the rock specimen to have lost all its load-bearing capability.

During the failure process, two different failure modes can be observed when the dip angle is varied. Figure 7.8 shows the failure modes and corresponding displacement vectors for the specimens with β varying from 0° to 90°. All the stress–strain curves in Figure 7.3 exhibit strong linear elasticity before the rock specimens reach their peak strengths; however, when the dip angle approaches 0° or 90°, the complete stress–strain curves exhibit more brittle type failure behaviour, and they show clear non-linearity and more ductility after the peak strength when the angle is between 30° and 75°. The nature of failure of the simulated rock specimens with low and high dip angles is caused mainly by splitting failure due to tensile stress; while the failure of specimens with medium dip angles is caused mainly by shear stress and the formation of faults.

Figure 7.8 shows that, for a low-dip specimen ($\beta = 0°$, $\beta = 15°$ or $\beta = 30°$) under axial compression, the increase in normal stress on the weaker layers is greater than the increase in shear stress; thus, cracks are unlikely to develop along the weaker layers of the B material. Instead the development of numerous major cracks occurs which initially are more or less parallel to the direction of the applied load, and finally coalesce and result in failure. The failure behaviour and patterns of crack development for the low-dip specimens are identical to isotropic rocks subjected to uniaxial loads (Tien *et al.*, 2006). This means that, for low dip specimens, the failure mode is not too influenced by the discontinuities in the rock. For specimens with medium-dip angles

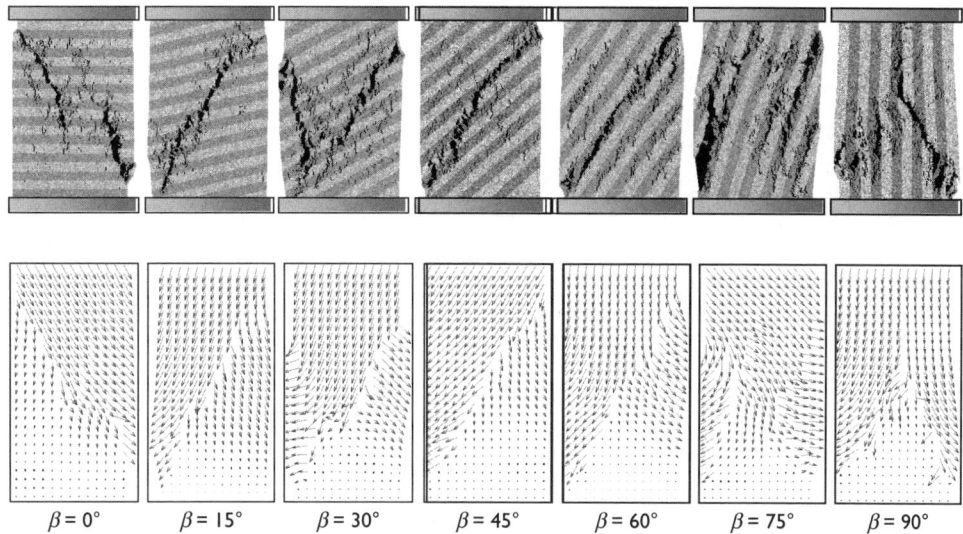

Figure 7.8 Failure configuration and displacement vectors at the collapse point for the seven specimens with different dip angles.

($\beta = 45°$, $60°$, $70°$), the failure is mainly caused by failure along the discontinuity, which in turn, leads to a delamination type of failure of the discontinuity.

For specimens with orientation angle ($\beta = 90°$), the failure occurs with the development of a major crack parallel to the direction of the applied load, and then the crack propagates as a shear zone. In such a situation, layers of material A and material B have the same axial deformation under an axial load but, since its stiffness is higher, the material A layer will carry more stress than material B. Since the specimen has no lateral confinement under a uniaxial loading condition and because the tensile strength of the weaker material B is less than that of material A, delamination occurs along the weaker layer B and is initiated by a 'macro-split' in the direction parallel to the applied load. The stronger material A is therefore further subjected to a moment (caused by distortion of the specimen), which explains the existence of a buckling failure mechanism.

The displacement vectors (Figure 7.8) show that the displacement along the fracture is larger and it causes sliding along the weak layers composed of material B. The numerical simulations reveal that the deformation mode remains symmetrical only for a specimen with vertical layers.

In order to validate the trends generated by the numerical model presented here, a series of numerical simulations were compared with the experimental laboratory results obtained by Tien and Kuo (2001). The mean value of elastic modulus, uniaxial compressive peak strength and Poisson's ratio of material A are taken as 9.1 GPa, 200 MPa, and 0.2 and of material B are taken as 8.24 GPa, 100 MPa, and 0.19, respectively. Both the homogeneity indices, m, are 4. It should be noted that the mechanical mean values of the elements in the numerical model are different from the values obtained from the macro-response of the whole specimens due to the heterogeneities at the mesoscopic scale.

128 Rock failure mechanisms

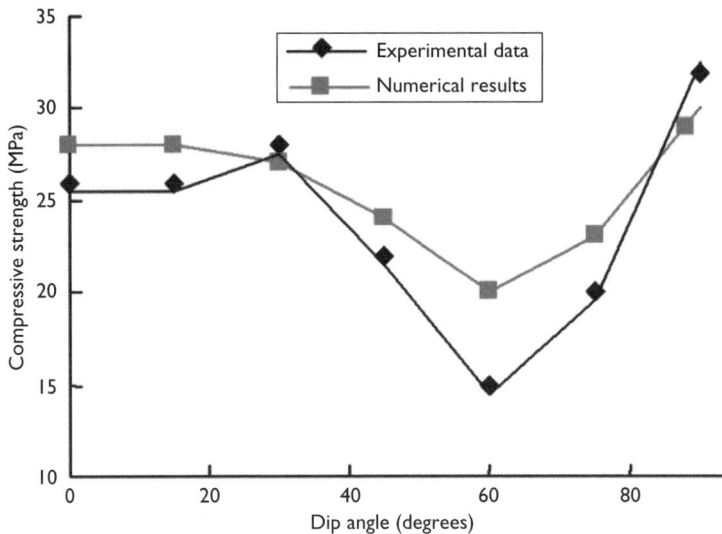

Figure 7.9 Experimental results and numerical simulation for the validation exercise.

Comparing the numerical results with the experimental data as shown in Figure 7.9, the basic trends are consistent. Also, the quantitative predictions seem to be quite reasonable. A close examination of the results, however, reveals that the data differences for 60° are quite significant, due to the fact that the numerical model is two-dimensional. Apparently, a more sophisticated approach could be employed by using three dimensional modelling but this would inevitably involve in a set of additional parameters regarding the layer geometry. In the paper by Cuisiat and Hudson (1993), an analytical approach concerning the stresses around an elliptical crack is combined with a statistical approach in order to study the influence of rock anisotropy on potential borehole breakouts. However, it is evident from the simulations in this Chapter and indeed throughout the book that using the finite element modelling to capture the mechanics together with the micro-structural characterisation using a statistical distribution is sufficient to provide a suitable approach for detailed research into the influence of anisotropy on rock failure.

Chapter 8

Loading, unloading and the Kaiser Effect

8.1 INTRODUCTION

The monitoring of acoustic emission activity produced during deformation is one of the well-known methods for observing material damage. Kaiser (1950) discovered that the acoustic emission, AE, from a stressed metal sample is zero if the applied stress is less than the previously applied maximum stress, a phenomenon now termed the Kaiser Effect. The potential value of this effect in rock mechanics is the possibility of establishing the *in situ* stresses in a rock mass by measuring the acoustic emission from site investigation cores loaded in uniaxial compression and establishing the applied stress at which significant acoustic emission begins to occur. Because the quantity 'stress' is a second order tensor with six independent components, it is essential that such tests for any specific core location are conducted on at least six sub-cores taken from the main core.

Kaiser Effect stress measurements on rock cores have been reported by Villaescusa (2002) and an overall review of the subject has been provided by Lavrov (2003). More recently, Zhang and Stephansson (2010) have also provided an overview of the subject. Hunt *et al.* (2003) used the PFC code to model the Kaiser Effect in rocks summarising as follows: "The numerical modelling package PFC^{2D} was employed for creation of a synthetic cored specimen and simulation of uniaxial compression tests. The ability of the numerical model to reproduce the Kaiser effect and the deformation rate analysis phenomenon was confirmed, and a direct comparison was made between laboratory and numerical observations. The link between the Kaiser effect/deformation rate analysis and development of micro-cracks was established."

Thus, the existence of the Kaiser Effect indicates that rock damage is irreversible and the consequential rock's memory of its stress history can be used to estimate the maximum previous stress applied to the rock. Nevertheless, this is still a controversial subject because of the difficulty of uniquely relating the rock damage to the most recent stress field experienced by the rock and determining the stress level at which the acoustic emission significantly increases. It is not possible to use an analytical model to study the Kaiser Effect due to the complications of the detailed material structure and the progressive micro-structural breakdown process that we have highlighted in previous Chapters.

Based on statistical strength theory and damage mechanics, Tang *et al.* (1997) proposed a one-dimensional statistical damage model to study the AE Kaiser Effect in rock failure. A relation between the damage variable, D (the proportion of failed elements), and the AE counts, N, has been derived for elements following the Weibull distribution. By using this model, Tang *et al.* (1997) provide an explanation for the mechanism of the Kaiser Effect in heterogeneous rock failure. One may assume that there exists a relation between the AE events and the micro-fracture forming events but the accuracy of this technique for assessing the damage will rest on the completeness of the AE record and the relation assumed for the AE data and the fracture geometry. In the simulation, every element failure is considered to be a source of an acoustic event since the failed element must release its elastic energy stored during deformation. Therefore, by recording the proportion of damaged elements, D, the changing proportion of surviving elements (1–D), and the associated AE event rate, magnitude and the locations and amounts of energy release, it is possible to provide an enhanced understanding of the Kaiser Effect.

8.2 NUMERICAL SIMULATION

A rectangular specimen is used to simulate the fracture process of rock subjected to uniaxial compression. The detailed mechanical and geometrical properties of the specimen are as follows: homogeneity index (m), 1.5; mean compressive strength (σ_0), 200 MPa; mean elastic modulus (E_0) 60 GPa; Poisson's ratio (v), 0.25; tension cut-off (λ), 10%; frictional angle (ϕ), 30°.

The numerical specimen is shown in Figure 8.1. Its low homogeneity index, $m = 1.5$, means that the rock material is significantly heterogeneous. The gray scale of the specimen indicates the values of the mechanical properties (strength and elastic modulus) of the rock material. In order to study the effect of loading history on the AE evolution during the rock fracture process, this specimen is loaded and unloaded six times before it fails. After the specimen reaches its peak strength, cyclical loading and unloading is conducted another six times. After reaching a specified maximum load in each loading cycle, the specimen is unloaded to a given non-zero prescribed value. All these cyclic loadings are carried out with a displacement rate of 0.0015 mm per step. The total number of loading and unloading steps is 720, which can be regarded as a rather slow loading process.

Figure 8.1 Numerical specimen with inhomogeneous elements, $m = 1.5$.

The AE counts for this cyclic loading are plotted in Figure 8.2 for the simulation and in Figure 8.3 for experimental results. The relation of the cumulative AE counts and stress (as a percentage of the strength) is depicted in Figure 8.4. The stress–strain curve under cyclic loading and unloading is shown in Figure 8.5. The AE spatial distribution in rock failure process is presented in Figure 8.6.

In Figure 8.2, during the first loading, AE occurs due to the local damage of the rock specimen when the stress reaches 32% of the peak strength of the specimen (see Figure 8.2, 1st cycle). No AE event was observed in the following unloading. During reloading, the number of broken elements is not increased until the loading reaches the maximum stress level of the last cycle (see Figure 8.2 2nd cycle). During the rest of the cyclic loading and unloading, this phenomenon is repeated again. In line with the analytical results (Tang et al., 1997) and the experimental results in Figure 8.3 (Li and Norlund, 1993), the numerical results reproduce the existence of the AE Kaiser Effect at a low damage stage of the rock failure process.

The Kaiser Effect is not exactly observed during all the cycles. During the sixth cycle, although the reloading does not reach the previous maximum stress, AE still occurs (see the 6th cycle in Figure 8.2). This phenomenon is known as the Felicity effect (i.e. when the AE recommences on reloading and hence how faithfully the Kaiser Effect is observed) and has been observed in experiments (Lavrov, 2001). The results of the cumulative AE events are plotted in Figure 8.4. It is clear that no AE event occurs during unloading and reloading to the maximum stress level of the last cyclic loading. This shows that the damaged numerical rock specimen can remember the stress history it has experienced.

Figure 8.5 shows the stress levels of the specimen at the different stages of loading and unloading throughout the complete stress–strain curve. Although the stress–strain

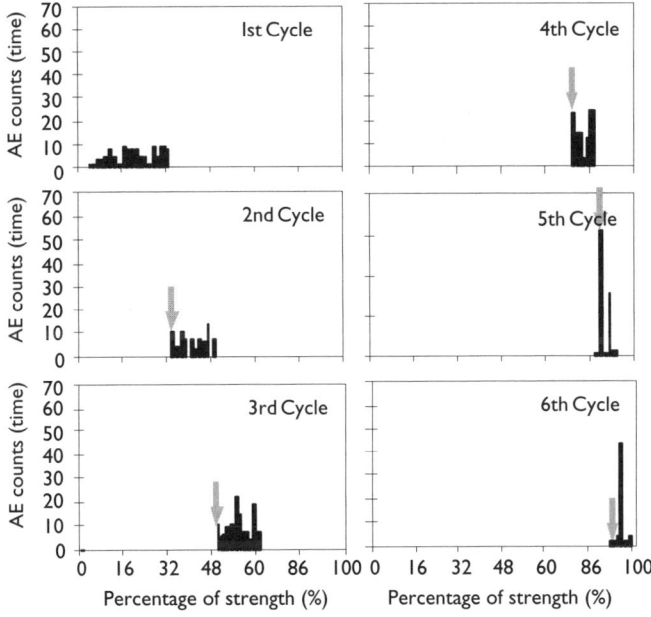

Figure 8.2 Numerical results of AE counts vs. percentage of strength.

132 Rock failure mechanisms

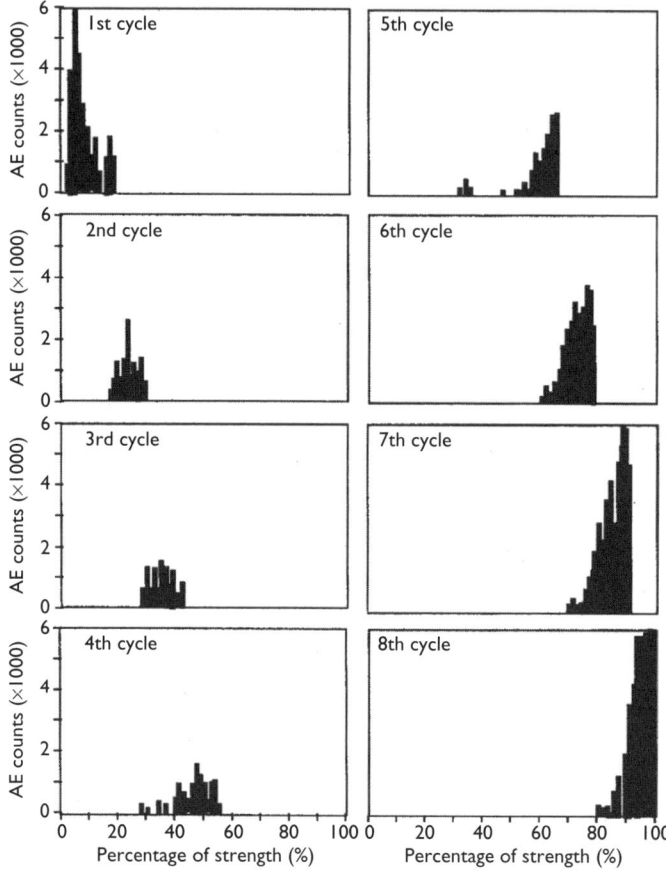

Figure 8.3 Experimental results of AE counts under cyclical loading (from Li and Norlund, 1993).

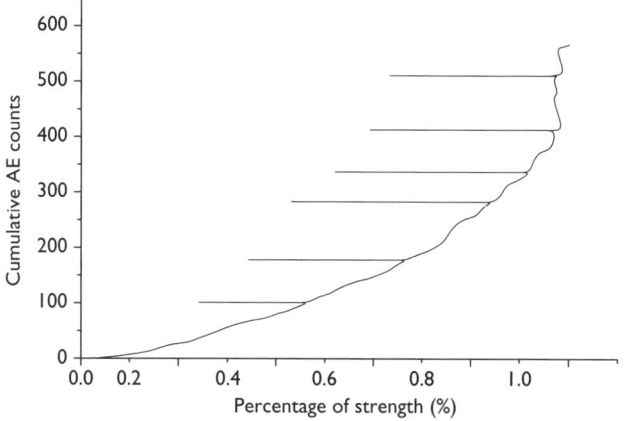

Figure 8.4 Curves of cumulative AE events *vs.* applied stress as a proportion of the compressive strength.

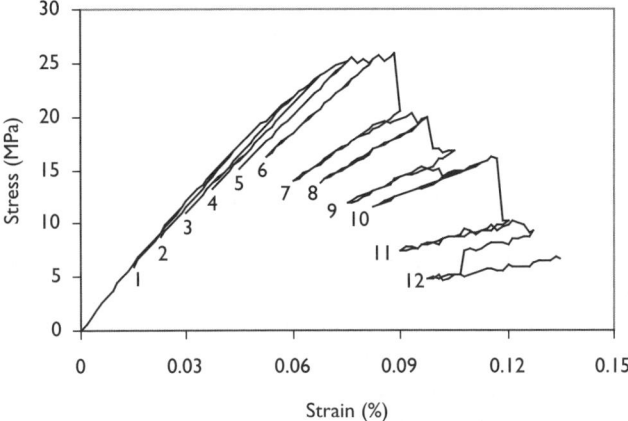

Figure 8.5 Complete stress–strain curve with loading and unloading cycles throughout.

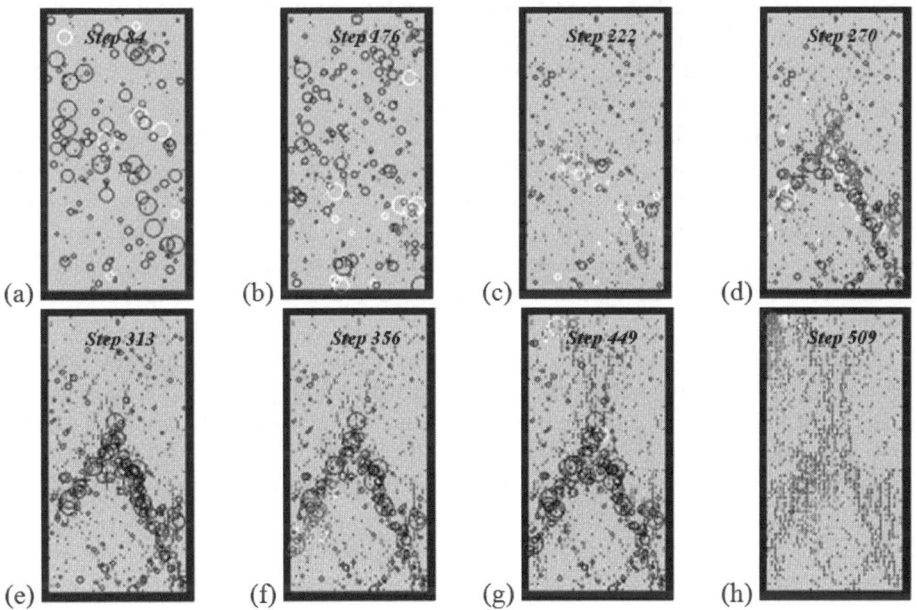

Figure 8.6 Spatial distribution of AE events during the rock failure process: Black circles are acoustic emission; gray circles are tensile failure; white circles are shear failure.

curve itself exhibits a strong non-linear characteristic, the unloading and reloading cycles appear linear, reflecting the local elastic modulus of the rock specimen. When the strain reaches 0.0885% (after the 6th cycle of loading), the stress reaches its maximum value of 25.5 MPa. However, before the peak-strength, the unloading and reloading paths are exactly the same but, after the peak-strength the unloading and reloading paths do not match exactly. During the 7th and 9th cyclic loadings, the stress drop occurs, although the loading does not reach the maximum stress level of the last cycle. During the final loading cycle (12th), an abrupt drop of stress is observed while unloading. Noting that the extensions of the paths of all the unloading lines tend to the origin, this reflects the linearly-elastic surviving elements in the model.

Although the mesoscopic elastic damage model used is simple, the non-linearity of deformation and the evolution of the material (such as its elastic modulus) during the fracture process can be simulated. At the same time, the loading path and the unloading path of the cycles are different—due to the fact that the change in mesostructure is irreversible. These results are in good agreement with the experimental results (Tang, 1988).

When viewed in colour on the computer screen, the modelling output distinguishes between tensile and shear failure events by colour coding the different results. In Figure 8.6 the red circles (reproduced here in gray) indicate tensile failure and the white ones indicate shear failure. The magnitude of the released energy after failure is represented by the diameter of the circle. These images show the entire evolutionary process of micro-structural breakdown. It is characterised by the initial randomly diffused damage and nucleation of the damage. During the first cycle, the damage, as tensile and shear failure, is randomly initiated throughout the whole specimen (see Figure 8.6a). With increasing loading, the AE counts are also increased in a random manner (see Figure 8.6b). However, while carrying out the third loading cycle, the AE events begin to nucleate at the middle and lower-right of the specimen (see Figure 8.6c). During the fourth loading cycle, a crack zone is developed due to the coalescence of broken elements in the specimen (see Figure 8.6d). When the loading becomes higher, another crack zone is formed at the lower-left (see Figure 8.6e). During the sixth cycle, the crack at the lower-right is developed further, and an inverted V-shaped crack zone is formed at the lower-middle of the specimen (see Figure 8.6f). For the rest of the loading cycles, the further AE events take place at the top of the specimen.

Even in this relatively straightforward case of the simulated uniaxial compressive loading of a heterogeneous rock specimen with loading cycles, the damage process is not simple. For *in situ* rock experiencing a succession of natural 3-D stress states over millions of years, needless to say, the situation is much more complicated. As mentioned, the use of the Kaiser Effect to measure the *in situ* rock stress is by uniaxially loading six small cores drilled out of a site investigation core at different angles and, by testing each of the small cores and detecting when significant acoustic emission begins, one can deconvolve the *in situ* stress state from the knowledge of the site investigation core's orientation, the orientations of the six mini-cores and the six values of AE initiation. As with all methods of *in situ* stress estimation, this Kaiser Effect approach is fraught with conceptual and practical difficulties. As one example, Lehtonen (2008) identifies the time-dependent decay of the Kaiser Effect which is directly related with the viscous stress relaxation behaviour of rock. However, continued numerical simulations and practical experiments will assist in the development of procedures to enhance the understanding of the Kaiser Effect and to improve the accuracy of the results when it is used for stress measurement.

Chapter 9

Time dependency of rock failure

9.1 INTRODUCTION

The time-dependent properties of brittle rocks are often ignored in the design of rock structures, due in part to the small additional strains that occur in the life of these structures. Although hard rock is usually not associated with large creep deformation, data collected from the tunnels and stopes of the deep South African gold mines do illustrate significant time-dependent behaviour (Malan, 2002). In fact, the design life of engineered structures is highly variable. The life of mine openings varies from a day for a longwall face opening to possibly more than a hundred years for the complete mine. However, the design life of civil engineering structures built on and in rock masses is of the order of 120 years. In the case of an underground repository for radioactive waste, the design life can be assumed to be hundreds of thousands of years. In this latter case, creep deformation may induce additional stress on the supporting structures of the repository and damage to rock subjected to long-term loading may create potential pathways for the migration of radionuclides to the biosphere. The time-dependent behaviour of rocks and rock masses is also of special interest for civil engineering, hydro-power engineering, and oil and gas exploitation. So, it is necessary to consider the time-dependent properties of rocks when designing excavations that must remain open for long periods of time.

Terms used when describing time-dependent behaviour of rocks are creep, stress relaxation, and fatigue. To avoid confusion in the use of these terms, they are defined via Figure 9.1 in which a rock specimen is illustrated being loaded in compression through an adjacent rock element represented by the spring. Creep is defined as increasing strain while the stress is held constant. Stress relaxation is defined as decreasing stress while the strain is held constant. In practice, and especially for the rock around an underground excavation, a rock element will be loaded via the stiffness of an adjacent element and so the time-dependent behaviour will be somewhere between the ideal conditions of creep and stress relaxation, as illustrated in Figure 9.1. Fatigue refers to an oscillation in the applied stress producing permanent strain. Of course, these definitions are similar when the force–displacement curve is being considered.

Disregarding time-dependent effects can lead to large discrepancies between predicted data and real field measurements. For instance when considering deep excavations in rocks, such as large diameter tunnels or drilling bore-holes, neglecting time effects may lead to incorrect evaluation of deformations at the walls and thus impact on the criteria for selection of the proper design. Over the last decades, many experimental and theoretical studies have been devoted to describing the time-dependent behaviour of

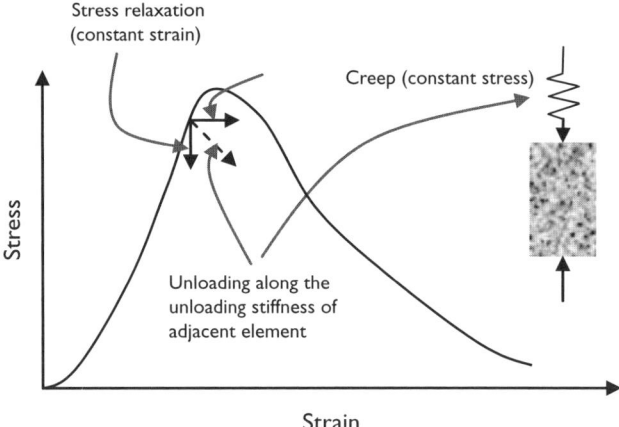

Figure 9.1 Definitions of creep, stress relaxation and time-dependent unloading (dashed line) along the stiffness of the adjacent rock element (the spring above the rock specimen).

rocks. In classical models, viscoelasticity or viscoplasticity have been used, e.g. Cristescu (1989) and Cristescu and Hunsche (1998) to study the time-dependent deformation and failure of rocks, i.e. the time-dependent deformation is entirely attributed to the viscous effects. Most authors have started from laboratory tests describing either creep or loading-rate effects and have afterwards tried to match the data with some empirical mathematical expressions. These models do provide useful information about the trends of time-dependent deformation. However, both these approaches depend on mathematical descriptions of the time-dependency and experimental results, but do not take into account the internal physical mechanisms and the field conditions. Moreover, some conditions are not attainable in laboratory studies so it is not sufficient to rely only on experimental results for a satisfactory description of the mechanical characteristics in the field. The extrapolation over long durations as well as to stress conditions which are not attainable in laboratory studies are possible only if the physical principles of the microscopic deformation mechanisms are known and can be integrated into the material laws.

For rock material, due to its brittleness and heterogeneity, a typical manifestation of creep or stress relaxation involves the progressive evolution of micro-cracks at constant stress or strain levels. In geomaterials, laboratory investigations suggest that the development of creep deformation is mainly associated with progressive evolution of the micro-structure (Shao *et al.*, 2003). Typical mechanisms can involve the sub-critical propagation of the micro-cracks in hard rocks (Atkinson, 1984; Atkinson and Meredith, 1987), pore collapse in highly porous rocks (Dahou *et al.*, 1995). For this reason, the creep model for rocks, particularly hard rocks, should be based on the physical processes governing the deformation and failure. The time-dependent failure of rocks is then considered as a macroscopic consequence of the progressive degradation of the rock structure at the microscopic scale. In the following sections, the time-dependent behaviour is numerically investigated via the initiation, propagation and linkage of micro-fractures during the creep process. The concluding section in this Chapter discusses the time-dependent degradation of building stones.

9.2 A CONSTITUTIVE MODEL FOR THE TIME-DEPENDENT BEHAVIOUR OF ROCKS

In this Section, we start with a brief presentation of the numerical constitutive model for time-dependent behaviour. The constitutive model for long term behaviour is formulated by extension of the short term model (Shao et al., 2003). In the work by Pietruszczak et al. (2002), a general methodology has been proposed for the description of creep in anisotropic rocks in terms of micro-structural evolution. Similarly to Pietruszczak et al. (2002) and based on the concept of damage mechanics, the numerical simulation assumes that time-dependent degradation affects the elastic modulus and failure strength of the elements. The time-dependent degradation of the elastic modulus and strength of rocks (i.e., of the individual elements in the finite element code) is assumed to be,

$$E_t = E_\infty + (E_0 - E_\infty) \cdot e^{-a_2 t} \tag{9.1}$$

where E_t is the time-dependent elastic modulus at time t, E_∞ is the long term elastic modulus at time t approaching infinity ($t \to \infty$), and E_0 is the initial elastic modulus.

$$f_t = f_\infty + (f_0 - f_\infty) \cdot e^{-a_1 t} \tag{9.2}$$

where f_t is the time-dependent strength at time t, f_∞ is the long term strength at time t approaching infinity ($t \to \infty$) and f_0 is the initial strength.

The parameters a_1 and a_2 in Equations (9.2) and (9.1) are two material parameters for the elements. Figure 9.2 illustrates the strength degradation law; the modulus degradation has a similar form.

Equations (9.1) and (9.2) can be easily implemented into a numerical integration algorithm using the finite element method, with nodal displacements as the principal unknowns. By introducing the material property time-dependent degradation into the

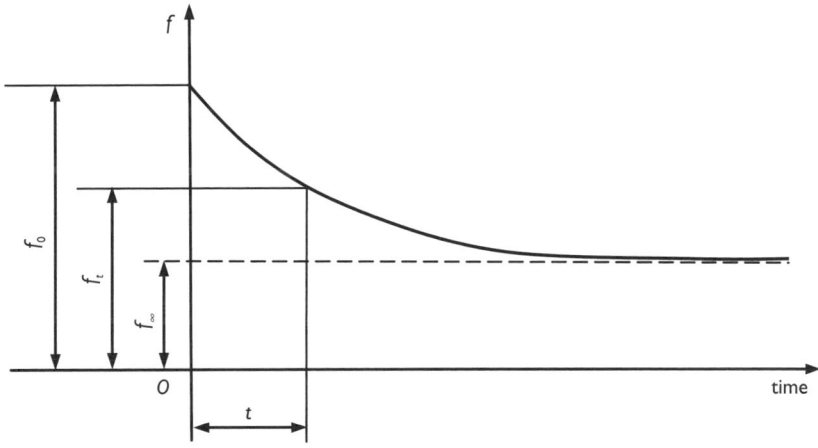

Figure 9.2 Strength degradation of an element following equation (9.2).

code, the damage behaviour with time can be obtained. In this model, the element may degrade and damage gradually with time according to the elastic damage constitutive relation. The combined action of time-dependent tensile damage and compressive-shear damage leads to the macroscopic failure of the rock material. In the following sections, the numerical simulation of a short term compression test, creep tests and relaxation tests under uniaxial loading conditions are illustrated.

9.3 ILLUSTRATIONS OF TIME-DEPENDENT MICRO-STRUCTURAL DAMAGE

9.3.1 The creep test

Following Figure 9.1, we start with a presentation of the simulation of the creep test, i.e. a constant load test. Corresponding to a laboratory investigation, the numerical simulation of the long-term loading test consists of two stages: initial loading and then constant loading. In the initial loading stage, the specimen is axially loaded to the pre-determined stress level. The constant loading stage is then initiated with the load maintained at the pre-determined stress level until the specimen fails.

Figure 9.3 shows the curves of strain and strain rate after loading to 80% of the instantaneous uniaxial compressive strength of rocks. The strain–time curve is of typical form with the 'instantaneous' response, transient deformation, stationary state, and the unstable stage. By calculating the slopes of the creep curve, the relation between the creep rate and the time can be obtained, as shown in Figure 9.3. During the whole loading process, the creep rates of the specimen undergo three stages, these are, decelerating, steady, and accelerating. The occurrence and duration of each stage depends on the behaviour of the rock material modelled and the load level applied, as discussed in the next section.

The deformation starts with a high deformation rate (transient creep) which decreases until it reaches a relatively constant creep rate (also termed secondary creep or steady-state creep). With the continuous degradation of the micro-structure, the

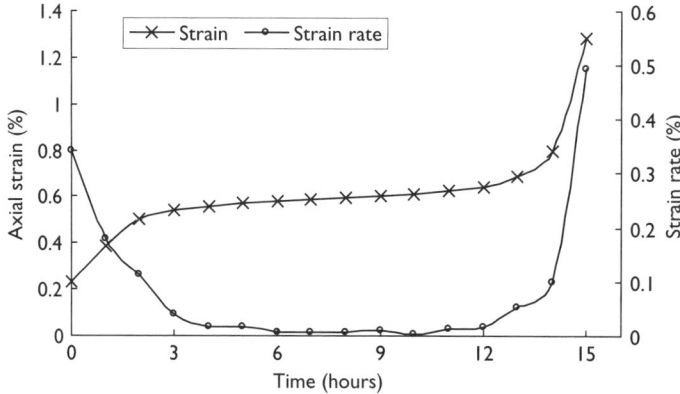

Figure 9.3 Numerically obtained strain vs. time and creep rate vs. time curves for the uniaxial creep test.

tertiary creep phase (also termed accelerating creep) is initiated. This begins with an increasing strain rate and finally ends in creep failure of the rock specimen.

Figure 9.4 shows the creep failure process of the numerical specimen. In these images, the gray degree indicates the magnitude of the shear stress (white is the highest) and in Figure 9.5, we present the location of the AE events occurring during the creep process. It can be seen that the onset of failure in the specimen is first indicated by the formation of a number of isolated micro-fractures (Figures 9.4a and 9.5a) with the micro-fractures influencing the macro-mechanical properties of the specimen. Moreover, the mechanical properties of the micro-element are reduced according to the time-dependent degradation law. More micro-fractures occur with time, as shown in Figures 9.4b and 9.5b. After a steady-state creep stage, a macroscopic fracture zone develops (Figure 9.4c) which results in a sudden flurry of AE activity (Figure 9.5c) and the micro-fractures around the macro-fracture begin to cluster together to form a new macro-fracture (Figures 9.4d, e and 9.5d, e). As the degradation of mechanical properties with time continues, two macroscopic fracture zones develop (Figures 9.4f and 9.5f). In Figure 9.4f there is a highly stressed area between the two macro-fracture zones. Finally, the interior macroscopic fracture zone forms and becomes inter-connected to form a V-shaped open fracture (Figures 9.4h and 9.5h).

Depending on the stress state, the volume of a rock increases (dilatancy) or decreases (compressibility) due to the overall micro-cracks opening or closing. During the tertiary creep phase, there is obvious dilatancy of the specimen. Acoustic emission is a good indicator of micro-cracking and volume increase. In Figure 9.6, the displacement vectors are presented. For the first creep phase, no obvious deformation is observed (Figure 9.6a). With the degradation of the mechanical properties, dilatancy comes into being gradually. For the tertiary creep phase, the dilatancy phenomenon becomes more obvious (Figure 9.6g, h).

During the creep tests illustrated in Figures 9.7 and 9.8, the stress was increased in successive steps and then kept constant for several hours. A schematic presentation of the loading path is shown in Figure 9.7. The mean stress was applied to the prescribed value and kept constant while the variation of strain with time was modelled. The stress steps chosen are 5, 10, 15 and 20 MPa. Creep strains versus time are shown in Figure 9.9 for various stress levels.

We expect that the time to failure will depend on the magnitude of the applied stress. The curves in Figure 9.8 have been obtained with specimens loaded at different stress levels: 12, 14, 16, 18 and 20 MPa. The short term uniaxial compressive strength of this rock specimen is $f_c = 100$ MPa. For lesser loading stresses, the transient creep phase ends with stabilisation after about 40 hours, as shown by Figure 9.7. If the loading stress is less than a certain limit, which depends on the short term compressive strength of the rock, only transient creep is apparently exhibited by the rock specimen, i.e., stabilisation is considered to be achieved if, after 5–10 additional hours under the same constant stress, no strain increase is exhibited. For these particular axial loading stresses, $f_1 < 0.6f_c$ will result more or less in transient creep only. For $f_1 > 0.7f_c$ the creep becomes unstable and failure is obtained. Tertiary creep was also observed several hours before failure.

During the time interval that creep takes place, the number of AE events is obtained (Figure 9.10). The last loading interval begins with active AE events and an accelerating creep (Figure 9.10). On each reloading, the AE suddenly increases immediately after the

140 Rock failure mechanisms

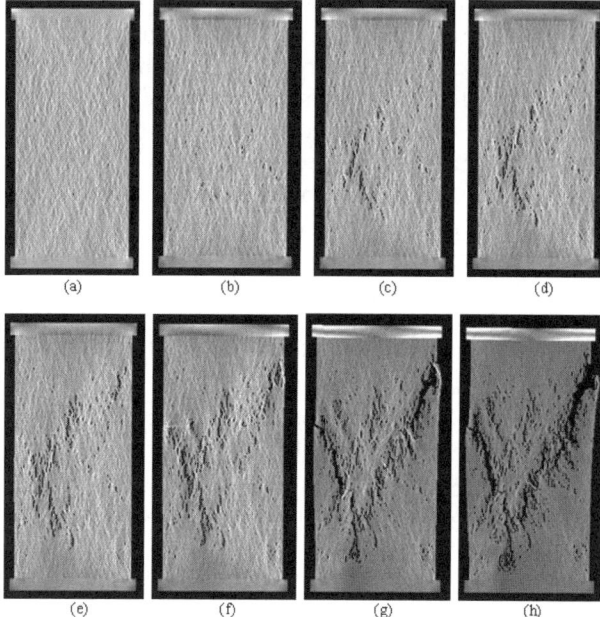

Figure 9.4 The shear stress field (gray shading) and fracturing (in black) during creep.

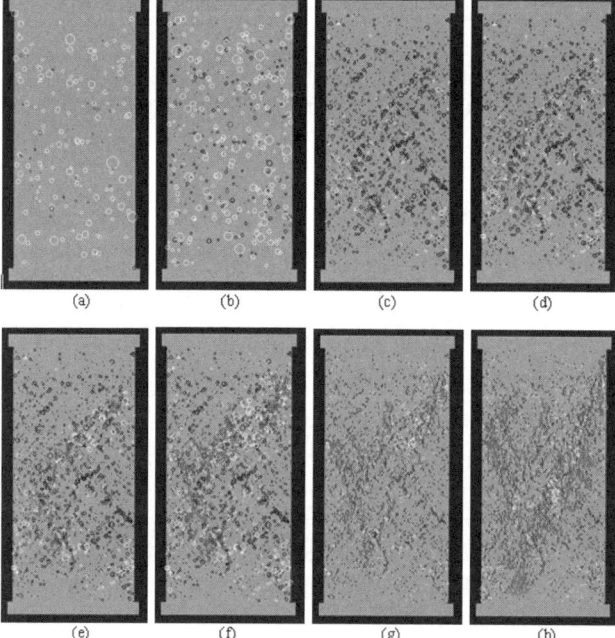

Figure 9.5 Numerically obtained AE locations in the rock specimen during the creep process.

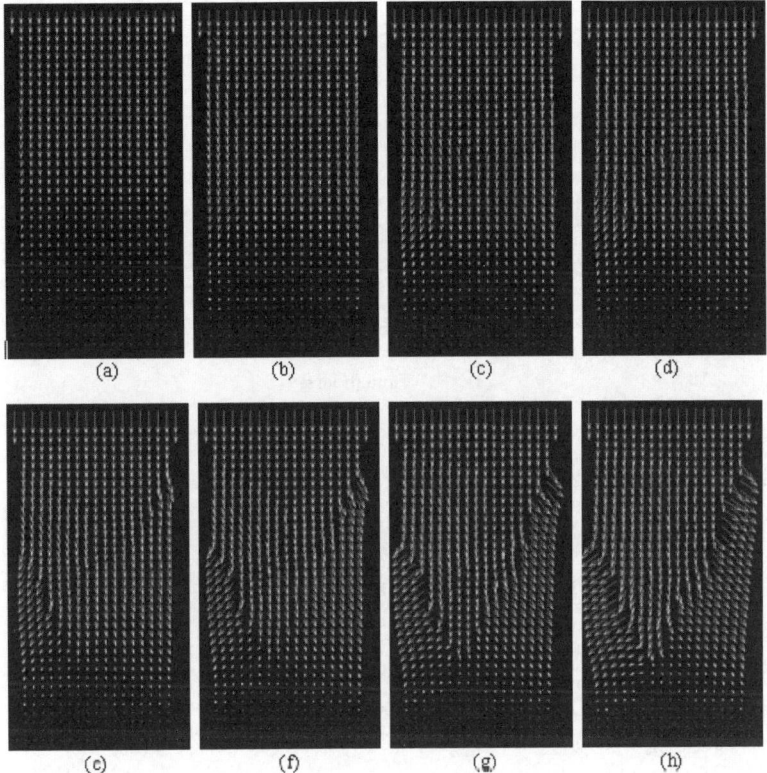

Figure 9.6 Displacement vectors corresponding to the creep failure process.

load increase followed by a quiescent period when the AE event number is very small but increasing gradually during the time interval in which stress is held constant.

9.3.2 The relaxation test

In terms of experimental results, creep observations are relatively easier to obtain than other time-dependent processes and so there are many results available in the literature. Also, the idea of a dead weight type of loading is directly applicable to many engineering circumstances, such as dam foundations. However, relaxation tests at a constant displacement or strain value or time-dependent unloading along a constant stiffness line (Figure 9.1) are less conceptually appealing and were more difficult to conduct in earlier decades because the displacement or the strain applied to the specimen has to be kept constant, or the stress and strain must follow a certain unloading stiffness. Consequently, experimental data for these tests are rare. Nowadays though, all cases in Figure 9.1 are easily followed in experimental tests using a closed-loop servo-controlled testing machine in which the feedback in the closed-loop can be arranged to achieve the required testing conditions.

In Figure 9.11, a simulation of a uniaxial relaxation test is presented. This test was carried out with an initial axial strain of 0.5% and the variation of the axial stress

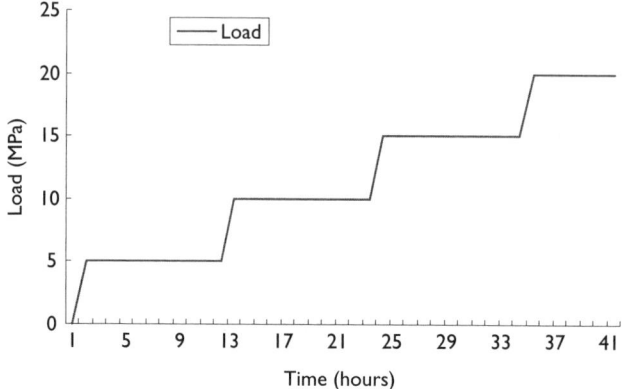

Figure 9.7 Stepped load increase with time intervals.

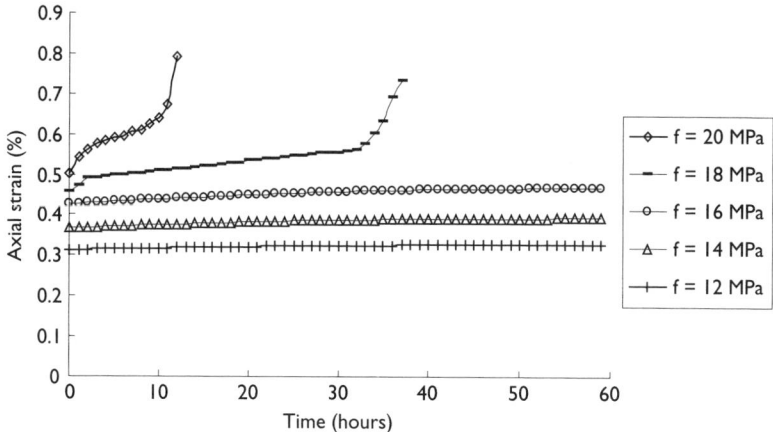

Figure 9.8 Creep deformation at different stress levels.

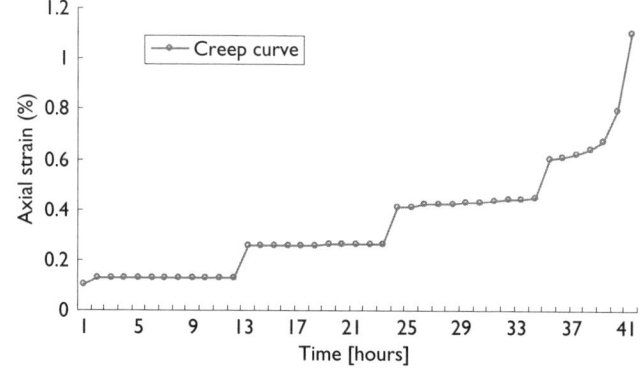

Figure 9.9 Creep deformation in the stress intervals.

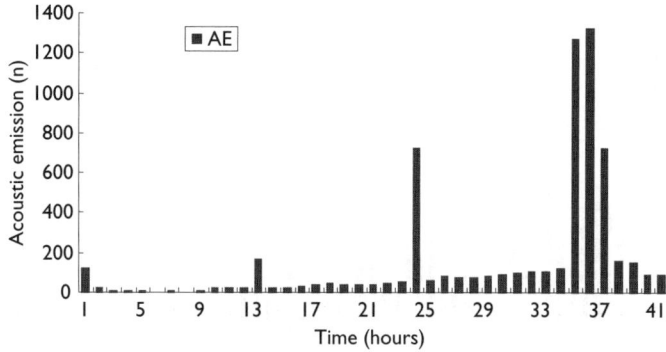

Figure 9.10 Acoustic emission sequence with stress intervals.

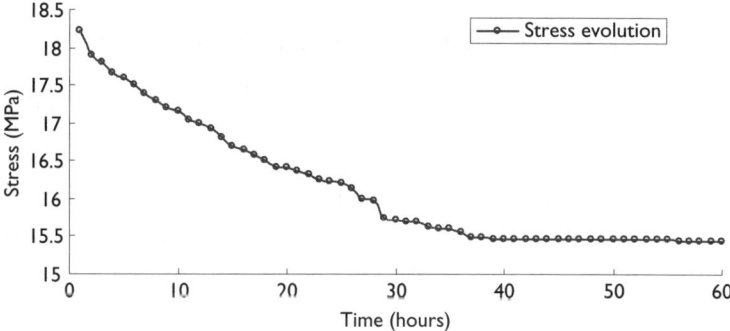

Figure 9.11 Stress relaxation curve.

was registered. The corresponding acoustic emission is shown in Figure 9.12. The model shows the progressive decrease of the axial stress due to the time-dependent process. As we might intuitively expect, the rate of stress relaxation of the rock is highest at the beginning, and gradually decreases with time and finally tends to be zero. The stress-time evolution curve can be divided into three intervals.

- In the first interval, the relaxation rate is higher when a large number of AE events is recorded.
- In the second portion of the curve, corresponding to the smooth descending part, the AE events are also active and the most stored elastic energy is released.
- Finally, after 37 hours, the stress evolution curve becomes smooth and only infrequent small AE events occur.

The influence of time is generally neglected in current rock engineering design but will be incorporated in future years as it becomes easier to characterise the time-dependent behaviour. A review of present knowledge in this subject for crystalline rocks is given in Hagros *et al.* (2008).

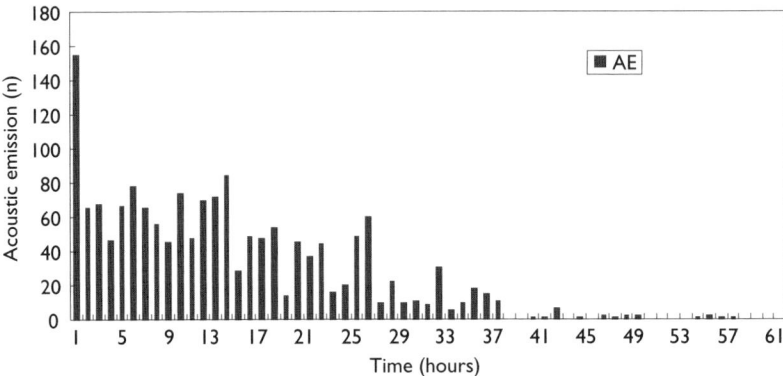

Figure 9.12 Acoustic emission *vs.* time in a uniaxial relaxation test.

9.4 DEGRADATION OF BUILDING STONES WITH TIME

In a similar way to the fact that geological history provides us with a multitude of examples of rock failure produced by past stresses in the Earth's crust, so inspection of old buildings provides a wealth of practical evidence demonstrating the degradation of building stones with time. Sometimes, such weathering is slight, as in the case of Stonehenge in the UK (the trilithons are sandstone with a silica matrix) which have resisted the British weather for thousands of years; in other cases, the degradation can be relatively rapid, as in the case of buildings constructed with weather-susceptible stone. Although, we are not presenting here any numerical simulation results for time-dependent degradation—because these would involve as yet undeveloped fully-coupled multi-variable models incorporating chemistry—the following examples of Carrara marble and the deterioration of historic buildings (highlighted by Caen limestone) are indicative of the complex nature of natural degradation processes.

The interesting case of Carrara marble is its susceptibility to strong temperature fluctuations (Williams, 2009). Thermal cycling and the consequent mismatch in expansion and contraction of the calcite crystals constituting the marble leads to deformation and bowing of the cladding panels on buildings. This is illustrated in Figure 9.13 for the case of Finlandia Hall in Helsinki, Finland. A detailed study of the susceptibility of such cladding panels has been conducted via an EU contract (SP Swedish National Testing and Research Institute, 2005).

The subject of the conservation of ancient monuments and sites is an especially interesting one in the context of rock failure mechanisms for two main reasons: the technical aspects of rock mechanics have to be considered in association with many other subjects, e.g. chemistry, archeology, long-term monitoring, management and artistic factors; and there are many issues associated with the subject that are highly controversial and often intractable as indicated by the following questions. To what extent is it necessary to conserve? Should the same stone be used? Who is going to pay for the conservation? Does the conservation have a long life? Should the same architectural style be used in the conservation?

Time dependency of rock failure 145

Figure 9.13 One face of the Finlandia Hall, Helsinki, Finland, showing convex bowing of the Carrara marble cladding panels due to large seasonal temperature changes.

 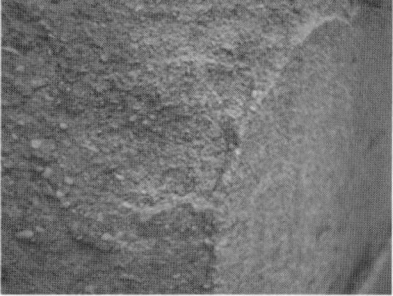

Figure 9.14 Deterioration of the Caen stone at Norwich cathedral in the UK.

Two of the seminal books on the subject are: "Conservation of Historic Buildings" by Sir Bernard M. Fielden, first published in 1982 with a 3rd edition in 2003 published by the Architectural Press, an imprint of Elsevier, and "Conservation of Building and Decorative Stone" by John Ashurst and Francis Dimes, first published in 1990 with a paperback edition in 1998 in the Butterworth-Heinemann Series in Conservation and Museology. These two books together provide background to the philosophy and technical aspects of such conservation, the content being applicable to all ancient sites and monuments.

In the UK, there are more than fifty cathedrals, many of which are medieval, having been built about 900 years ago and are jewels of Britain's architectural heritage. The cathedrals were built with different types of stone, some even with stone imported from Caen in France (a Jurassic limestone). However, the stone has deteriorated—mainly because of the weather: sun, rain, and frost. In Figure 9.14 are examples of the surface spalling and general deterioration of the Caen stone at Norwich cathedral in the UK. It is a mammoth, and indeed in some cases impractical, task to attempt to replace all the deteriorating stone with fresh stone of the same origin and exactly the same geometry. At Norwich Cathedral, wood has been used in places as a cheaper and more easily worked stone substitute during restoration, but now, despite a mortar skim coating, the wood itself is deteriorating.

Further discussion on the degradation of building stones is beyond the scope of this book but let us finish this Chapter with an extract from an 1842 paper "Observations on stone used for building" by C.H. Smith published in the Transactions of the Royal Institute of British Architects, and as quoted in the "Conservation of Building and Decorative Stone" book mentioned earlier: "Whoever expects to find a stone that will stand from century to century, deriding alike the frigid rains and scorching solar rays, without need of reparation, will indeed search for the philosopher's stone."

Chapter 10

Coalescence of fractures

10.1 INTRODUCTION

As discussed in Chapter 4, the dominant mode of failure of rock specimens under uniaxial compression is splitting where the fracture surface is approximately parallel to the direction of applied loading. This type of splitting failure normally involves a sequence of progressive micro-fracturing. The micro-cracking results from a high tensile stress concentrated at inhomogeneities, such as crack-like flaws, pore-like flaws, or soft or hard inclusions. For example, a circular hole, i.e. with zero elastic modulus, will create tensile stresses in the micro-structure above and below the hole potentially causing splitting parallel to the applied compressive load; whereas, a circular grain of high elastic modulus can be split in half similarly to the Brazilian test described in Chapter 3 but with the micro-fracture also parallel to the applied load. These initiated micro-cracks will propagate with increased loading, and eventually extend to the specimen surface to form a macroscopic splitting failure. Because of the importance of the coalescence of the developing fractures, in this Chapter, we analyse and describe further the growth of axial cracks from pre-existing flaws (including crack-like flaws and pore-like flaws) and their coalescence.

10.2 MODELLING OF CRACK GROWTH FROM CRACK-LIKE FLAWS IN COMPRESSION

Although many experimental results for specimens containing a pre-existing crack-like flaw under uniaxial and biaxial compression have been published, there is no existing study which investigates systematically the effect of progressive failure in brittle solids with the changing of the size of specimen, the flaw length, the inclination of the flaw, an array of flaws, and multi-flaws with a random distribution. Nemat-Nasser and Horii (1982) and Horii and Nemat-Nasser (1985,1986) studied the mechanism of crack coalescence in specimens made of Columbia resin CR39 under uniaxial, as well as biaxial, compression. Their specimens contained a series of flaws which consist of different flaw lengths and orientations. In general, large flaws control the mechanism of coalescence in the form of axial splitting under uniaxial compression (with no or little crack growth in the small flaws) while, under biaxial compression, the growth of large cracks is followed by growth of smaller cracks and the final failure is a coalescence of the smaller cracks in the form of a shear zone or fault.

Ashby and Hallam (1986) used both experimental and analytical approaches to study the crack problem. In their experiments, they tested PMMA plate specimens, 10 mm thick, 170 mm high and 25–100 mm wide, containing a single crack-like flaw with flaw length $2a$ of 16 mm, inclined at angles ψ at 20°, 30°, 45°, and 60° to the compression axis. They also studied crack growth in an array of flaws (three flaws with a diagonal arrangement). The experimental results showed that cracks initiated from the flaw and interacted with the surfaces of the specimen in a way that caused them to grow faster than they would in an infinite medium. For the crack study with multi-flaws, they used the observations from the experimental study of multi-holes (Sammis and Ashby, 1986) and developed a simple analysis based on beam theory to describe this interaction in their work (Ashby and Hallam, 1986).

Wong and Chau (1998) and Bobet and Einstein (1998) and Bobet (2000) conducted research on crack initiation, propagation and coalescence of flaws in a synthetic rock containing two through-going flaws under uniaxial compression. The size of the specimen was fixed and the angle of the flaw, the position of flaws (from overlapping to non-overlapping) and the frictional coefficient of the flaw were changed. For overlapping flaws, linkage of the tensile cracks appears to be dominating the coalescence mechanism. For non-overlapping flaws, coalescence through both shear and mixed (tensile and shear) cracks has been observed.

The particular problem studied here is similar to the experimental work of Ashby and Hallam, (1986) with the changing of the size of specimen, the flaw length, and the inclination of the flaw. In addition, the crack growth study for three array flaws with the changing position of diagonal, vertical and horizontal, and the crack growth study for multi-flaws with a random distribution will also be investigated. The Chapter will summarise the numerical analysis of the growth of wing cracks from pre-existing flaws in specimens subjected to a uniaxial compression stress. The flaw lies at an angle ψ to the axial loading direction. The initiation and the growth of wing cracks from a single flaw are considered first; then the interaction between the growing wing cracks from three flaws is simulated; finally, the more complex multi-flaws are studied. In all the cases, plane strain is used and all the specimens are subjected to displacement controlled axial compression. Other similar numerical work has been conducted recently by Ord and Hobbs (2010) who studied the mechanics that leads to the formation of joint sets in rock masses using the Itasca code PFC2D. They also studied the detailed formation of fractures during the generation of the complete stress–strain curves for virtual specimens.

10.2.1 An angled crack-like flaw

The numerical specimens for studying the growth of wing cracks from a pre-existing crack-like flaw are 170 mm long. Three parameters are considered in this study for investigating the crack growth behaviour: (1) flaw length, (2) flaw angle, and (3) specimen width. The mechanical properties for the specimens used in this investigation are as follows: homogeneity index, $m = 4$; mean compressive strength, σ_0, = 200 MPa; mean elastic modulus, $E_0 = 50$ GPa; Poisson's ratio, $v = 0.25$; tension cutoff, $\lambda = 10\%$; frictional angle, $\phi = 30°$.

As indicated in Figure 10.1, a pre-existing single flaw with length $2a$ of 10, 20 and 30 mm is numerically simulated to examine the influence of flaw length on the

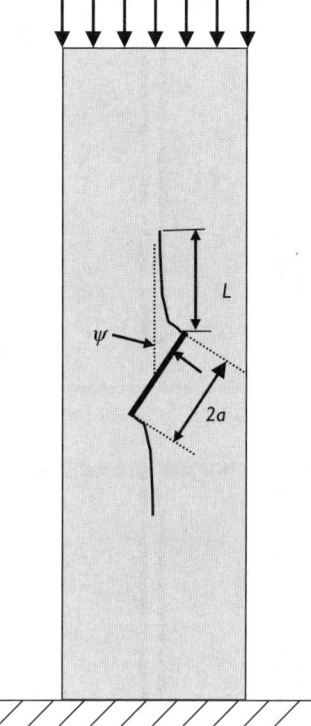

Figure 10.1 Schematic of numerical specimen geometry.

growth of the wing crack. The specimens are 50 mm wide, with flaw angle ψ inclined at 45° to the axial compression direction. The numerical specimens are 170 mm high and 50 mm wide $255 \times 75 = 19{,}125$ elements. The sequence of wing crack growth for these three specimens is shown in Figure 10.2a & b. The gray colour in Figure 10.2b shows the shear stress distribution in the specimens obtained in the same simulations (the lighter the gray, the higher the shear stress). Due to the heterogeneity of the specimens, the crack path is not smooth when compared to the experimental results with Ashby and Hallam (1986), but then we would not expect the crack path to be smooth in a real rock.

The influence of flaw length on the initiation and propagation of the wing crack is plotted in Figure 10.3 by using the normalised length of the wing cracks (L/a) versus the applied axial stress. It is shown that the wing crack is easier to propagate from the larger flaw than from the smaller flaw. The overall crack growth rate (length of the wing crack L per unit stress) from the longer flaw is faster (about twice) than that from the shorter-flaw specimen. The crack growth rates for flaw lengths of $2a = 10$, 20 and 30 mm are 1.8, 3.6 and 10.4 mm/MPa, respectively.

Figure 10.4 shows the ultimate failure modes of the specimens. For the specimen containing the shorter flaw ($2a = 10$ mm), the failure is in splitting mode. For the specimen containing a longer flaw ($2a \geq 20$ mm), it is observed that several cracks initiated in the form of a shear zone at the peak stress level. The cracks grew unstably

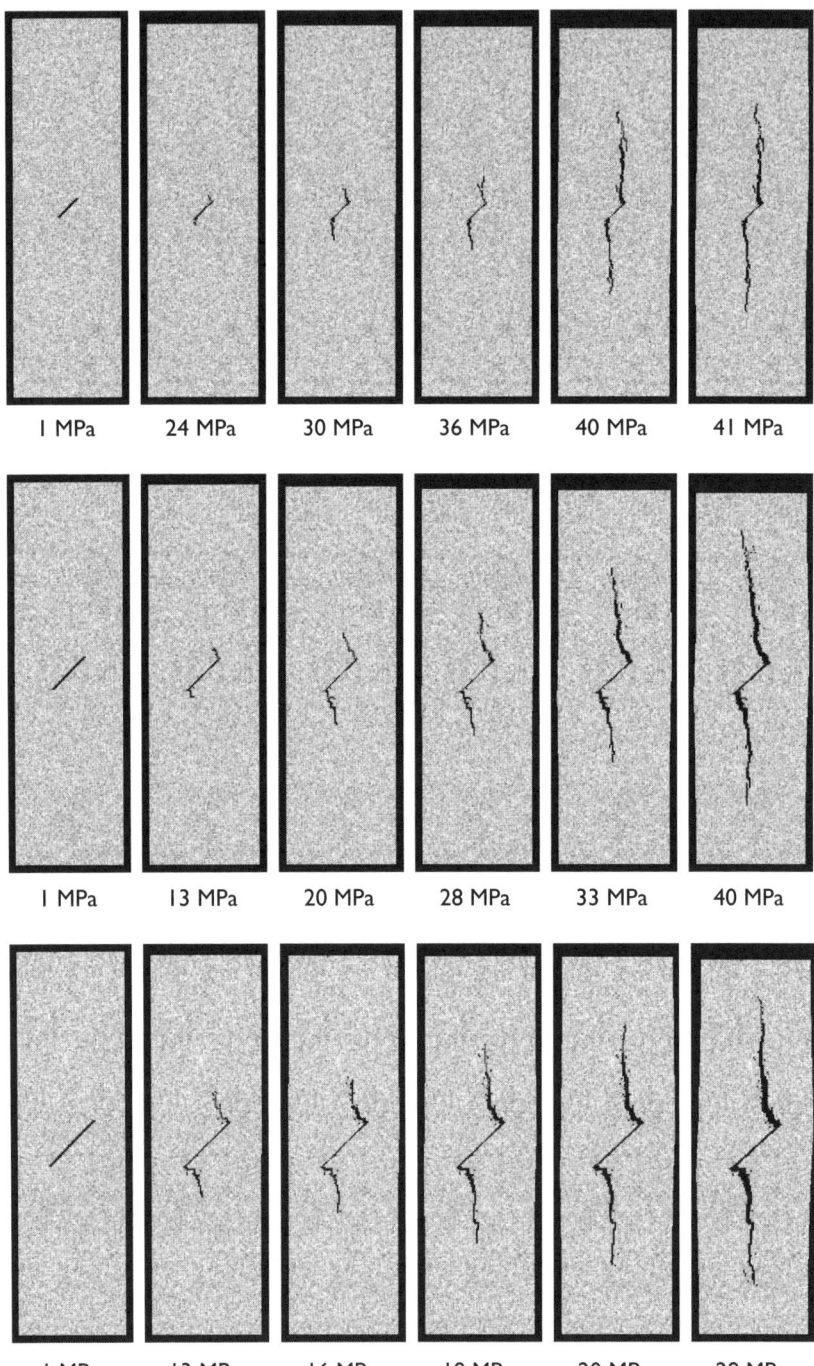

Figure 10.2a Sequences of the wing crack growth under uniaxial compression for specimens containing crack-like flaws of lengths 10, 20 and 30 mm.

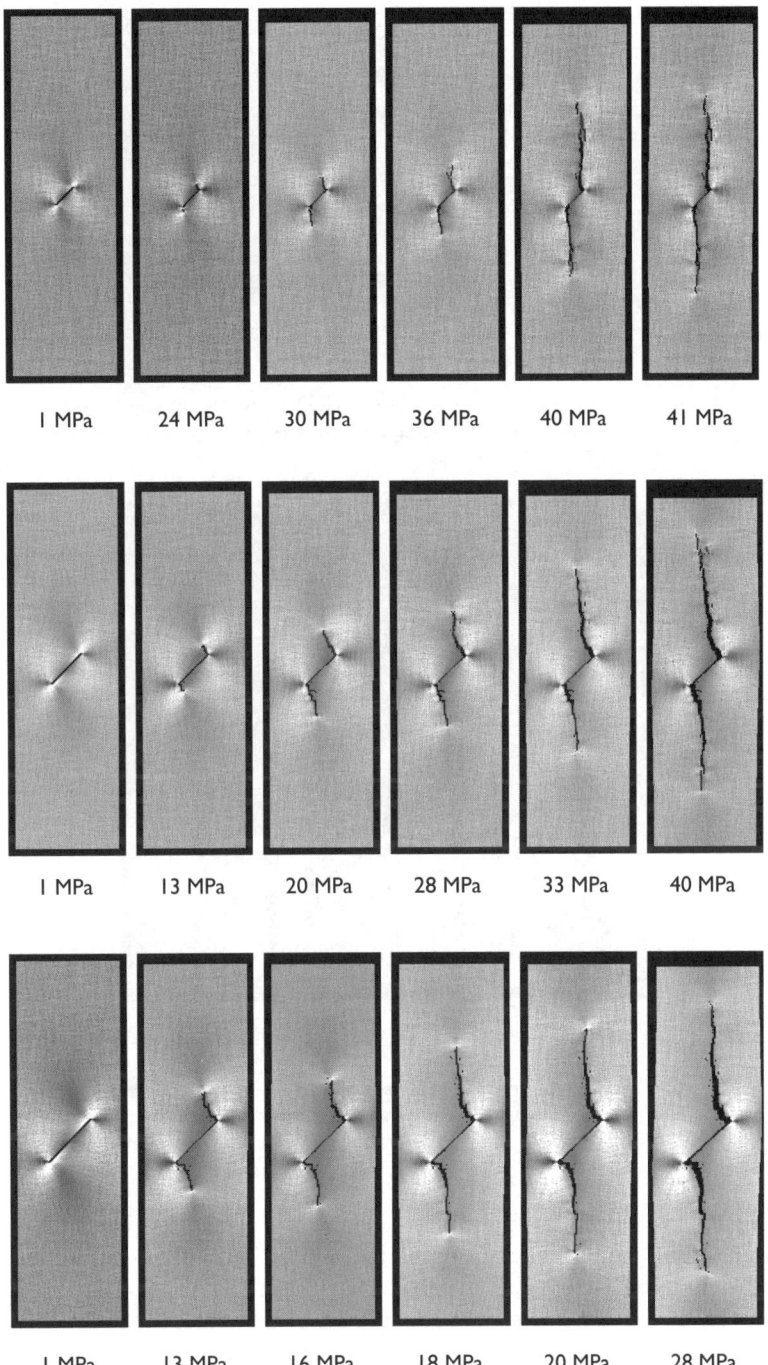

Figure 10.2b Shear stress distribution during the wing crack growth under axial load for specimens containing crack-like flaws of length 10, 20 and 30 mm.

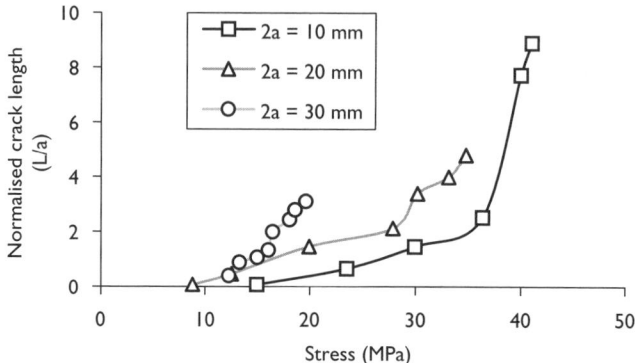

Figure 10.3 The influence of the initial flaw length (10, 20 and 30 mm) on the growth of wing crack under uniaxial compression.

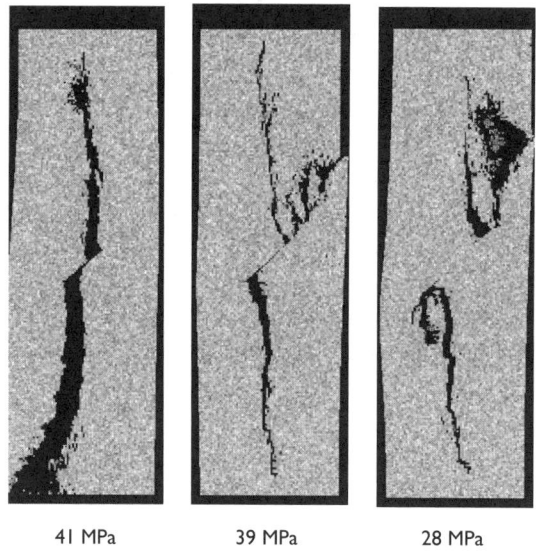

Figure 10.4 Ultimate failure modes for specimens containing pre-existing crack-like flaw lengths of 10, 20 and 30 mm (in these images the flaw is not so visible but is present between the main developed cracks).

towards to the tip of the flaw and the free surface of the specimen (Figure 10.4). The size of the larger flaw compared to the size of specimen is about 0.4–0.6. As observed from Figure 10.4, it is found that when the cracks approached the upper and lower surfaces of the specimen, the propagation of the crack slowed down or sometimes stopped. Based on the above observations from the numerical simulations, two conclusions can be drawn. Firstly, for the same size of specimen, cracks nucleate more easily and grow faster from the tips of a longer flaw than that for a shorter flaw. Horii and Nemet-Nasser (1985) observed that, for a specimen containing different lengths of flaws, the longer flaw is more susceptible to wing crack initiation and propagation than the shorter flaw. Secondly, the boundary effect is one of the factors influencing the mode of failure and growth of the crack. For a distance between the flaw tip and free boundary of the specimen less than twice the flaw length, shear failure may occur if the shear stress around the flaw tips is high. This mode of failure is observed when there is no lateral boundary restraint and the end boundary is friction free. If end friction exists under the applied loading of the specimen, splitting failure may occur instead of the shear failure.

Figure 10.5 shows the stress distributions during the growth of wing cracks from the pre-existing flaw at angles ψ of 25°, 45° and 60° under gradually increased axial compression. The specimens are 50 mm wide with flaw length, $2a$, of 20 mm. Figure 10.6 shows the influence of the angles, ψ, of the pre-existing flaw on the crack initiation and propagation behaviour. It is found that, for flaws with orientation near 60° to the axial loading direction, wing crack propagation occurs more easily. The crack initiated from the flaw inclined at 25° is the most difficult to propagate. According to the analysis of Ashby and Hallam (1986), the most favourable flaw angle for crack nucleation is equal to $\psi = 1/2 \tan^{-1}(1/\mu)$, where μ is the frictional coefficient of the flaw surface. They inserted two brass shims into the slot to control the friction between the flaw surfaces. The frictional coefficient in their study is about 0.25 (by back calculation from the experiments). For the experimental study of Wong and Chau (1998), the frictional coefficients are in the range of 0.6–0.9 (the calculated flaw angle of ψ is at about 24°–29.5°). Their experiments showed that the most favourable flaw angles ψ for crack nucleation were in the range of 25°–30° from the axial load. Therefore, it seems that the favourable flaw angle for crack nucleation depends on the frictional coefficient μ of the flaw surface.

Figure 10.7 shows the ultimate failure modes of three specimens. It is found that only the specimen having a flaw inclined at 25° fails in the splitting mode; the other two failed in a combined mode of splitting and shear. The reason for this failure is more or less the same as the discussion where the distance between the flaw tip and the side boundary is less than twice the flaw length.

Figures 10.8 and 10.9 show the influence of the specimen width (25, 50, and 100 mm) on the growth of the wing crack. The specimens contain a flaw inclined at 45° with length $2a = 20$ mm. The illustrations in Figure 10.8 show that the wing cracks grow more easily in narrow specimens, and the specimens are split into two thin columns where the width is about half the flaw length. This indicates that there exists a strong interaction between the growing crack and the side boundary of the specimen. From Figure 10.8, it can be seen that, when the narrow specimen is loaded, an outward bending of the ligaments appears on either side of the growing wing cracks. This phenomenon was also observed in the experiments of Ashby and Hallam (1986) and explained by them with a beam model. They studied the crack growth behaviour of inclined flaws within silicone rubber plates (in which large elastic deflections are possible) with detailed measurements of the shape change of the plates during loading. Their observations and measurements indicated that the speed of crack propagation would increase if additional bending displacements appear in the plate. They reported that, when a narrow specimen is loaded, the growth of wing cracks splits the specimen into vertical columns. These narrow columns bend outward and cause a larger sliding displacement in the finite plate. This extra displacement increases the stress intensity

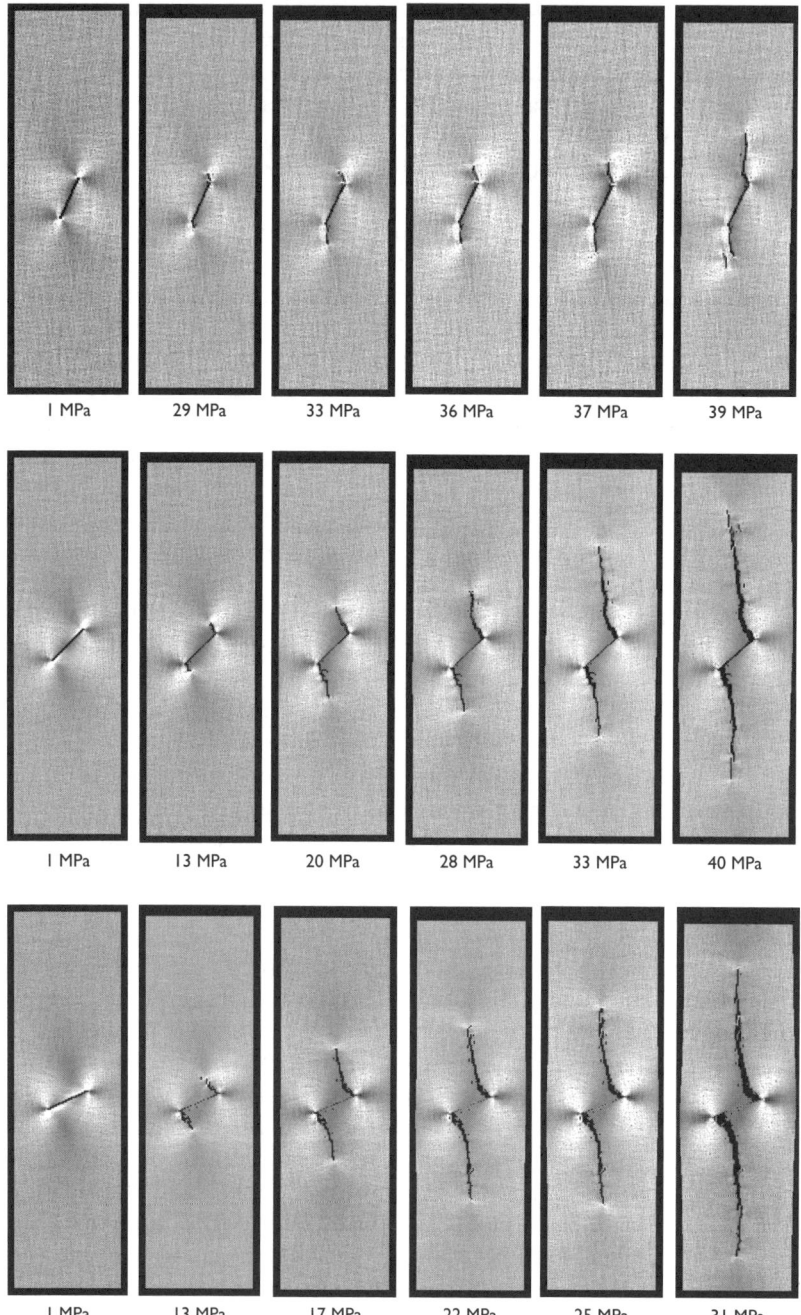

Figure 10.5 Stress distribution during stages of the growth of the wing cracks during uniaxial compression for specimens containing a crack-like flaw at ψ angles of 25°, 45° and 60°.

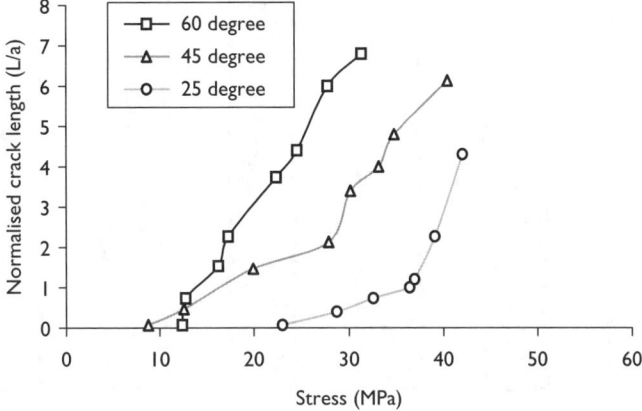

Figure 10.6 The influence of the varied initial flaw angle ψ (25°, 45° and 60° from the axial load direction) on wing crack growth in specimens under uniaxial compression.

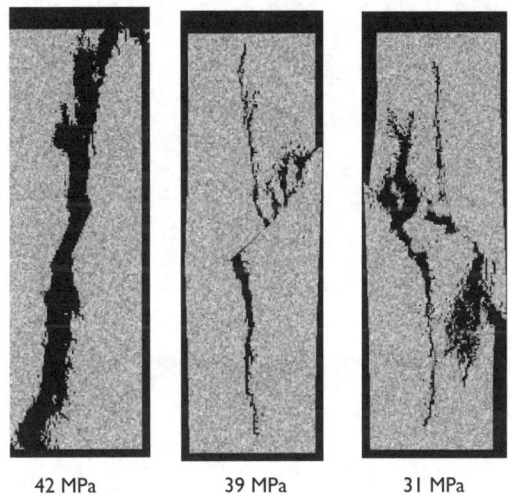

Figure 10.7 Ultimate failure modes for specimens containing crack-like flaws at angles ψ of 25°, 45° and 60° to the applied load.

in the crack tips and makes the crack grow further. The final failure of the narrow columns is caused by buckling. However, if the dimensions, H (height) and W (width), of the specimen are much larger than the length of the pre-existing flaw, a slight confinement may exist at the tip area of the wing cracks. This may allow wing crack growth in a stable manner, as shown in Figure 10.8 for a specimen of 100 mm width. The crack growth behaviour shown in Figure 10.9 also confirms this observation.

Figure 10.10 shows the failure modes of the specimens with different widths (W = 25, 50 and 100 mm). For the narrow specimen (W = 25 mm), buckling occurs

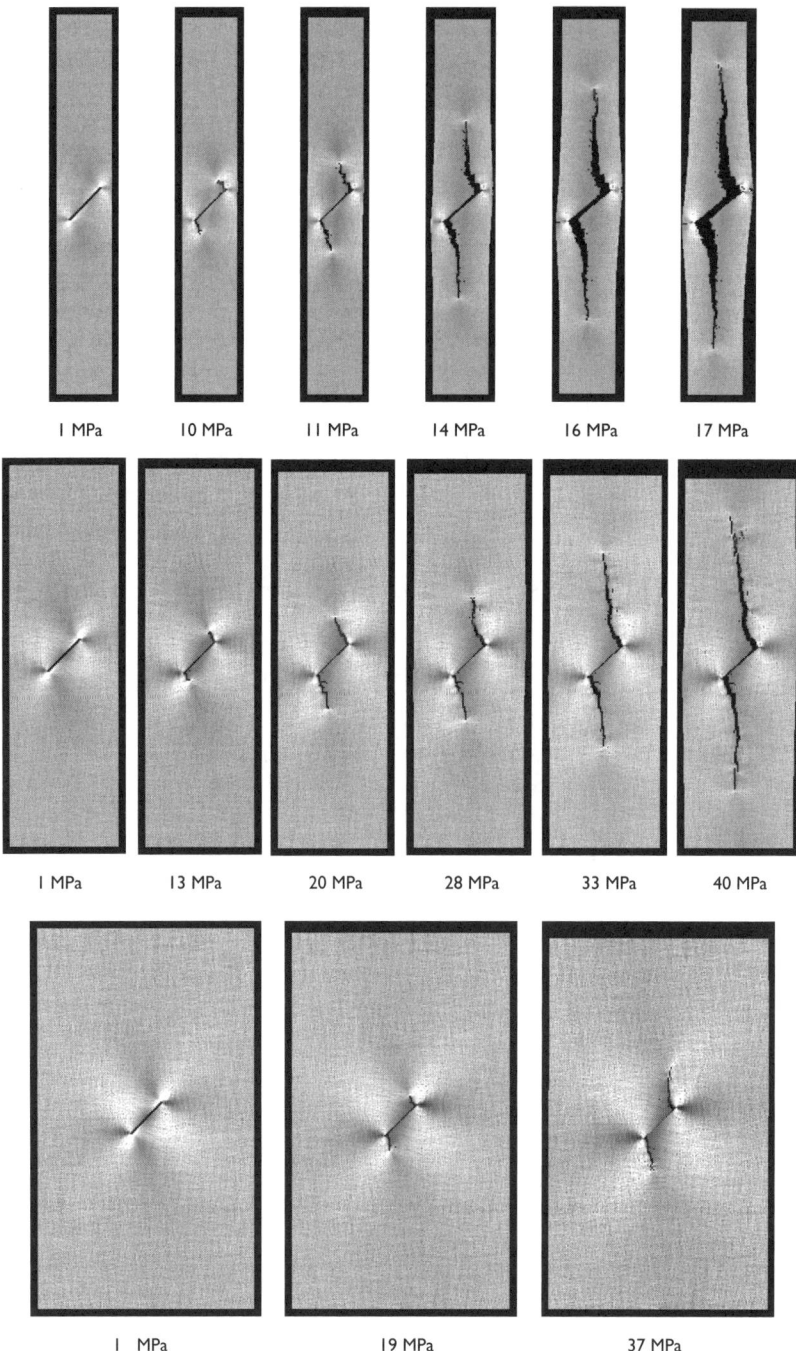

Figure 10.8 Stress distribution during crack growth under uniaxial compression for specimens with widths 25, 50 and 100 mm.

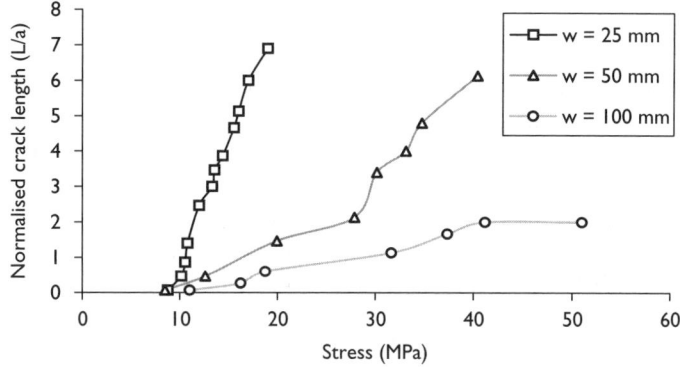

Figure 10.9 The influence of the specimen width (25, 50 and 100 mm) on the growth of wing cracks under uniaxial compression.

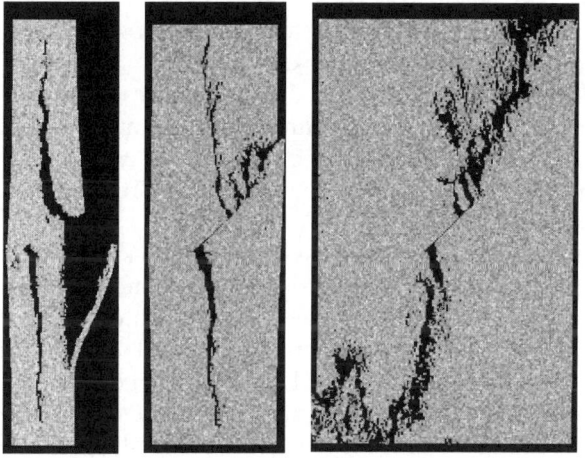

Figure 10.10 Ultimate failure modes for specimens with widths 25 mm, 50 mm and 100 mm.

when the cracks are close to the two ends of the specimen. Buckling is due to the strong interaction between the cracks and the free boundary of the specimen. For the specimen with a width of 50 mm, this interaction is reduced. The failure occurs by inducing shear cracks instead of the buckling tensile crack near the flaw tip. In the specimen with a width of 100 mm, however, the failure occurs by unstable *en echelon* fractures, a failure mode more complicated than simple splitting. From the above observations, it can be concluded that the growth rate of wing cracks in a wide specimen is much slower than that in a narrow specimen. Furthermore, for the narrow specimen, if the splitting column is thinner than the half-length of the flaw, strong interaction between the cracks and the free boundary of the specimen occurs and this interaction causes buckling failure. A similar phenomenon can also be observed in the multi-flaw-specimens.

158 Rock failure mechanisms

10.2.2 Crack growth from an array of crack-like flaws

10.2.2.1 Wing crack growth from three flaw arrays

According to previous experimental studies on specimens containing multi-flaws (Nemat-Nasser and Horii, 1982; Horii and Nemat-Nasser, 1986; Ashby and Hallam, 1986), wing cracks grow from each flaw in a stable manner when the length of the wing cracks is short. However, when the wing cracks become longer, they start to interact with each other in a way that will slow down the growth of the crack, or even stop it until a sudden coalescence occurs between cracks.

In order to study the influence of the geometrical configuration of crack-like flaws and the interaction between flaws, specimens 170 mm high and 100 mm wide containing three-flaw arrays with the same length $2a$ of 20 mm were numerically tested. The corresponding number of elements of this specimen is $255 \times 150 = 38,250$. The arrangement of the flaw arrays is of three types: diagonal (D Model), vertical (V Model) and horizontal (H Model), as shown in Figure 10.11 in which the stress distribution of the three model specimens before crack initiation is indicated (the lighter the grayscale, the higher the stress concentration).

Figure 10.12 shows the numerical results of crack growth in the associated specimens. It can be seen that crack growth is easier in the D Model specimen than in the V Model and H Model specimens. The data for the crack growth in the D Model specimen are plotted in Figure 10.13, noting that the numbering of the flaws I, II and III in each flaw group has been given in the Figure 10.11 caption. For comparison, the data for an isolated flaw in the specimen with the same size (Figure 10.9 for $W = 100$) are also plotted in Figure 10.13.

It can be seen that, although the stress levels for crack nucleation in each flaw are more or less the same, the crack growth from flaws I, II and III extend rapidly at a load well below that for an isolated crack. Referring to Figures 10.8 and 10.11, the free boundary away from the tips of the pre-existing flaws of I and III of Figure 10.11 is less than that of an isolated flaw (Figure 10.8). Thus, the tips of the flaw near the free boundary may allow the crack to grow more easily. For the middle flaw (II) of Figure 10.11, the crack growth is still easier than for the isolated flaw, although the boundary condition around the tips of the flaw is the same. This numerical result agrees well with the experimental observation made by Ashby and Hallam (1986). They concluded that the wing cracks grow more easily when other cracks are nearby. Although the numerical simulation for the D Model specimen also shows that cracks may grow more easily when another crack is nearby, this is not always the case. To accentuate the point, a comparison between the three simulations of the D Model, V Model and H Model is made.

From the results shown in Figures. 10.11 and 10.12, there are two main observations.

- Firstly, as seen from the second column of Figure 10.11, a high stress concentration (the white colour) occurs at the flaw tips, and stress interaction occurs inbetween in the nearby flaw area. The numerical simulation for the D Model specimen shows that cracks may initiate more easily when the growth direction of the wing cracks is away from the interaction stress field (see the first and second column of Figure 10.12). However, for the V Model, the zone of stress interaction

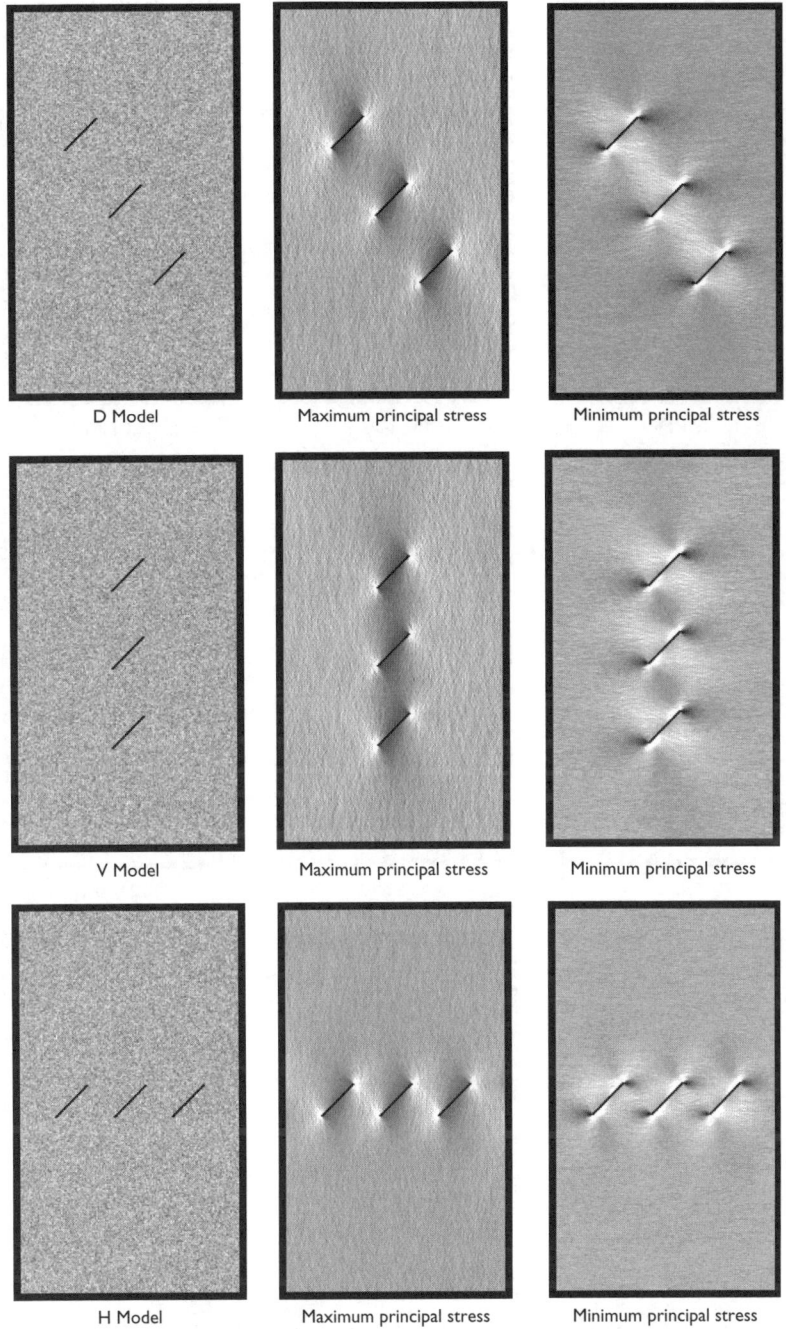

Figure 10.11 Stress fields showing the interaction between flaws for the three types of three-flaw arrays: Diagonal (D), Vertical (V) and Horizontal (H). The flaw numbers are I, II and III read from the top down or left to right.

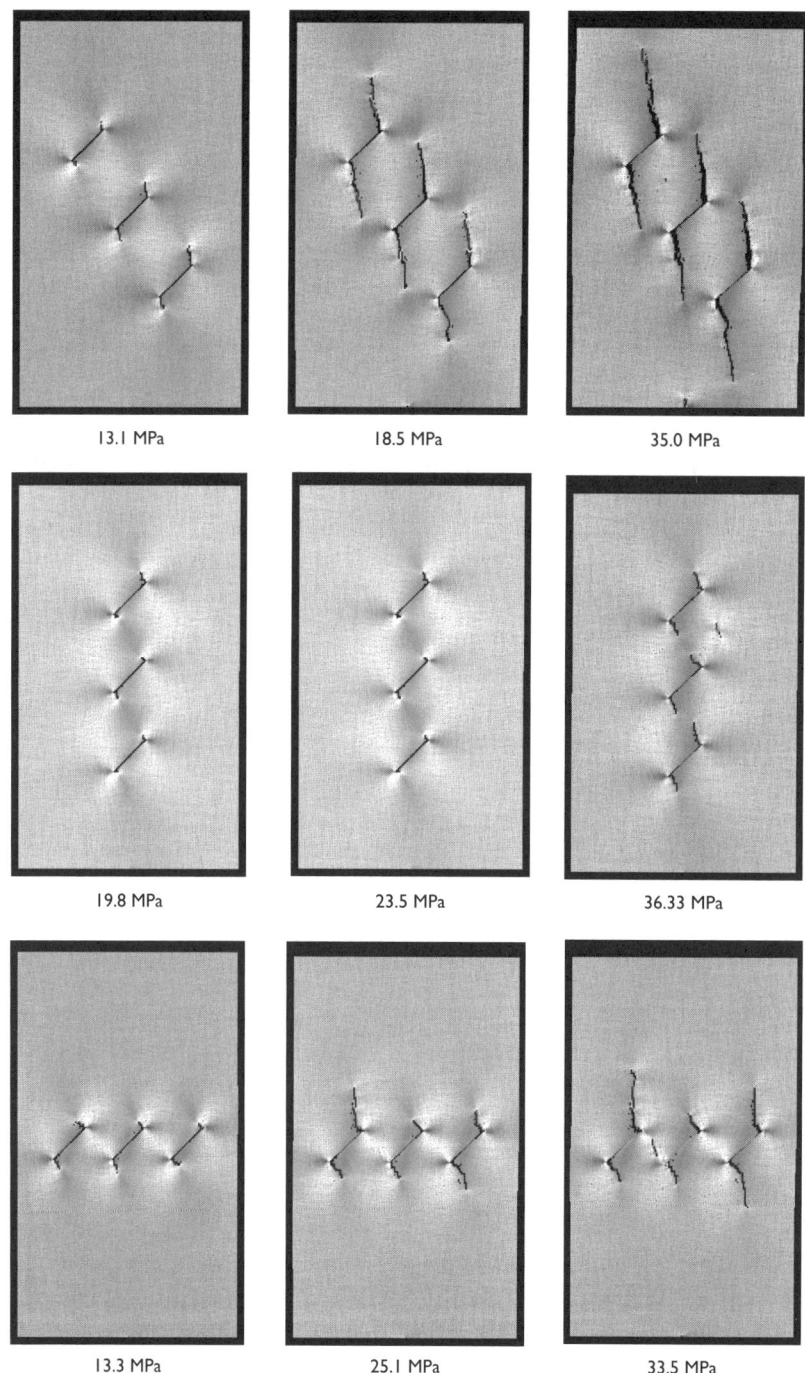

Figure 10.12 Stress distribution of crack growth under uniaxial compression for specimen containing three crack-like flaws (D Model, V Model and H Model).

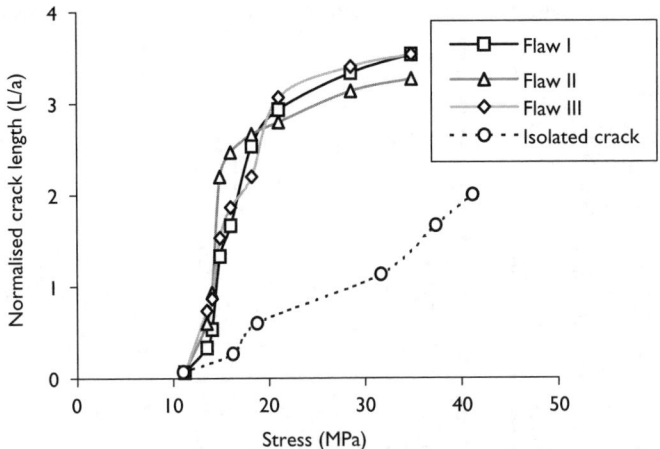

Figure 10.13 Crack growth under uniaxial compression in the specimen containing three pre-existing crack-like flaws arranged in the D Model configuration. For the arrangement of flaws I, II and III, refer to Figure 10.11, and the data for the isolated crack are obtained from Figure 10.9.

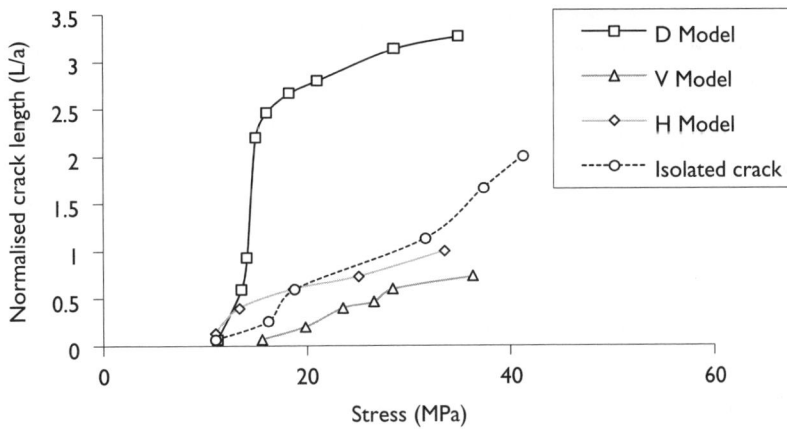

Figure 10.14 The influence of the flaw arrangement (D model, V model and H model) on the growth of cracks under uniaxial compression. The normalised crack length is taken as the mean value from the three flaws I, II and III. The growth of the isolated crack is plotted for comparison.

between the nearby flaws is located vertically around the tips of the flaws and this is also the location for tensile crack initiation and propagation. With this stress interaction, a higher stress is required for crack initiation (about 20 MPa).

- Secondly, for the H Model, numerical simulations show that crack growth from the central flaw is much slower than from the cracks of the two side flaws. This is because the growth of tensile cracks in the central part is suppressed by the stress interaction from the two side flaws. Figure 10.14 shows more clearly that the crack

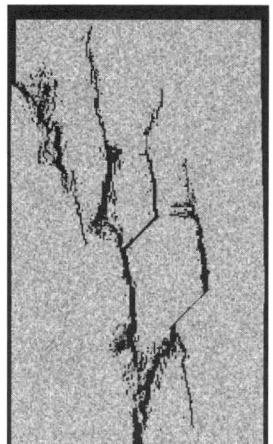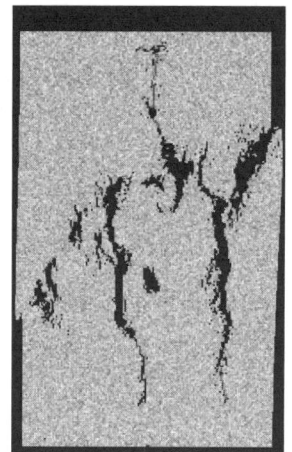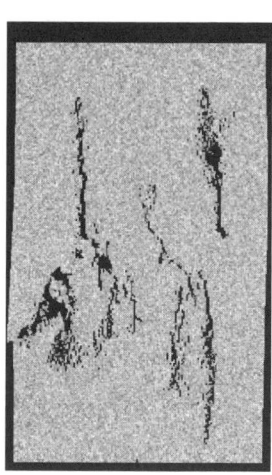

Figure 10.15 The failure modes of the specimens containing three crack-like flaws with the different arrangements of D model, V model and H model.

growth behaviour is strongly affected by stress interaction between flaws; the crack growth rate in the V and H Models is slower than that of an isolated flaw. Therefore, it can be concluded that stress interaction is one of the important factors that determines the initiation and propagation of a crack from the tips of the pre-existing flaw, and stress interaction depends strongly on the locations of the initial flaws.

The conclusion from Ashby and Hallam (1986), that cracks grow more easily when another crack is nearby, is not always applicable. Similar phenomena will be shown again in this Chapter regarding crack growth in specimens containing multi-crack-like flaws.

Figure 10.15 shows the ultimate failure patterns for the three model specimens. Although the crack propagation path in the D Model is longer, the coalescence stresses of the three different models are more or less the same (this is also seen from Figure 10.14).

10.2.2.2 Wing crack growth from randomly distributed multi-flaws

The complexity of crack initiation, propagation, interaction and coalescence can be demonstrated further in the following numerical simulation for a specimen containing a number of randomly distributed crack-like flaws with different lengths and angles. Figure 10.16a shows the sequence of progressive failure and Figure 10.16b shows the stress distribution for a multi-flaw specimen. The lightest colour shows the maximum stress concentration.

From these Figures, there are four main observations.

- Firstly—crack nucleation. Tensile cracks nucleated first from a 45° large flaw, which is located near the free boundary at the right side of the specimen (see Figure 10.16a, stage b). The distance between the tip of the flaw and the free

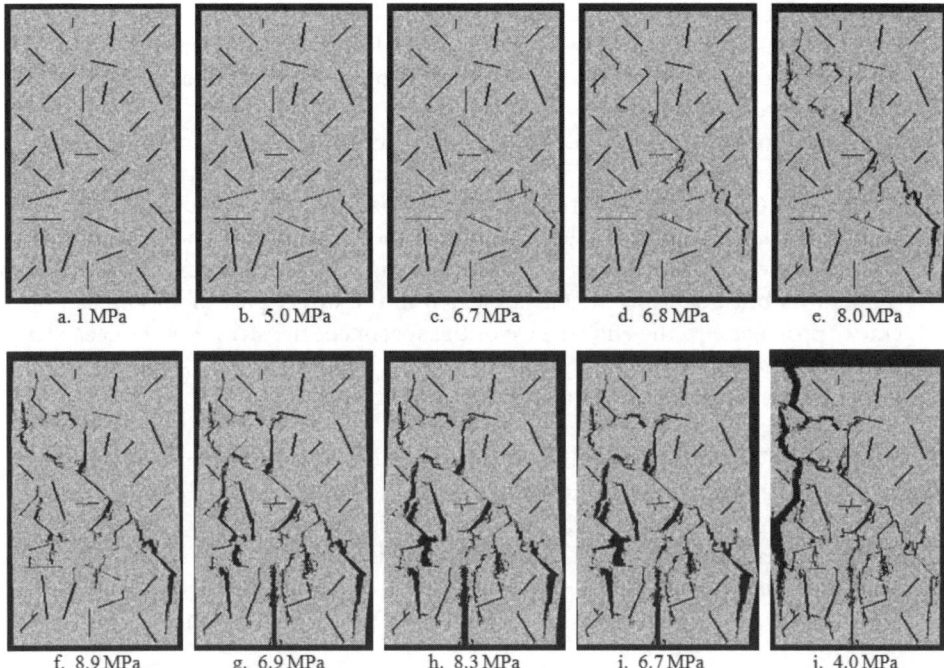

Figure 10.16a Sequence of progressive failure during growth of cracks under uniaxial compression for a specimen containing multi-flaws (background is variation in elemental stiffnesses).

surface is less than half of the flaw length. However, the longer flaw located at the centre of the specimen has experienced no crack nucleation at this stress level. Thus, confinement around the flaw tips is clearly one of the factors that influences the crack nucleation. After stage b, tensile cracks initiated from those flaws located along the sub-diagonal of the specimen (see Figure 10.16a, stages c–d). All these flaws are of different lengths and at different angles to the axial load direction. In general, cracks initiate from the larger flaws first (see Figure 10.16a, stage c), and then initiate from the shorter flaws (Figure 10.16a, stage d). Therefore, for a multi-flaw-specimen, flaw length and the location of the flaw are the important factors in determining the nucleation of a crack.

- Secondly—crack propagation. It is observed that, although cracks initiated from the flaw near the free surface of the specimen (see Figure 10.16a, stage b), cracks propagate in the direction of the axial compression only, and do not grow towards the free boundary (Figure 10.16a, stages c–e). The final failure of the specimen is caused by lateral buckling (Figure 10.16a, stages f and i). The same phenomenon was also observed in the experiment by Nemat-Nasser and Horii (1982). They generated a pre-existing flaw close to a free boundary (about half of the flaw length) with different boundary profiles: dome shaped, concave shaped and straight line. Cracks occurred at the tip of the flaw, propagating along the direction of the axial load and sub-parallel to the free boundary. This observation confirmed our earlier discussion that, if the splitting column is thinner than

the half-length of the flaw, strong interaction between the cracks and the free boundary of the specimen occurs and causes buckling failure.
- Thirdly—unstable crack growth. It is observed that, after the peak stress, cracks still grow continuously, even if the stress drops to half of the maximum load (Figure 10.16a, stages f–j).
- Fourthly—stress distribution. The initially uniform stress distribution at the initial stage (Figure 10.16b, stage a) changes to a highly non-uniform pattern (Figure 10.16b, stages e–f) with both crack initiation and propagation. High stress concentration (the lightest colour) occurs at the lower part of the specimen when the stress level reaches the peak value. That is why more cracks initiated from these flaws but not from the flaws located in the upper part of specimen. When cracks propagate to the end surface of the specimen, the stresses are released and concentrate only along the undamaged areas (Figure 10.16b, stage j). There is difficulty in observing this in a physical experiment.

Figure 10.17 shows the stress–strain curve for the multi-flaw-specimens. In Figure 10.17, the black curve shows the axial stress plotted against the axial strain for the specimen shown in Figure 10.16, and the gray curve shows the corresponding values of the cumulative crack length that is calculated by summing the steps in the crack propagation. The points a–j marked on the stress–strain curve correspond to those in Figure 10.16 at the different stages a–j. From Figure 10.16a and b, it is known

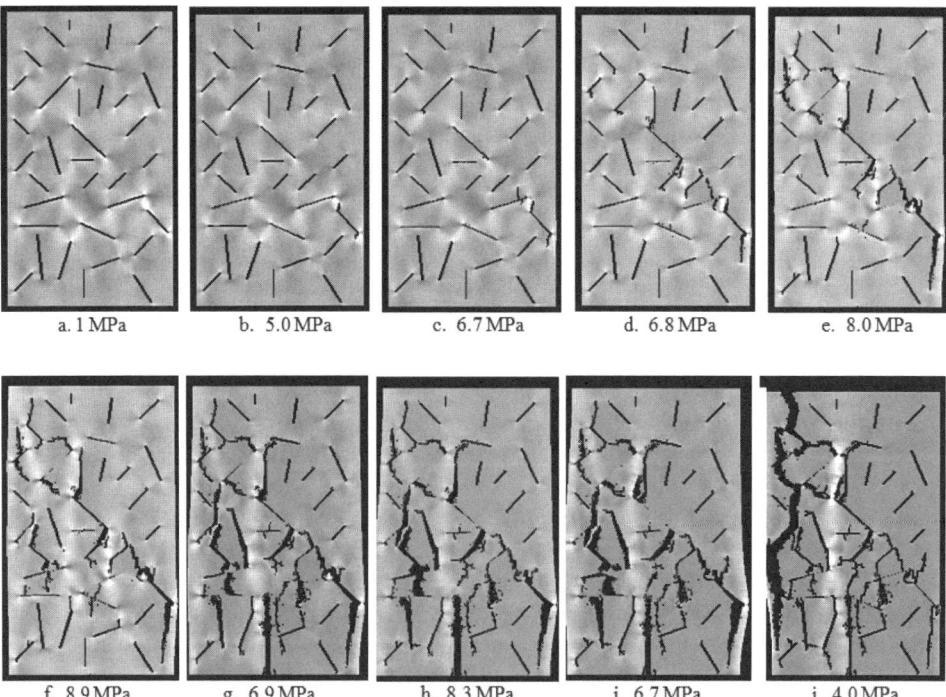

Figure 10.16b Stress distribution during the growth of cracks under uniaxial compression for specimen containing multi-flaws (background is shaded to show the shear stress variation).

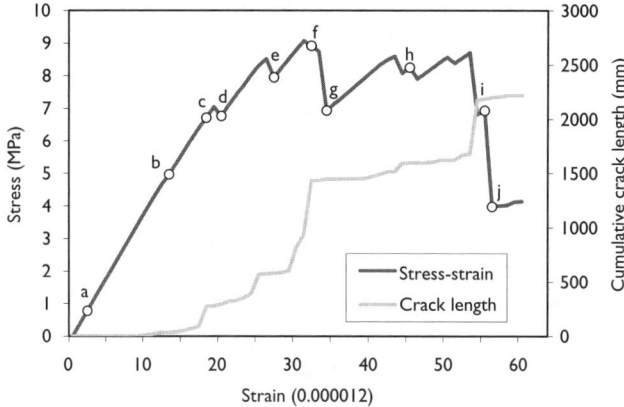

Figure 10.17 Stress–strain curve for specimen containing multi-flaws with the cumulative crack length value. The points a to j correspond to the stages in Figure 10.16 with the same letters.

that the first tensile crack nucleates at the point marked b, followed by the first crack coalescence at point d. After that, the stress–strain curve shows that the overall elastic modulus reduces—due to the continuous nucleation of tensile cracks. According to the curve for the cumulative crack length, every large increase of the cumulative crack propagation length results in an associated considerable stress drop. The largest stress drop indicates the axial splitting of the specimen (stages i–j).

10.3 CRACK GROWTH FROM A PORE-LIKE FLAW IN COMPRESSION

The pre-existing flaws in rocks may be crack-like (as in Section 10.2) or pore-like, such as soft inclusions or grain boundary cavities, which we address in this section. The problems considered here have been studied by Sammis and Ashby (1986) both experimentally and analytically. In their experiments, plate specimens containing a single hole of the same size or an array of holes with various diameters were tested. The purpose of their experiments was to investigate the interaction of growing cracks with the surfaces of the specimen and with each other. Figure 10.18a shows photographs of the study of Sammis and Ashby (1986) where the cracks extended in a polymethylmethacrylate (PMMA or Perspex/Plexiglas) plate containing a central hole were photographed as the compressive load was increased. Typical crack growth from the array of holes is shown in Figure 10.18b.

The experimental results show that cracks initiating from the holes interact with the surfaces of the specimen in a way that causes them to grow more than they would in an infinite medium. It is also seen that cracks grow more easily in narrow plates than in wide ones. A simple analysis based on beam theory was developed to describe this interaction. For solids containing multiple holes, the appearance of neighbouring holes and cracks may produce an amplifying or shielding effect on the stress state around the crack tips. So we address these questions using numerical modelling.

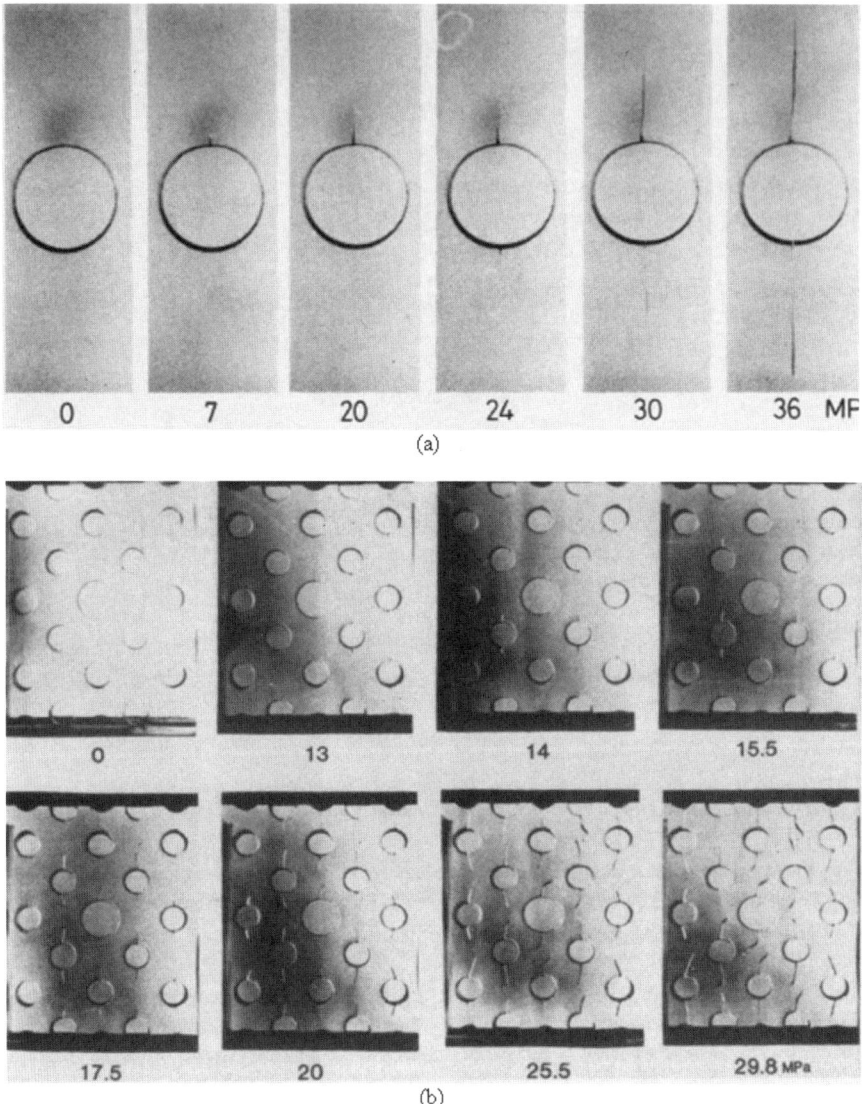

Figure 10.18 Crack growth from single hole a) and an array of holes b) in PMMA plates, after Sammis and Ashby (1986).

10.3.1 Modelling crack growth from a single hole in specimens under compression

10.3.1.1 Crack growth from a single hole in specimens of different width

Crack growth from a single hole is similar to the ring test simulations that we discussed in Chapter 3 except that here the specimen is uniformly loaded at the boundary, rather than as diametral compression in the ring test. We first study crack growth from a circular hole by loading specimens with different widths under uniaxial compression. The numerical specimens are originally 170 mm high and 25, 50 or 100 mm wide, and the corresponding numbers of simulation elements are $255 \times 38 = 9{,}690$, $255 \times 75 = 19{,}125$ and $255 \times 150 = 38{,}250$, respectively. In all the cases, the tensile stresses are always greatest at the top and bottom of the hole (or close to these locations—because we are modelling a heterogeneous rock). Therefore, a pair of cracks of length, L, emanating from these points is created in tension.

Figure 10.19 shows the processes of crack initiation and propagation in the specimens. The cracks grow in a stable manner, and a progressively greater load is required to drive them further away from the hole at which they nucleate. Due to the heterogeneity of the rock material, the crack path is not smooth.

Typical data for crack growth obtained from the simulations are plotted in Figure 10.20 on normalised axes: the crack length L is plotted as L/a where a is the radius of the hole. The crack growth is stable in the sense that a further stress increment is needed to drive the cracks further from the hole. It can be seen that the cracks initiate at an earlier stage and grow more easily if the specimen is narrower (i.e. $W = 25$ mm) compared to wider specimens (e.g. $W = 100$ mm). This can be seen clearly in Figure 10.19. When the narrow specimen is numerically loaded, there is a clear outward bending deformation of the ligaments on either side of the growing cracks. This result, shown in Figure 10.19 for narrow specimens, is important in understanding the interaction of cracks with surfaces and other cracks (to be simulated later). In fact, the simulations presented in the following sections all show that bending displacements appear in specimens whenever the crack length becomes comparable with the specimen width. The bending moment is caused by the built-in effect at the ends of the specimen when the crack propagates toward them.

The failure modes for the three specimens having different widths are shown in Figure 10.21. It can be seen that, for the narrow specimen, buckling occurs when the cracks reach the two ends of the specimen, which indicates a strong interaction between the crack and the specimen surface parallel to the cracks. For the wide specimen, however, this interaction is less apparent. Instead, failure occurs by a new crack initiating in shear zones around the hole. The specimen of medium width fails by two shear cracks connecting the hole with the specimen surfaces. This result differs from the theoretical prediction for a hole in an infinite domain—because of the interaction between the growing cracks and the upper and lower surfaces, which appears to retard the crack growth as the crack is approaching the end surfaces (as seen in Figure 10.20 for $W = 50$ mm). Thus, the shear cracks zones are more likely to develop.

10.3.1.2 Crack growth from a single hole with different diameters

The growth of cracks from a pre-existing single hole of diameter $2a$ of 10, 15 and 20 mm is studied to examine the influence of hole diameter on the crack growth behaviour. The size of the specimens is 170×50 mm. The simulated sequence of crack growth and associated stress distribution for these three specimens is shown in Figure 10.22. It is

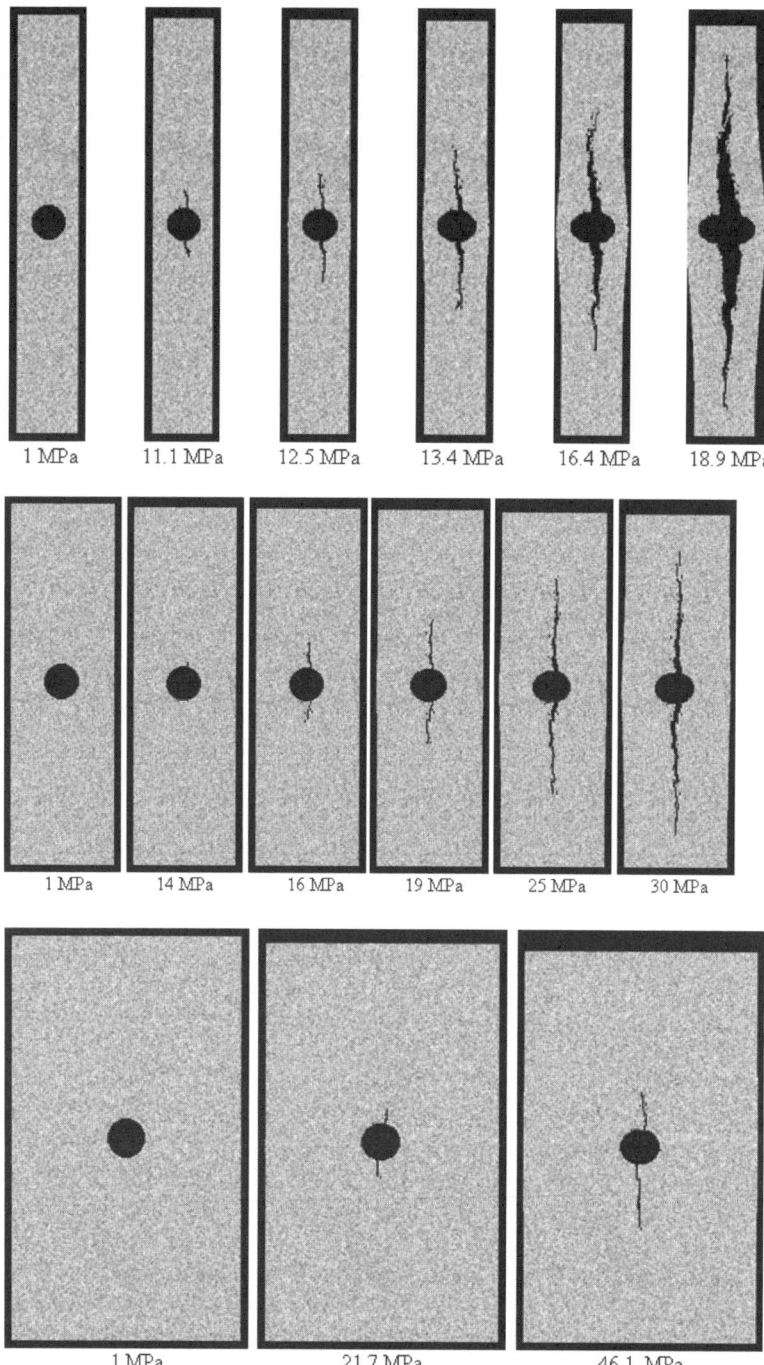

Figure 10.19 Sequence of crack growth from single holes under uniaxial compression for numerical specimens with widths 25, 50 and 100 mm.

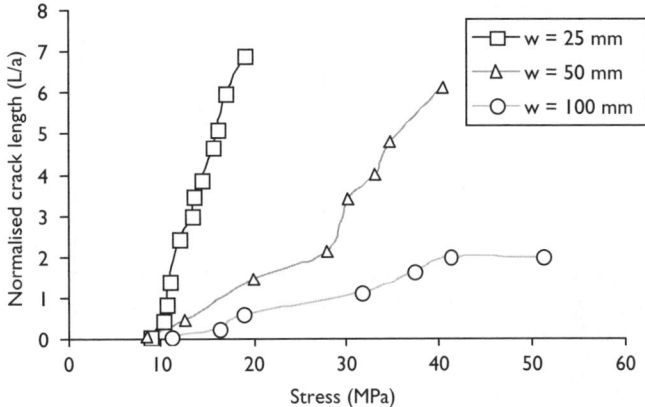

Figure 10.20 Crack growth in specimens under uniaxial compression showing the influence of the specimen width.

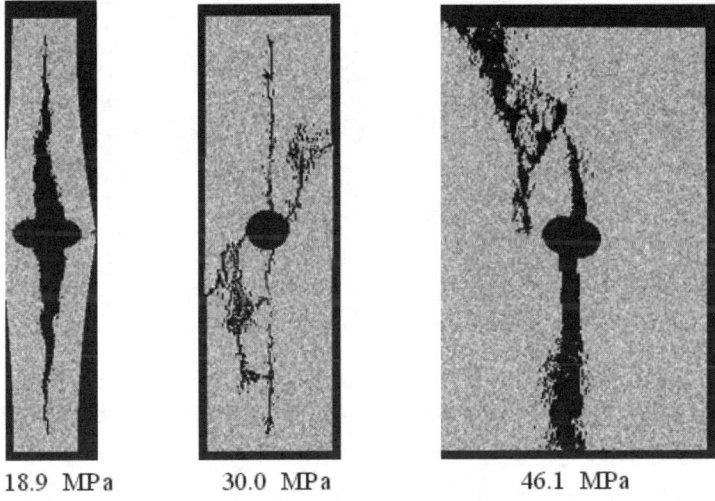

Figure 10.21 The failure modes of the specimens with different widths of 25, 50 and 100 mm.

seen that the crack initiation and growth in the specimen containing the smaller hole ($2a = 10$ mm) is more difficult in the early stage of loading. However, when the axial stress reaches about 35 MPa, the crack growth dramatically increases until global failure of the specimen. For the specimens containing larger holes, however, the cracks initiate earlier and have a higher growth rate initially, and then slow down later. As mentioned earlier, the slowing down of the crack growth rate is mainly caused by the boundary effect (i.e., the interaction between the cracks and the upper or lower surfaces).

Figure 10.23 shows the influence of the hole diameter on the crack initiation and propagation behaviour. It is expected that specimens containing isolated, pre-existing holes of a larger size are more susceptible to crack nucleation in the early stages than

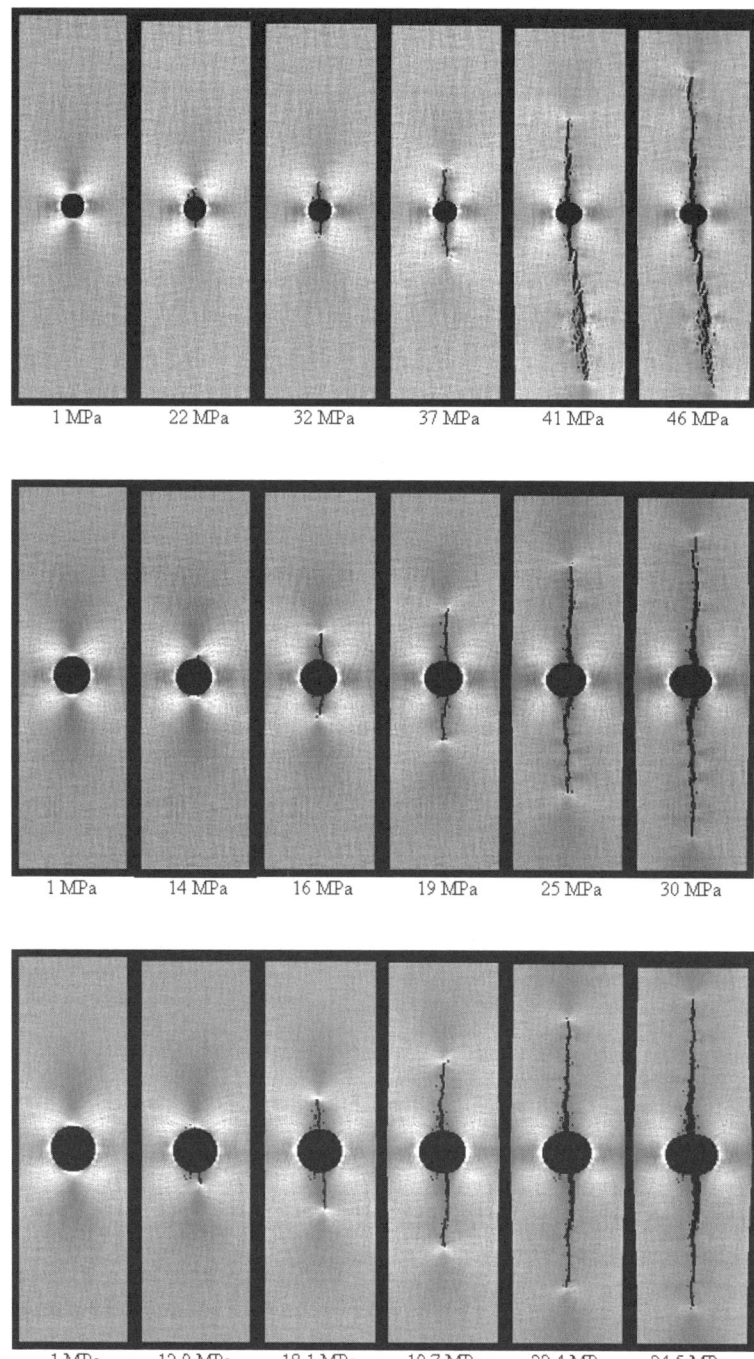

Figure 10.22 Stress distribution during crack growth under uniaxial compression for specimens containing a hole of diameter 10, 15 and 20 mm.

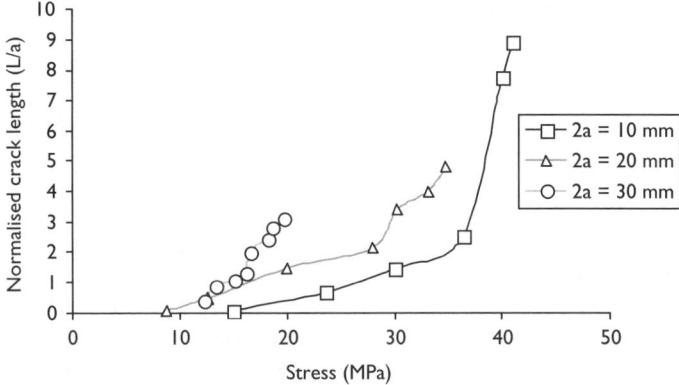

Figure 10.23 Increase in normalised crack length in specimens under uniaxial compression showing the influence of the hole radius.

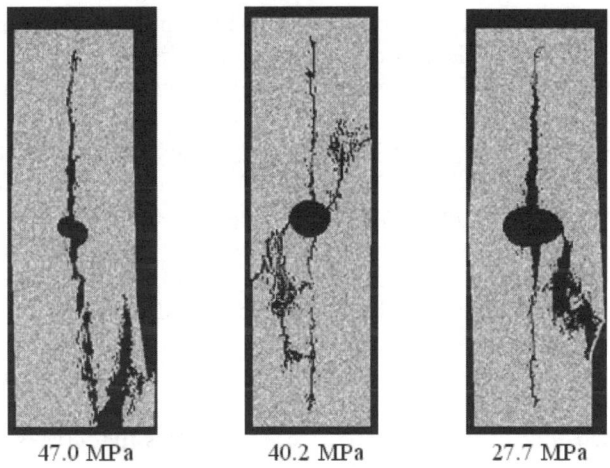

Figure 10.24 The failure modes of the specimens containing a single hole but with different diameters of 10, 15 and 20 mm.

those specimens containing smaller holes. This result agrees well with the experimental observations made by Sammis and Ashby (1986).

The failure modes presented in Figure 10.24 show that only the specimen containing the smaller hole fails in a splitting mode; whereas, the other two fail by cracks initiated in a shear zone connecting the hole and the free surface, due to the proximity of the hole to the free boundary of the specimen. This failure is caused by the following sequence: at first the cracks grow fast, and then, when the tips of the cracks are near the ends of the specimen, the propagation slows down, or sometimes stops, due to the boundary effect. If the shear stress in the area between the hole and the free boundary of the specimen is high enough, shear failure may occur between the hole and the free boundary, as shown in Figure 10.24.

10.3.1.3 Modelling of crack growth from an array of holes in a specimen under compression

10.3.1.3.1 Crack growth from an array of three holes

In order to study the interaction of adjacent holes, specimens containing an array of three holes arranged in diagonal, vertical and horizontal lines were established (referred to as the D Model, V Model and H Model, similar to Figure 10.11 for crack-type flaws). Figure 10.25 shows the three model specimens and the stress distribution before crack growth occurs. The specimens are 170 mm high and 100 mm wide, and the corresponding number of elements is $255 \times 150 = 38{,}250$. All the holes have the same diameter of 15 mm. The specimens are loaded in uniaxial compression under displacement rate control and in plane stress conditions.

Figure 10.26 shows how the tensile stress before crack growth falls rapidly with distance outward from the surface of the central hole, becoming compressive at certain distances for the three models. For comparison, the result of the specimen containing a single hole is also plotted in Figure 10.26 (the dotted line). The tensile stresses at the top and bottom of the holes are 1.14, 0.49 and 0.35 times the axial stress for the D, V and H models, respectively. Therefore, it can be expected from Figure 10.26 that a crack will be easier to initiate in the D model than in the V or H model. This can be seen clearly in Figure 10.27 showing the results of crack growth in the associated specimens.

From Figure 10.27, it is seen that, for the D Model, the crack extension initially nucleated from the poles of the pre-existing hole, then coalescence occurred between the holes along the diagonal line (which is in a high shear stress zone). The data relating to the crack growth in the D Model specimen are plotted in Figure 10.28. For comparison, the data for an isolated flaw in the specimen of the same size (shown in Figure 10.20 for the specimen width of 100 mm) are also plotted in Figure 10.28. It can be seen that, in all cases for holes I, II and III, the cracks extended rapidly at a load well below that for the case of the isolated hole. In this case, cracks grow more easily when other cracks are nearby.

Although the result for the D Model specimen shows that cracks may grow more easily when another crack is nearby, this is not always true for the other models. To emphasise this point, two other simulations of the V and H model specimens containing three similar flaws but with different arrangements of the flaw locations are shown in Figures 10.25 and 10.26. No crack nucleates from the central hole in the H Model, even if the axial stress reaches the maximum, indicating that the existence of the nearby holes arranged in the horizontal line decreases the tensile stress around the central hole. This may be caused by the presence of the free specimen boundaries and can be seen more clearly in the following sections regarding crack growth in specimens containing multiple hole-like flaws. Figure 10.29 shows more clearly that the crack growth behaviour depends strongly on the locations of the initial hole flaws. In this case, the normalised crack length is the length of the crack located near the central hole.

Figure 10.30 shows the ultimate failure patterns for the three model specimens. The coalescence stress, 17.2 MPa, of the D Model is the lowest among the three model specimens. Figure 10.30 also shows that global failure of the specimen does not necessarily result in the complete loss of the strength.

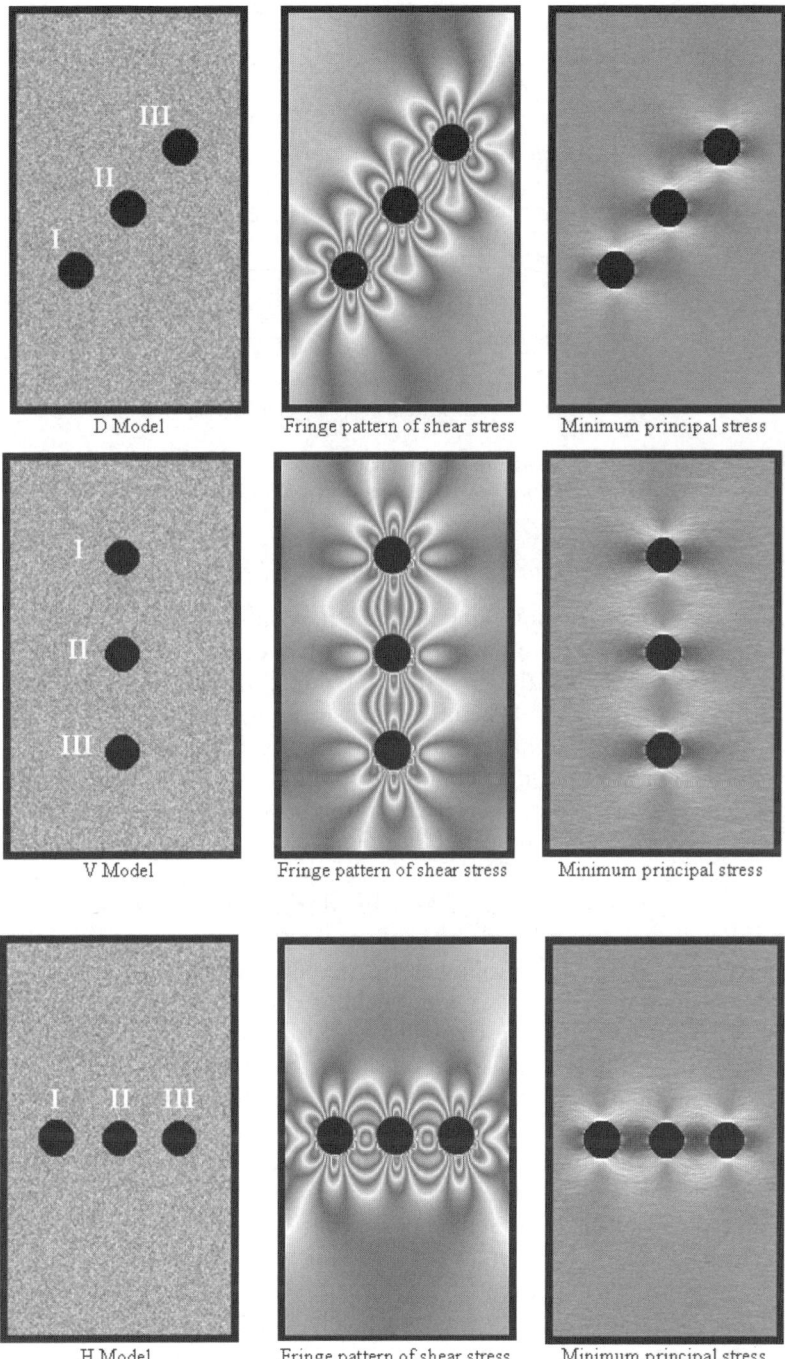

Figure 10.25 Stress field showing the interaction between holes. The fringe pattern in the central images has the same appearance as a photoelastic pattern because both are generated by the magnitudes of the shear stresses.

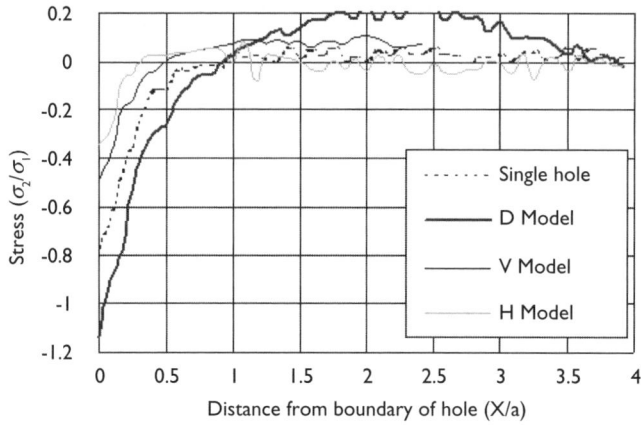

Figure 10.26 The stress σ_z along the longitudinal axis of the central hole in the specimens under compression showing the influence of the three-hole orientation.

10.3.1.3.2 Crack growth from multi-holes uniformly distributed

Experiments on specimens containing multiple flaws also show that, when loaded in compression, cracks grow from flaws. The individual cracks, when short, grow in the same way as isolated cracks, the growth being dominated by the stress field around the flaw from which it grows. When the cracks become longer, they start to interact with each other in a way that makes their growth easier. Figure 10.31 is a typical result of the crack growth in a specimen containing pre-existing hole-like flaws distributed uniformly. It can be seen that the onset of fractures in this specimen subjected to uniaxial compression is first indicated by the nucleation of cracks from some of the holes, as indicated in stage b. With these cracks propagating, more cracks nucleate from other holes, as seen in stage c. At stage d, the cracks become longer and the interaction between them induced the coalescence between some of the propagated cracks (stage e). This is quickly followed by the development of macroscopic failure immediately beyond the peak stress with buckling occurring in stage j.

Figure 10.31 shows that the changing stress distribution is highly non-uniform compared with the initial stress distribution shown in stage a. From this simulation, we have a better understanding of the mechanics of material strength. In Figure 10.32, the stress–strain curve and the cumulative crack length are plotted. Comparison between the stress–strain curve shown in Figure 10.32 and the failure processes shown in Figure 10.31 reveals once again that reaching the maximum stress or strength of a specimen does not necessarily mean abrupt failure of the specimen (see stage f). The stress–strain curve becomes non-linear when cracks occur in the specimen (stages b, c, d and e). At stage f, the specimen reaches its maximum strength. Although the load bearing capacity drops, it is not necessarily a stage of specimen collapse. The specimen continues losing its load bearing capacity, and yet the cracks do not propagate thoroughly until stage j.

Also, the comparison between the stress–strain and cumulative crack length plots shown in Figure 10.32 indicate a close relation between them. Note that every large stress drop (points e, f, g, h, i, j) in the stress–strain plot corresponds to a high rate of cumulative

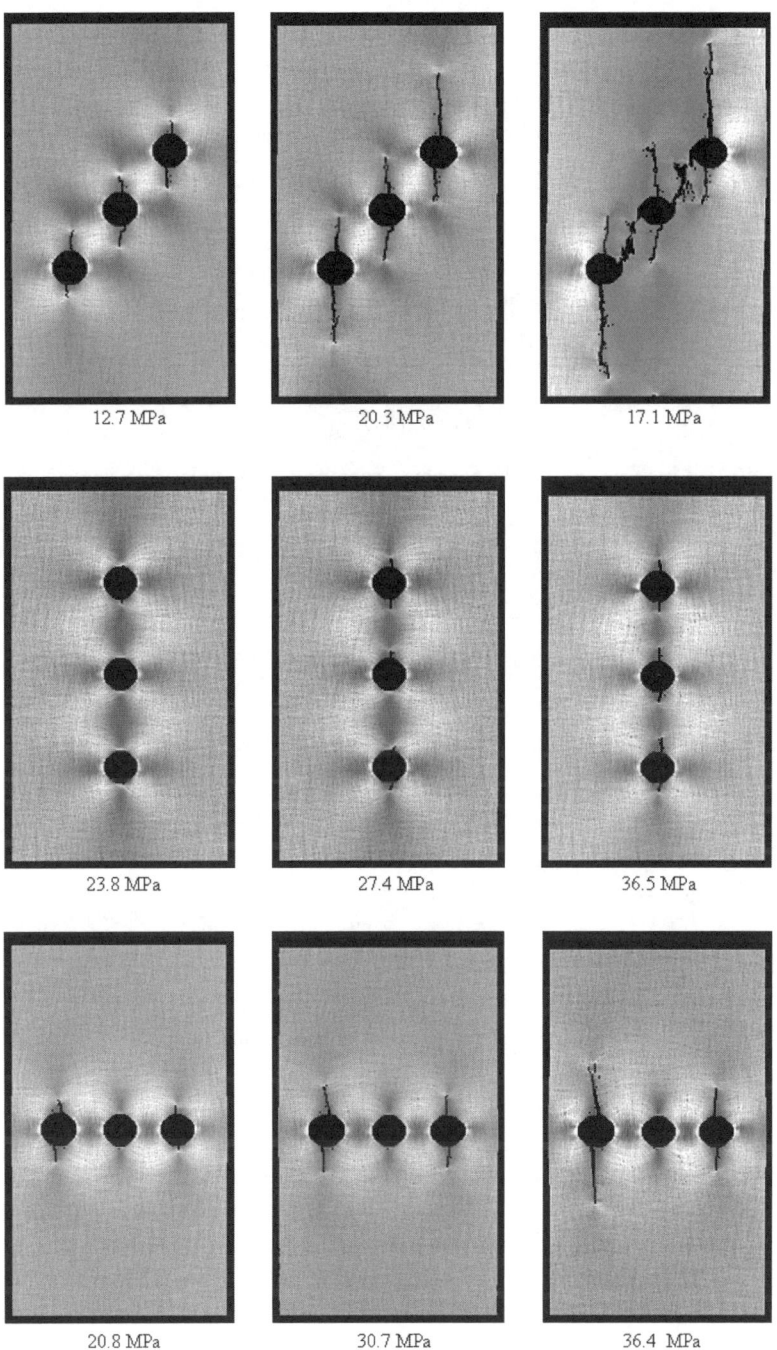

Figure 10.27 Stress distribution during crack growth from three holes under uniaxial compression, showing the influence of the nearby holes on the central hole (from top to bottom, D, V and H models).

176 Rock failure mechanisms

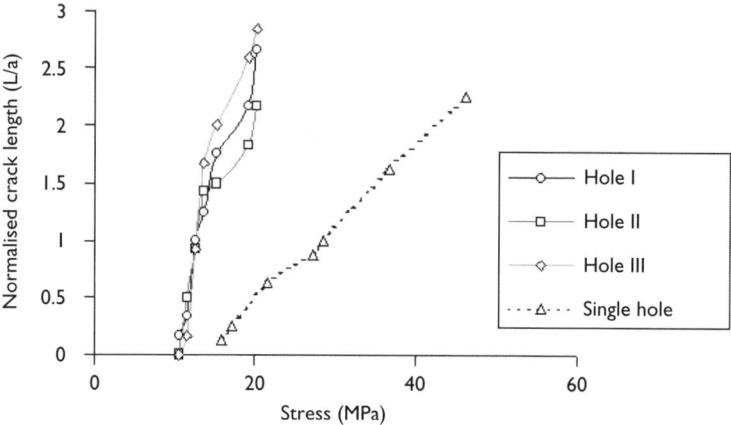

Figure 10.28 Crack growth for the specimen containing three holes arranged in a diagonal line (D Model) showing the influence of nearby holes on the central hole.

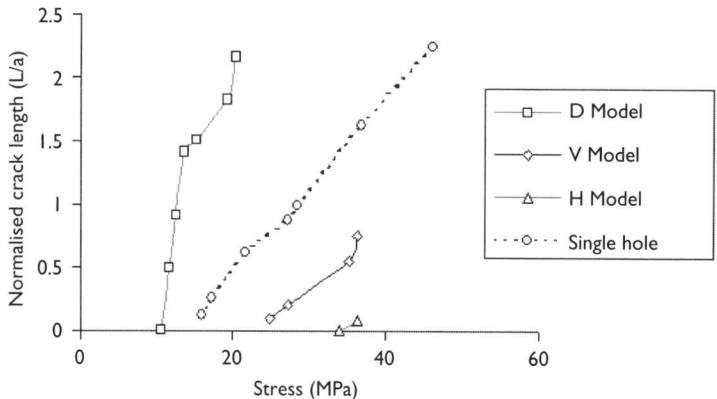

Figure 10.29 Crack growth in specimens containing three holes, showing the influence of hole arrangement.

crack length. The reason for these stress drops is in fact the jump phenomenon of the crack propagation. After a few cracks have propagated, stress relaxation occurs and the crack propagation stops. With further increase in the external displacement loading, the concentration of stress at the crack tips reaches the strength and the crack propagation starts again. The process continues until interaction with other cracks commences and a more complex failure patterns forms, which results in a large stress drop (as shown in Figure 10.32, in which at least four notable stress drops are observed).

10.3.1.3.3 Crack growth from multiple holes distributed randomly

The complexities of crack initiation, propagation, interaction and coalescence can be demonstrated once again for the following specimen containing a number of

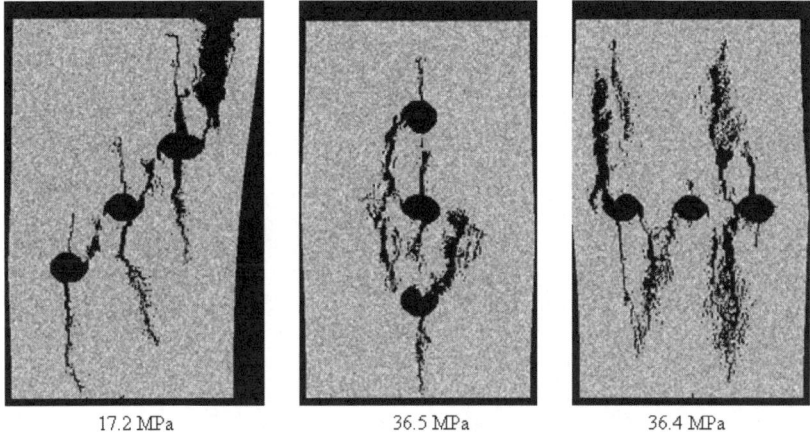

Figure 10.30 The failure modes of the specimens containing three holes in different arrangements.

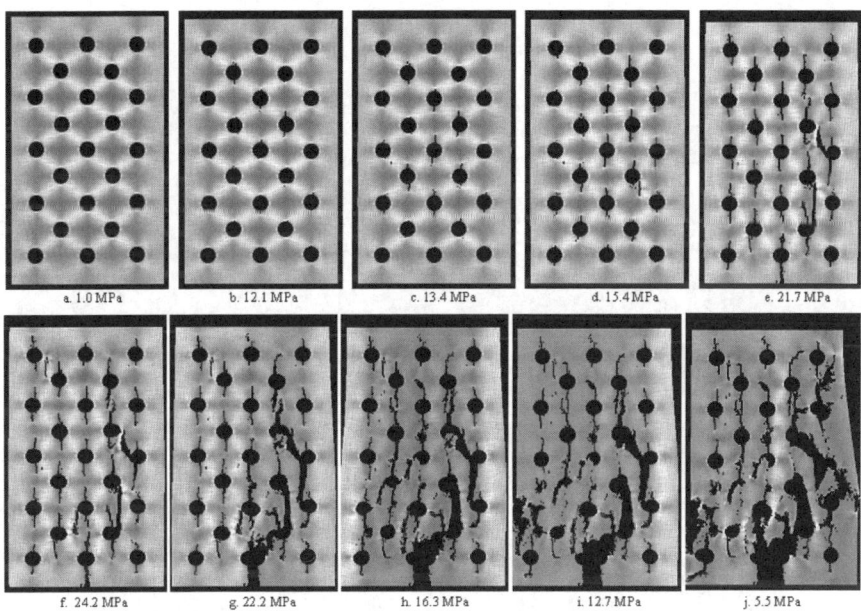

Figure 10.31 Sequence and stress distribution of crack growth from multiple holes uniformly arranged in a specimen under uniaxial compression.

randomly distributed pre-existing hole-like flaws with different diameters, as shown in Figure 10.33 for the sequence of failure. As predicted, based on the studies above, under overall uniaxial compression, tension cracks first nucleate at the larger hole flaws (stage a), or holes close to the boundary (stage b), or holes located on a diagonal line (stages c, d, h), extend in the axial direction (stages b–d), link with other cracks

178 Rock failure mechanisms

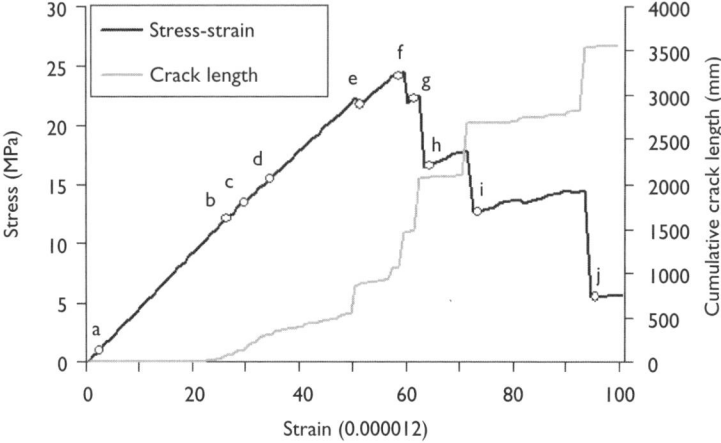

Figure 10.32 Complete stress–strain plot and cumulative crack length plot for specimen containing uniformly distributed holes.

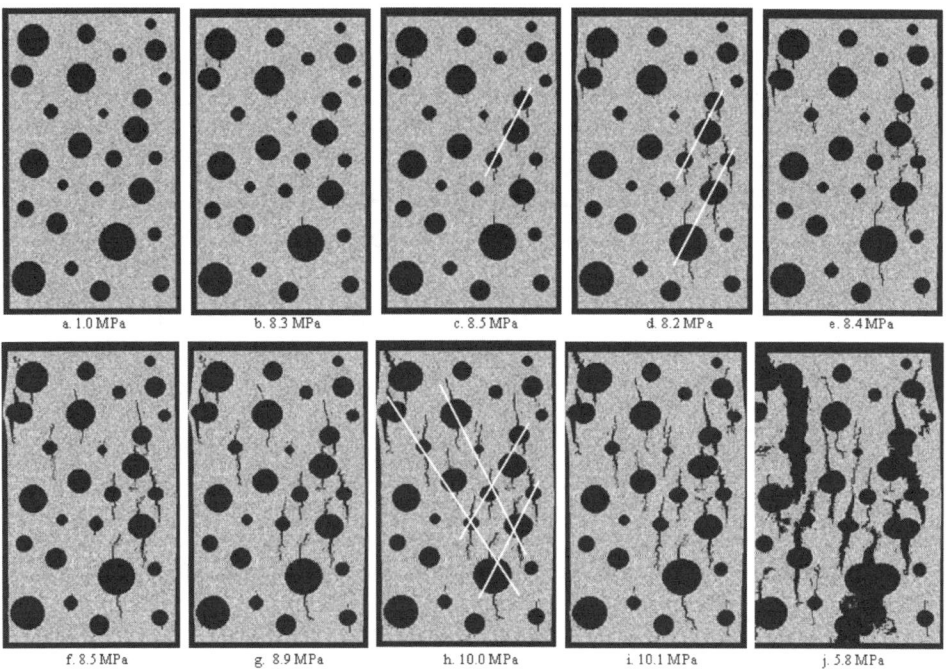

Figure 10.33 Sequence of crack growth from multi-holes randomly arranged in a specimen under uniaxial compression.

and flaws (stages e and i), and the specimen fails by axial splitting (stage j). Note that many holes have not caused any crack in the failed specimen.

The interaction between crack and specimen surface is clearly demonstrated by the holes in the upper left corner of the specimen (Figure 10.33), in which two cracks that initiate from a hole near the specimen surface seem to have no desire to move toward the free boundary (stages d to i). On the contrary, and as observed in experiments performed by Nemat-Nasser and Horii (1982), cracks nucleate from the hole, propagate away from the hole parallel to the free boundary, and then more or less follow the contour of the free boundary. Even after splitting, the specimen can withstand almost half the maximum load (5.8 MPa as indicated in stage j in Figure 10.33). In the specimen containing uniformly distributed holes, cracks nucleate from most of the holes before the peak stress is reached (Figure 10.31); whereas, in the specimen containing randomly distributed holes, there are still many holes left inactive prior to the peak stress.

10.3.1.3.4 Plaster tests on blocks containing cylindrical holes

Jespersen *et al.* (2008) report physical and numerical (finite difference and finite element) tests on plaster blocks containing cylindrical holes, see Figure 10.34.

Although the geometry of the holes in Figure 10.34 is not exactly the same as in the numerical models presented in this Chapter, it is similar to the diagonal set of holes in Figure 10.30 (left-hand side) and the uniform multi-hole example shown in Figure 10.31 and the shearing between the diagonal holes is well matched in the respective physical and numerical models.

Figure 10.34 Uniaxial compression of a plaster specimen containing cylindrical holes (from Jespersen et al., 2008).

Chapter 11

Dynamic loading of rock

11.1 INTRODUCTION

The time over which a rock is subjected to stresses and strains varies over a large spectrum, from the milliseconds of blasting to the millions of years during natural tectonic processes. In Chapter 9, we considered time dependency through considerations of creep and stress relaxation. In this Chapter, we will describe the dynamic response of rock, i.e. the loading of rock over fractions of a second.

An example of high strain rate natural fracturing of rock is shown in Figure 11.1 in which there have been two rapid fracture events as the fracture has moved from the right to the left across the rock mass. The morphological study of such dynamically created fracture surfaces is known as tectonofractography (see Bahat, 1991) and similar surfaces occur both on the large scale in nature and on the small scale in laboratory testing. The scientific research in this field has focused on the determination of the properties of rocks and the development of accurate constitutive models and failure criteria at high strain rates (Zhao et al., 1999), often in a military context, in order to provide sufficient information for the design of the excavation and support of rock engineered structures.

An example of a laboratory experiment for studying the effect of pre-existing fractures on stress waves in the context of pre-split blasting is shown in Figure 11.2. In each of the three photographs in Figure 11.2a, three detonators in a line are simultaneously initiated in a polymethylmethacrylate (PMMA, perspex, plexiglas) block in which there are two pre-existing fractures. In all three photographs, the star cracks around the detonators are clearly visible. In photograph (a), some damage caused by reflection of the compressive wave as a tensile wave can be seen. However, in photographs (b) and (c) and because of the closer proximity to the pre-existing fractures in the PMMA, plus the almost total reflection of the stress waves, there is denser fracturing. This experiment demonstrates why pre-split blasting (where the intention is to create a pre-split fracture surface through a rock mass) is such a robust technique, even when the pre-split basting boreholes are not exactly parallel and/or when there are natural pre-existing fractures in the rock mass, see Figure 11.2b.

Although there is a good elasticity foundation for the subject of rock dynamics through the theory of stress waves in solids (Kolsky, 2003), dynamic loading has been somewhat neglected in the history of rock mechanics research, probably because of the difficulty of conducting experiments in the laboratory and in the field. Also, and despite this excellent elasticity foundation, unfortunately the rock mass is not elastic and so refraction and reflection of stress waves at fracture boundaries (Figure 11.2)

Figure 11.1 Two dynamic natural fracture events as the fracture, with its curved front, moves to the left. Note the surface irregularities on the fracture surfaces which are typical of dynamic fracturing (scale: about 4 m across the picture, Carboniferous strata, South Wales, UK).

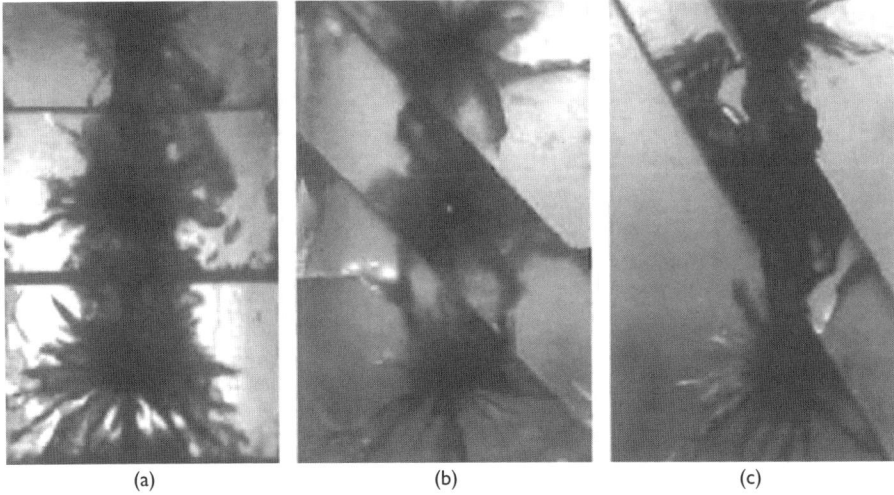

Figure 11.2a Laboratory experiment to evaluate the influence of fractures on pre-split blasting (from Worsey, 1981).

and attenuation have to be taken into account as well. Moreover, there is a wide spectrum of strain rates when we consider rock dynamics *sensu lato*. Indeed, one of the important features recognised for rocks and other brittle materials is their rate-dependence, i.e., the fact that their properties (ultimate strength, Young's modulus, fracture energy) are highly dependent on the loading rate. The general trend is an increase in dynamic strength as the loading rate increases. Efforts have been made to improve the knowledge of the constitutive relation for a wide range of strain rates by developing more realistic material characteristics.

Dynamic loading of rock 183

Figure 11.2b Example of the half-barrels of pre-split blastholes in a fractured rock mass in Scotland.

In this Chapter, we will focus on the laboratory testing scale. Readers interested in the larger scale aspects of rock dynamics are referred to the recent book by Barton (2007). Experimental investigations have provided good opportunities for examining the ultimate failure patterns of rock samples; however, we believe that analysing the complete history of dynamic failure is better than simply examining the final outcome. The post-experimental observations do not provide sufficient information about the evolution of the stress field nor the micro-fracture development. Continuing the theme of Chapter 3, but moving into the dynamic field, the focus of the current Chapter is to numerically investigate the failure mechanisms and the fracture patterns of rocks using Brazilian samples under different stress wave amplitudes. This includes the way in which the stress wave in heterogeneous rocks affects the dynamic fracture propagation and patterns.

11.2 THE SIMULATION MODELS

To evaluate the influence of heterogeneity on stress wave propagation, four samples with different homogeneity indices, $m = 2, 5, 10, 100$, subjected to a stress wave input are used for the simulations. The model, with sample and bars, is divided into 400×15 elements with 5 mm as the length scale of the element, as shown in Figure 11.3.

184 Rock failure mechanisms

The rock sample was positioned between two transmitter bars as an analogue to the Split Hopkinson Pressure Bar (SHPB) used in laboratory experiments (ISRM News Journal, 2009). The parameters and calculation conditions are as follows:

	Specimen	Bar
Homogeneity index (m)	2, 5, 10, 100	200
Mean elastic modulus (E_0)	60 (GPa)	210 (GPa)
Poisson's ratio (v)	0.25	0.2
Density	2.5E-6 (kg/m³)	7.8E-6 (kg/m³)

To investigate the influence of stress waveforms in terms of their peak value on the fracture process and failure pattern of heterogeneous rocks, 2-D Brazilian disc samples are used for the simulations, as shown in Figure 11.4. The radius of the disc is 80 mm. A magnified portion of the finite element layout for the disc is also illustrated in Figure 11.4. The model is divided into 25,600 square elements.

The parameters and calculation conditions for these samples are listed below:

	Sample	Bar
Homogeneity index	3	200
Young's modulus	37.5 (GPa)	210 (GPa)
Compressive strength	205 (MPa)	
Tensile strength	18 (MPa)	
Poisson's ratio	0.25	0.20
Density	2.5E-6 (kg/m³)	7.8E-6 (kg/m³)

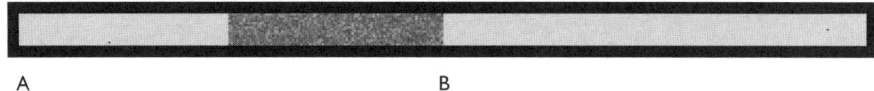

A   B

Figure 11.3 Numerical model of the sample (left centre) and transmitter bars, (400 × 15 elements with 5 mm length scale for the elements in the sample).

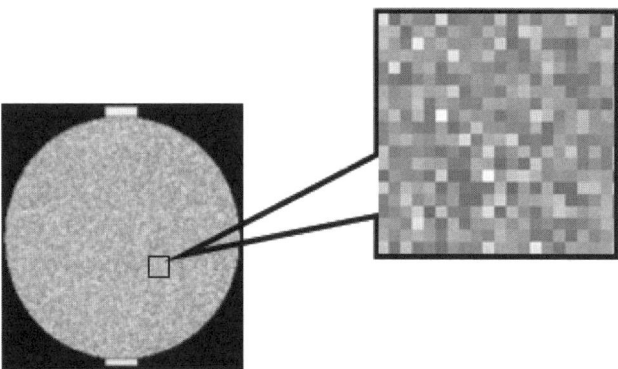

Figure 11.4 The numerical Brazilian disc sample and the diametral loading conditions (the sample with 160 × 160 elements).

Dynamic loading of rock

The modelling system consists of an incident boundary and a transmitter platen to transfer the incident compressive pulse that propagates towards the sample. The pulse is partially reflected at the location of the transmitter platen and is partially transmitted through the sample.

11.3 SIMULATION DEMONSTRATION

11.3.1 Influence of heterogeneity on stress wave propagation

As discussed in the previous Chapters, rock is a heterogeneous material, and the heterogeneity plays a significant role in the fracture process and the failure pattern. To demonstrate this influence, the stress wave in samples that did not fracture was simulated using the model shown in Figure 11.3. Figure 11.5 shows the stress wave propagation along the bars and through the sample. The samples with $m = 2, 5, 10$ and 100, with the values corresponding to relatively heterogeneous (2, 5), medium homogeneous (10) and relatively homogeneous (100) rocks, are used for the simulations.

The stress waves at points A and B along the bars in Figure 11.3 are shown in Figure 11.6. These compressive stress waves reached point A in the first bar at 5 μs and reached the B point in the second bar at 200 μs. Figure 11.6 shows that, after passing through the heterogeneous samples, the four stress waveforms for the specimens with different homogeneity indices differ significantly. After the stress wave travels through the sample, the peak value becomes lower and the pulse length becomes longer for heterogeneous rock than for homogeneous rock. This influence will surely cause the

Figure 11.5 Stress wave propagation along the bars and the sample for $m = 2$.

186 Rock failure mechanisms

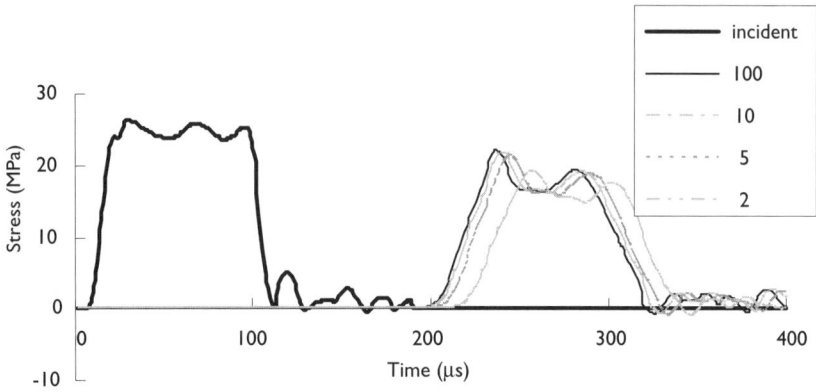

Figure 11.6 Geometry of the stress wave vs. stress–time curves at the A (black curve) and B points (gray curves) along the two bars (see Figure 11.3 for the locations of the A and B points).

failure pattern in a rock specimen to be different when sample failure modelling is conducted.

11.3.2 Influence of stress wave amplitude on the fracture process and failure pattern

The Brazilian sample shown in Figure 11.4 is used to investigate the influence of stress wave amplitude on the fracture process and failure pattern. Figure 11.7 shows the applied stress waveforms with three peak values of stress, with the rise time t_0 up to the peak stress being constant. Figure 11.8 shows the stress history obtained in the transmitter platen during one of the sample simulations. The dashed line averages the oscillations on the plateau of the incident pulse.

Results for the numerically obtained fracture processes and failure patterns at selected time intervals are presented in Figure 11.9. In order to aid visualisation, displacements have been magnified by a factor of 5. Also shown in the images are the contours of stress magnitude, taken as the fractional values of maximum shear stress. A fully fractured surface is shaded in black; whereas the zones that are intact or failed, but not fully fractured, remain in the shade of the stress magnitude contours. The comparison between the three cases shown in Figure 11.9 demonstrates the dependency of the dynamic fracture patterns on the transmitted stress wave amplitude.

Snapshots along the first row in Figure 11.9 show the failure pattern and shear stress evolution for the case of the input stress waveform with the lower peak amplitude of 75 MPa. The fractures start nucleating and propagating at about 120 μs when the peak value of the compressive pulse is reflected from the lower platen and reaches the centre of the sample. Double major fractures and branching are observed. Due to the rock heterogeneity, the fractures are well developed in the bottom area on the right side when the sample transmits the maximum load. The development of these fractures generates relief waves which temporarily halt the failure process. The stress

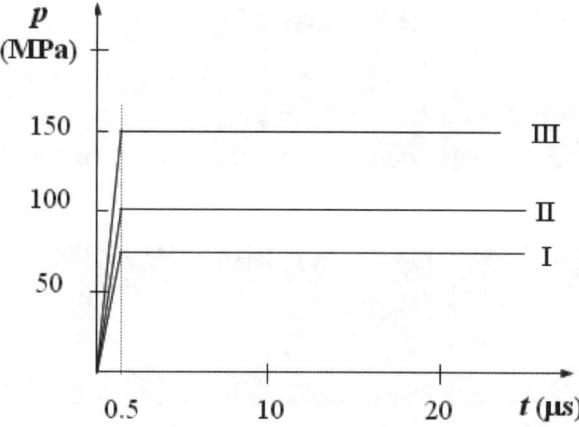

Figure 11.7 Applied stress waveforms with three peak values of stress.

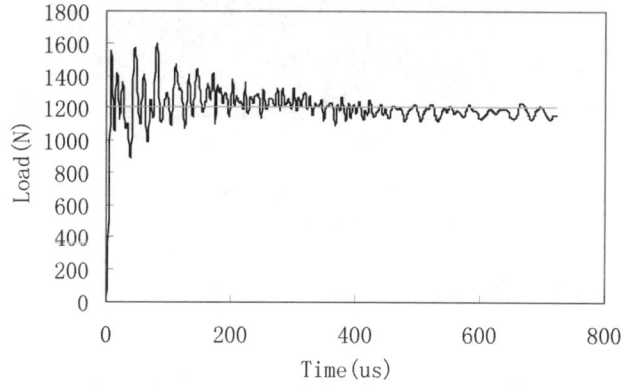

Figure 11.8 Stress history obtained in the transmitter platen during the simulation of a sample with input peak stress of 150 MPa.

waves subsequently travel from the left side towards the centre of the disc, inducing further fracture growth, as well as some micro-fracturing in the central area, which finally forms a main fracture. This through-going fracture is clearly seen in the snapshots in the first row in Figure 11.9.

Contrasting with the first row in Figure 11.9, the third row reveals that, for the case of the stress waveform with the highest amplitude of 150 MPa, the fractures start nucleating and propagating at about 20 μs, which is much earlier than for the previously discussed case. The fractures around the loading areas nucleate due to the incident wave, not the reflected wave as in the first case. On the two sides of the lower support areas, however, the compressive stress waves are reflected as tension waves when they reach the free surface, and are then superimposed upon the tail of the compressive waves. These superimposed stress waves develop an increasing amount of tension until, when the tensile stresses are high enough, they induce opening fractures

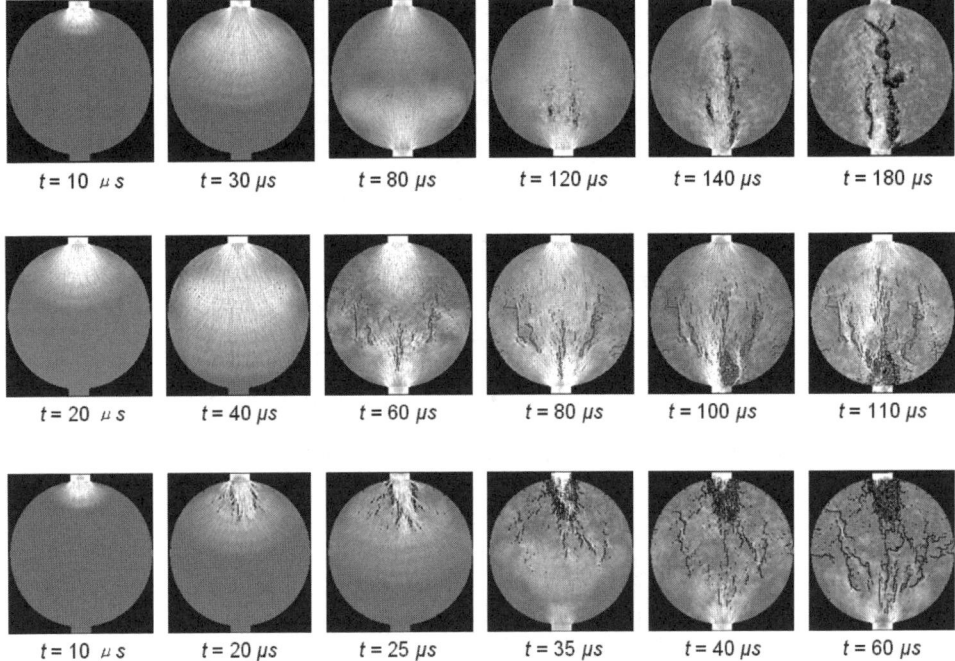

Figure 11.9 Fracture sequence and failure patterns for different stress wave amplitudes, the peak stresses being 75 MPa, 100 MPa and 150 MPa.

parallel to the surfaces. The process is also accompanied by the formation of a new wedge-shaped inclined fracture zone due to the intensity of the pressure stress waves near the loading area.

These results show that the fracture processes and the failure patterns are markedly affected by the stress wave amplitudes that are applied to the samples. The nucleation of fractures and failure patterns for peak stresses 75 MPa and 150 MPa revealed in the first and third row of images in Figure 11.9 differ significantly. For lower peak stress, the fractures start from the vicinity of the bottom compressive zone after about $t = 120$ μs; whereas, for higher peak stress, the fractures occur almost immediately after the stress wave front enters into the sample, which is 100 μs earlier than that for the lower peak stress. It is also seen from these snapshots that the failure patterns are enriched by the random distribution of the elemental stiffnesses and strengths.

The snapshots in the third row in Figure 11.9 further reveal the development of profuse fracturing at the loading area, which even leads to some fragmentation. Secondary fractures, parallel to the main diametral fracture, also appear as the load decreases, leading to the typical columnar failure of Brazilian tests (Yu *et al.*, 2004). For smaller stress wave amplitude, the model predicts the formation of the principal fracture that nucleates into the centre of the sample and grows towards the loading areas, as well as some secondary fracturing parallel to the main fracture and near the loading areas. For a higher amplitude stress wave, the simulation illustrates the formation of radial fractures starting in the circular border and growing to the centre.

Chapter 12

Rock failure and water flow

12.1 INTRODUCTION

Among the problems that are faced during rock engineering design, one of the most challenging is the prediction of fluid flow through fractured rocks. In addition to the need to characterise existing fractures, during the excavation of tunnels and underground chambers, new fractures are formed. This rock failure can cause significant local changes in the permeability of the rock mass. Consequently, and because an excavation acts as a sink, the rate of water flowing into the tunnels and caverns will increase. In particular, stress-induced changes in the flow properties can potentially affect the performance of one particular rock engineering project: underground vaults and deposition holes for the storage of nuclear waste—because the objective is to avoid unacceptable quantities of radionuclides reaching the biosphere. Thus, predicting the behaviour of fluid flow through fractured and fracturing rocks, especially in highly stressed rocks, is a formidable task, mainly because of the difficulty of specifying the fracture geometry and the transmissivities of the individual fractures. The book by Zhang and Sanderson (2002) and the recent book by Franciss (2010) provide detailed examples of numerical and analytical approaches respectively.

The two major characterisation problems when modelling water flow through fractured rock masses are specifying the overall geometry of the fractures and specifying the transmissivity of the individual fractures. Needless to say, it is not possible to obtain complete information concerning the fractures, even if there is good access to borehole, outcrop and larger scale data. Furthermore, it may be necessary to conduct thermo-hydro-mechanical (THM) studies to also account for the influence of temperature and rock stress on the rock mass system. Given the complexity of the problems, there is increasing emphasis on numerical models to study the influence of the various factors in benchmark example cases and to simulate real rock engineering projects. In the radioactive waste disposal area, the DECOVALEX project (DEmonstration of COupled Models and their VALidation against EXperiment), which is an international consortium of implementers and regulators, has been exploring such models since 1992 (e.g. Stephansson *et al.*, 1996; Stephansson *et al.*, 2004; DECOVALEX, 2009).

However, most of these models/simulations consider a given set of circumstances and do not incorporate the effects of the extension of existing fractures, the initiation of new fractures, and the coupled effects of flow, stresses and damage on the extension of existing/new fractures and the permeability of the rock mass. To improve the performance and upgrade the safety of engineering projects, it is necessary to accurately predict the behaviour of fluid flow in fractured and fracturing rocks, and particularly the effects that arise from the damage to the rocks (that is, the initiation, development and coalescence of fractures). To this end, a coupled finite-element strategy is developed, in which the problem is formulated in the context of the theories of fluid-saturated porous media and damage mechanics. The theory of fluid-saturated porous media can be derived from the classical phenomenological approach of Biot using the effective stress concept of Terzaghi. The theory of damage mechanics that is applied in this strategy is based on the elastic damage model.

To solve the coupled flow–damage problems, a numerical method is needed to directly simulate fracture initiation, propagation and coalescence in stressed rocks. Fortunately, many numerical techniques, such as the finite element, boundary element, finite difference and discrete element methods, have been developed to simulate the damage behaviour of rocks or rock masses. Recognising the dominant roles played by heterogeneity and fractures in the deformation and collapse behaviour of rock structures, Li and Zimmermann (1997) simulated fracture propagation using a laminate model. They introduced an additional set of lamina, which was orientated according to the maximum principal stress at the onset of fracturing, and associated it with a softening stress–strain law. Van Mier (1997) used the lattice model to simulate concrete and sandstone laboratory scale specimens. Yuan and Harrison (2006) provide a review of the state-of-the-art in modelling progressive mechanical breakdown and associated fluid flow in intact heterogeneous rocks. Here, we illustrate a flow–stress–damage (FSD) coupling model of saturated rocks to simulate the overall response of rock masses arising from the fracture process under hydraulic and boundary loadings.

12.2 ROCK FAILURE UNDER HYDRAULIC PRESSURE

Many experimental studies have been conducted in order to quantify the change in permeability of rocks during laboratory tests by measuring permeability in the axial load direction. Furthermore, it is reasonable to assume that permeability will increase, particularly for slightly impermeable rock, when the stress–state induces material damage by micro-crack opening and growth. To provide evidence for and explain the principle of the FSD model in the numerical model, we first present experimental results for permeability variation as a function of strain and damage for rock specimens under biaxial compressive loads with water pressure. These results include measurements of the flow in the pre- and post-failure ranges of the complete stress–strain curve. Specimens of sandstone, limestone, and conglomerate from the China Dongtan Coal Mine were used in the study. Each specimen had a diameter of 50 mm and a height of 80 mm.

Yang *et al.* (2001) reported the physical and mechanical properties of these rocks. In the experiments, all of the specimens were tested under fully saturated conditions. The experiments were conducted in an electro-hydraulic servo-controlled testing machine. The test system included a triaxial cell and a water injection pump. The detailed structure of the triaxial cell and the specimen installation are illustrated in Figure 12.1. The cell had a hollow piston so that water under pressure could flow

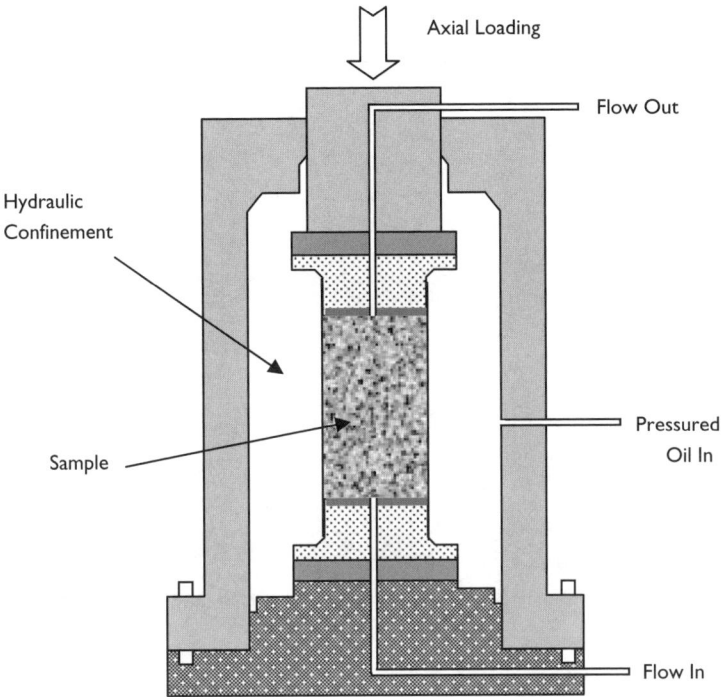

Figure 12.1 Experimental configuration for the FSD tests.

axially through the specimen while it was subjected to triaxial loading. The procedure for conducting the coupled mechanical/flow tests was similar to that for a normal triaxial compression test. A constant water pressure difference, 1.5 MPa, between the two ends of the specimen was applied with an injection pump. When the axial load was applied, flow input and output rates from the specimen could be measured by using a flow meter and a scale in the injection pump. The permeability at different loads was then calculated using the flow rate and cross-sectional area of the specimen. The test was terminated when the specimen failed in the post-failure portion of the stress–strain curve.

A total of 14 specimens were tested, and the results indicate a strong coupling between mechanical loading and axial flow. Figure 12.2 shows one of the typical results of the complete stress–strain curve and the associated permeability variation during the loading of sandstone.

It is generally accepted that, as evidenced in Figure 12.2, the stress–strain curves for brittle rock samples subjected to compression tests display four distinct regions (I–IV): closure of pre-existing micro-cracks (region I); elastic behaviour zone (region II); stable crack growth (region III); and unstable crack growth (region IV). As stated by Souley *et al.* (2001), in relation to these regions, compression tests with permeability measurements performed in the laboratory on two different granites indicate the following results.

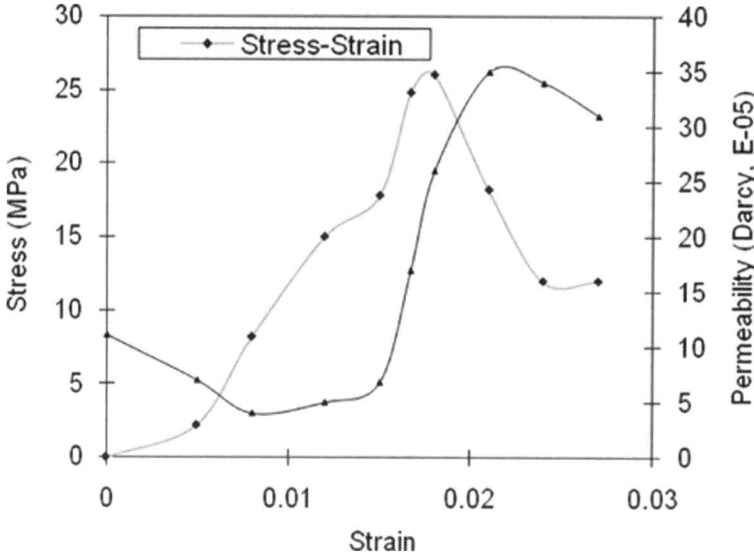

Figure 12.2 Experimentally obtained relations among stress, strain and permeability for sandstone.

1. Due to the closure of pre-existing pores and micro-cracks, permeability decreases in region I.
2. Permeability remains approximately constant both in the elastic region II and at the start of the stable crack growth region III.
3. Permeability increases as a function of deviatoric stress in regions III and IV representing the onset of unstable crack growth. The change in permeability is non-linear and can be significant in region IV. This is related to macroscopic failure produced by the coalescence of microcracks.

Based on these general observations, a model for coupling between flow, stress and damage is introduced into the simulation code. The formulation of the model is based on the following assumptions.

- The rock is fully saturated.
- The flow of the fluid (water) is governed by Biot's consolidation theory.
- The rock is assumed to be an elastic-brittle material with residual strength, and its loading and unloading behaviour is described by elastic damage mechanics.
- An element is considered to have failed in the tension mode when its minimum principal stress reaches the tensile strength of the element, and to have failed in the shear mode when the shear stress satisfies the strength criterion defined by the Mohr–Coulomb failure envelope.
- The permeability varies as a function of the stress state during elastic deformation, and increases dramatically during the elemental failure process.
- The local heterogeneity in the properties of the rock is characterised by Weibull's probability density function.

Rock failure and water flow

For rock materials, when the stress of the element satisfies the strength criterion (such as the Coulomb criterion), the element begins to fail and the damage to the rock will cause a dramatic increase in permeability. During elastic deformations, rock permeability decreases when the rock is compressed, and increases when the rock extends. However, the variation of permeability in these situations is limited. On reaching the peak load, a dramatic increase in rock permeability can be expected as a result of the generation of numerous micro-fractures. In other words, the damage to the rock causes a significant increase in permeability.

12.3 ILLUSTRATIONS OF FLUID FLOW IN HETEROGENEOUS INITIALLY INTACT ROCK

In this simulation, the parameter values are as follows: homogeneity index $(m) = 1.5, 2, 3, 6, 20$; Young's modulus $(E) = 33.8$ GPa; internal friction angle $(\phi) = 30°$; compressive strength $(f_c) = 220$ MPa; ratio of compressive and tensile strengths $(f_c/f_t) = 10$; Poisson's ratio $(v) = 0.25$; residual intensity coefficient $\lambda' = f_{cr}/f_c = f_{tr}/f_t = 0.2$, i.e. the residual compressive and tensile strengths are 0.2 of the initial strengths; coefficient of pore-water pressure $(a) = 0.1$; permeability coefficient $(K) = 0.001$ mD; mutation coefficient of permeability $(\zeta) = 20$; coupling coefficient $(\beta) = 0.05$. Figure 12.3 shows the numerically obtained relation between load, permeability and loading step, and the associated acoustic emissions (AE events) for the specimen with the homogeneity index $m = 1.5$. The overall permeability is obtained from the following Equation:

$$K = LQ/\Delta P = L\Sigma q/\Delta P \qquad (12.1)$$

where $Q = \Sigma q$ is the total flux flowing through the specimen, L is the length of specimen, and ΔP is the hydraulic pressure difference between the two ends of the specimen.

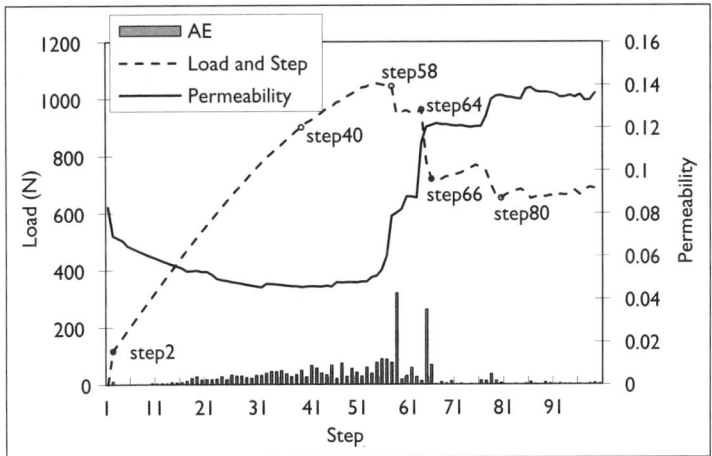

Figure 12.3 Numerically obtained relations between load, permeability and loading step and the associated acoustic emissions (AE events, relative counts) of the specimen with homogeneity index $m = 1.5$.

Figure 12.3 shows that there is a relation between the loading and the flow properties along various portions of the complete stress–strain curve. Axial permeability decreases with the axial load during the initial portion of the stress–strain curve, which is due to the axial load increase. In the second portion of the stress–strain curve before the maximum stress, the axial permeability decrease rate slows and then gradually increases due to the formation and propagation of the micro-fractures. In the post-peak portion, the axial permeability increases dramatically due to the formation of macro-fractures.

A comparison of Figures 12.2 and 12.3 shows general agreement between the experimental and numerical results. Both results show that, during the elastic deformation, rock permeability decreases. As the specimen is loaded to the non-linear deformation stage, the permeability slowly increases until yielding, at which stage there is a significant increase in the permeability with every permeability increase corresponding to a stress drop. As the specimen is loaded in a displacement controlled manner, sudden failure with a large number of acoustic emissions results in a rapid stress drop. Note that the change in permeability due to the closure of pre-existing micro-cracks (e.g. decrease of permeability in region I) is not modelled since the closure of cracks is not considered in this simulation.

Figures 12.4a-i and 12.5a-i show the simulated progressive failure process and the formation of macro-fractures during the hydraulic and boundary loading, Figure 12.4 for the inhomogeneity and failure and Figure 12.5 for the shear stress. The stress levels in Figure 12.5 are represented by the brightnesses of the gray scale: a higher stress level is indicated by a lighter gray.

Figure 12.6 shows the sequential plots with maps of the AE source locations during the failure process. The stress interval for each plot is shown in Figure 12.3. Each circle represents one fracture event, and the diameter represents the relative magnitude of energy release. The gray background circles give all of the event locations before the current step. The black circles represent the events during the current step that are caused by shear and tensile failure.

The AE locations for events that occur during loading to 85% peak stress are plotted in Figure 12.6a. Note that the events are distributed throughout the specimen, which indicates that a uniform deformation occurs during this portion of the loading. The AE pattern changes dramatically in Figure12.6b, in which the AE events that occurred during step 58 following macro-fracture nucleation are plotted. While a few events are still occurring throughout the volume of the specimen, most are clustered in the nucleation zone. This nucleation zone is the site for the macro-fracture plane. Consequently, in step 64, a distinct AE event zone crossing the specimen in the diagonal direction has developed from the nucleation site. Figure 12.6c and Figure 12.4 g-i show that the AE event zone coincides well with the fracture plane or fault.

The flow velocity field, with arrows indicating the velocity vectors, is shown in Figure 12.7. As the nucleated macro-fracture provides the pathway for significant fluid flow, the overall permeability of the post-peak specimen is three times higher than that of the pre-peak specimen, as shown in Figure 12.3.

Studying Figures 12.3, 12.6 and 12.7 shows that, although crack growth can be initiated at low stress levels due to the existence of inhomogeneities, large changes in permeability occur only when the stress level creates a number of connected flow paths (the onset of percolation flow). This change in permeability due to fracture initiation, propagation and coalescence in a compressive stress state has been observed by

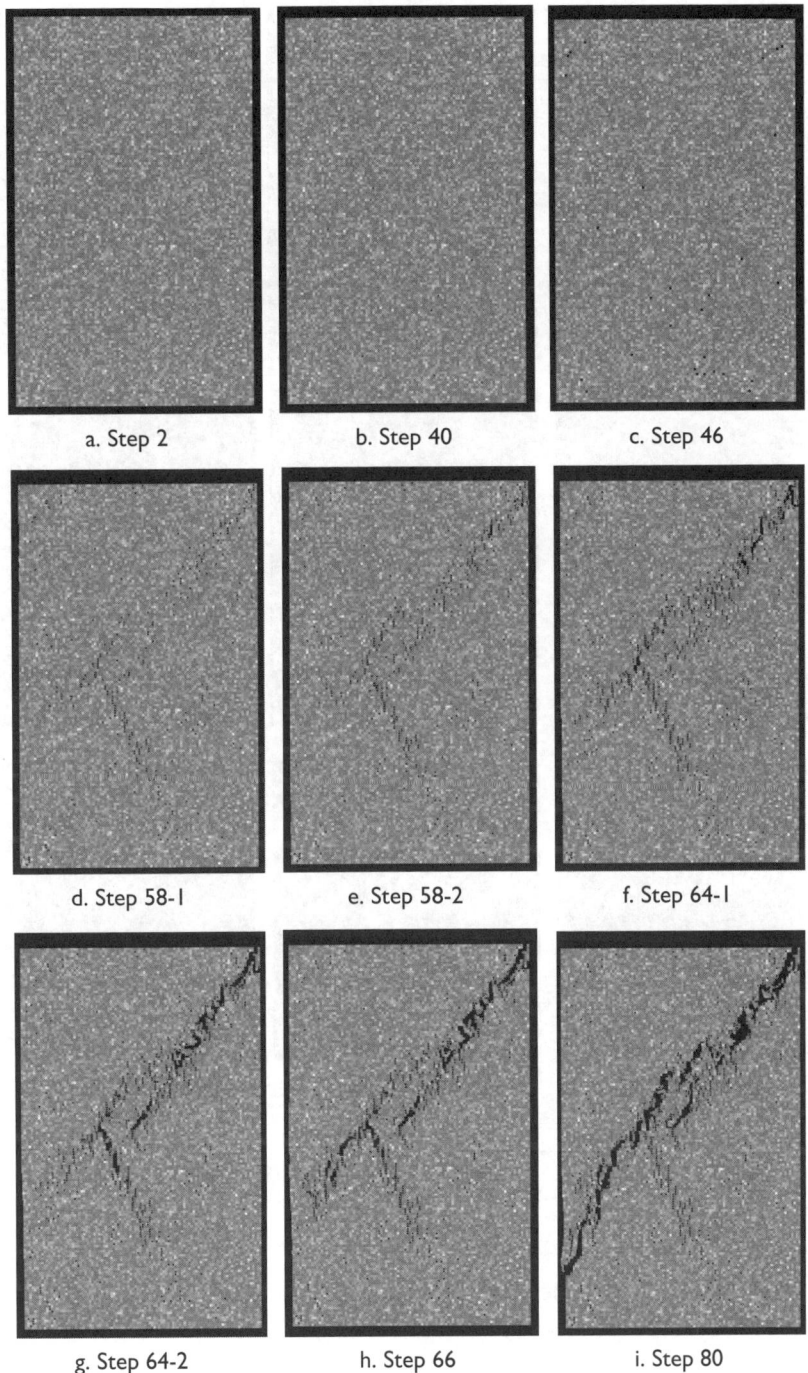

Figure 12.4 Failure process of specimen ($m = 1.5$) containing hydraulic pressure. Plot shows the inhomogeneity and failure of the rock elements.

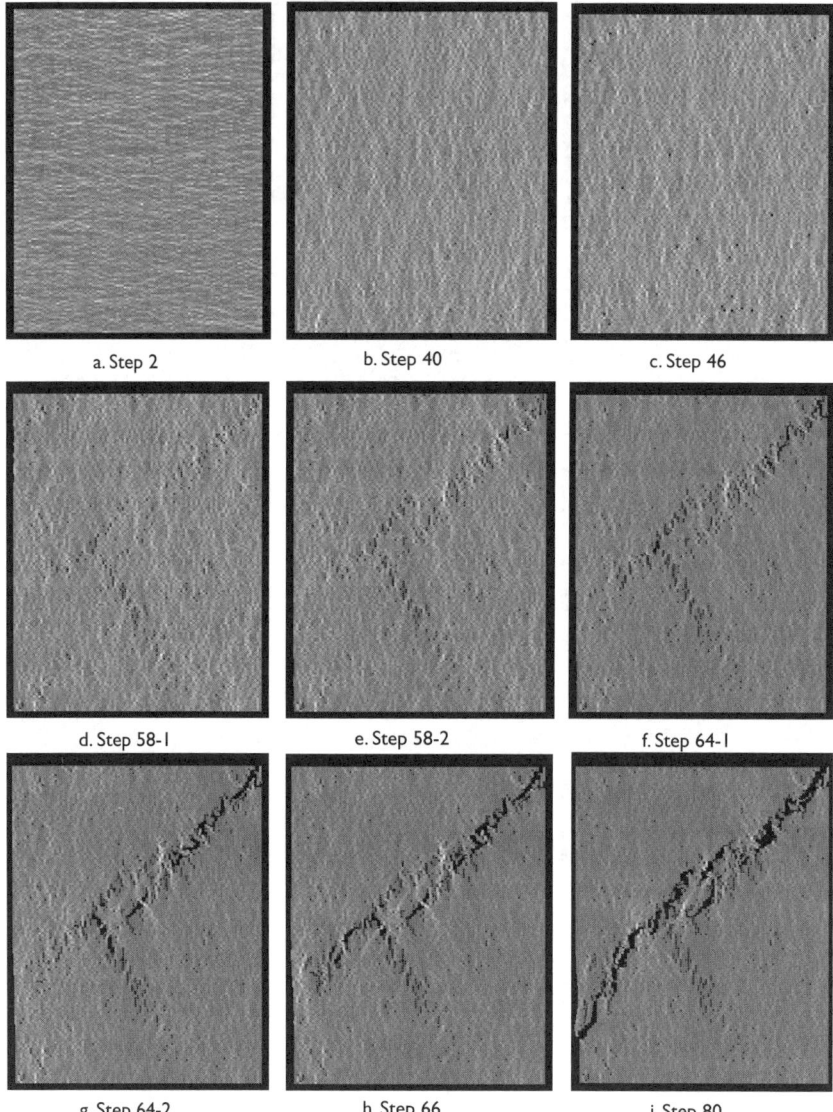

Figure 12.5 Shear stress evolution during failure of the specimen ($m = 1.5$) containing hydraulic pressure.

many investigators and is in accordance with the well-known curves of the mechanical properties of rock during compression tests.

12.3.1 Evolution of flow paths

Preferential flow, which is the concentration of flow into narrow channels, has long been observed to occur in saturated fractures. Fracture initiation and propagation play

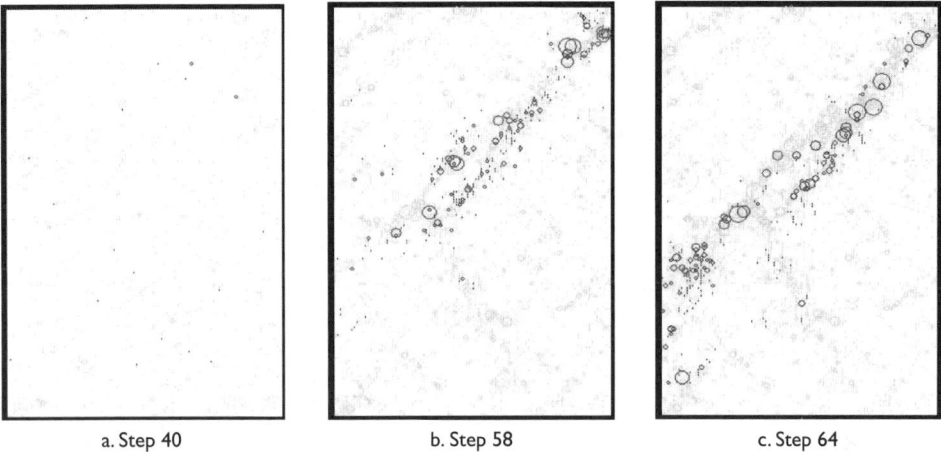

a. Step 40 b. Step 58 c. Step 64

Figure 12.6 Sequential plots showing the maps of AE source locations during failure of specimen ($m = 1.5$) containing hydraulic pressure (for Step locations, see Figure 12.3).

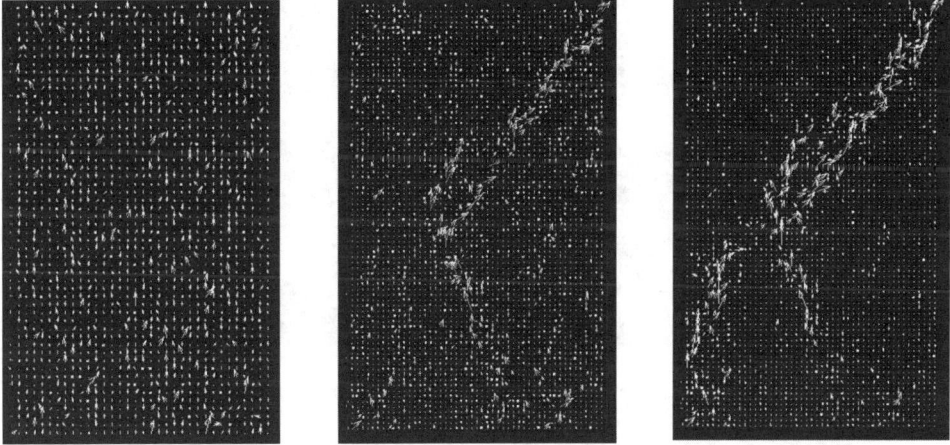

Figure 12.7 Flow velocity field with arrows indicating the velocity vectors during failure of the specimen ($m = 1.5$).

important roles in fluid flow behaviour by providing fast pathways for the migration of fluid. The conventional approach for describing the seepage of fluid through fractured rock is based on macro-scale continuum concepts and volume averaging over a large scale. This approach predicts that the fluid proceeds with a spatially uniform infiltration front. However, experimental and field studies have provided strong evidence that water proceeds non-uniformly along fast flow paths through fractures in rock masses. Especially when located in highly stressed rock masses, existing fractures can propagate and new fractures can nucleate. Consequently, there can be considerable changes in the fluid flow path, from distributed flow paths to localised flow paths.

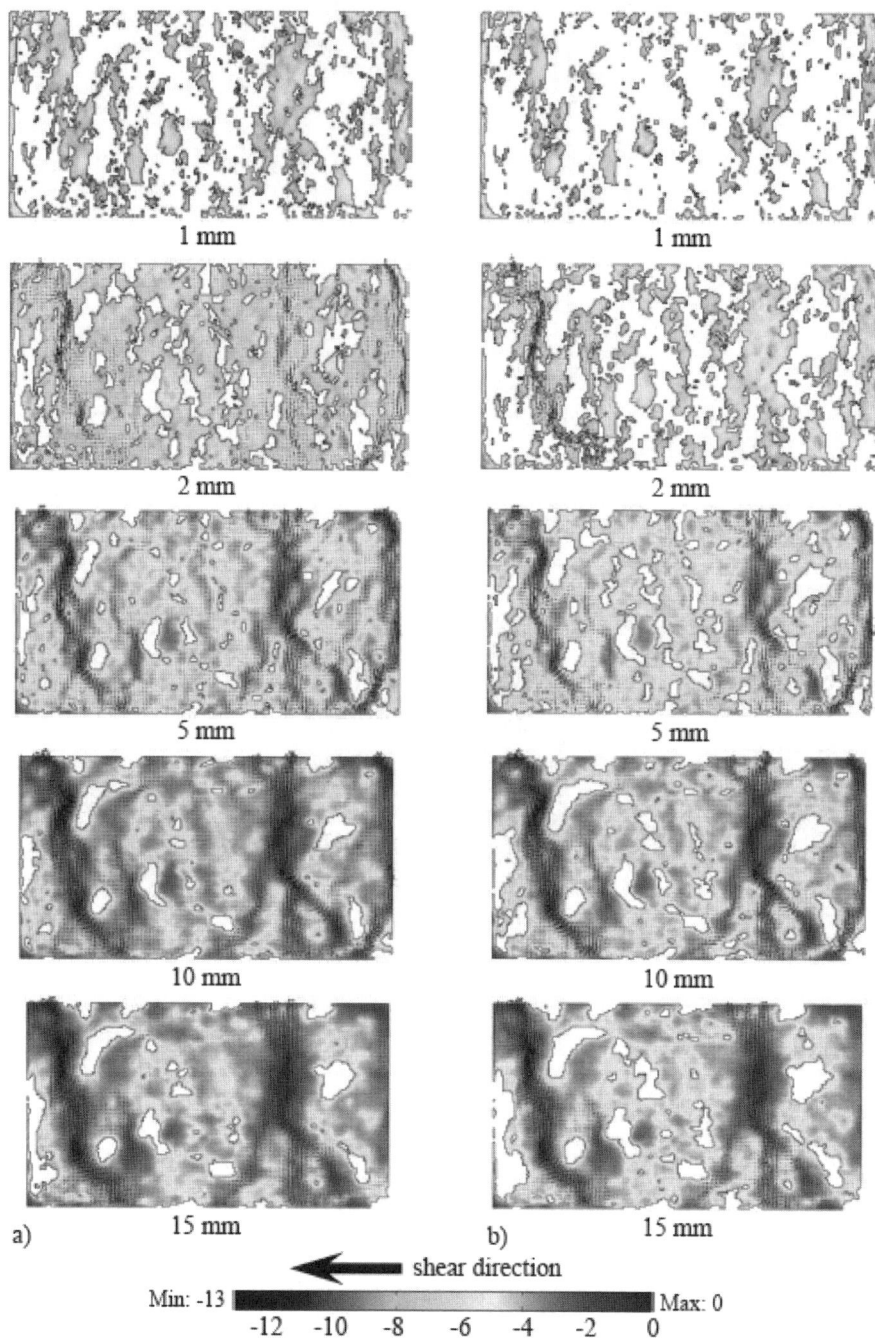

Figure 12.8 The change of flow paths with transmissivity evolution for overall flow parallel with the shear direction at different displacements of 2, 3, 5, 10 and 15 mm with constant normal loading. The legend is the order of magnitude of transmissivity (m^2/sec). From Koyama (2007).

The flow rate arrows that are shown in Figure 12.7 suggest that due to localisation during the fracture nucleation, fractures with non-uniform apertures in the failed specimen cannot be replaced with a system of fractures with equivalent uniform apertures. Due to the heterogeneity of the permeability, a non-uniform fluid flow field is obtained. The results reveal that the fluid flows rapidly along localised preferential flow paths through the newly formed fractures. When fluid moves through a specimen with heterogeneous random properties, the elements with lower permeability act as barriers to flow. As a result, the fluid flows in a random fashion. It is important to note that these random-looking fluctuations in pressure and flow rate are not in fact directly random, but are the deterministic results of the random distribution in the mechanical properties of the specimen—that is, the specimen's heterogeneity.

As soon as the fractures nucleate, the fluid will by-pass the specimen matrix and enter them. As the macro-fractures that are formed during the post-failure of a specimen are much larger than the micro-fractures in the matrix, the rate of infiltration into the macro-fractures is greater than that for the micro-fractures in the matrix. From Figure 12.7 it can be seen that the preferential flow along fractures will accelerate fluid transport to a rate well above that predicted by conventional models. In addition, the tortuous route of the flow paths through the fractures in the failed specimen is dependent on the heterogeneity of the material.

This type of phenomenon is also expected to occur in a rock fracture that is being sheared, as illustrated from the numerical results of Koyama (2007)in Figure 12.8.

We can also study the effect of the degree of heterogeneity on the permeability as the rock is taken through the complete stress–strain curve using four numerical specimens with different homogeneity indices. Figures 12.9 and 12.10 depict the load–displacement (loading step) curves and the associated permeability for the five simulated specimens. It is clear that the load–displacement relation and the permeability characterisation depend strongly on the heterogeneity of the specimens.

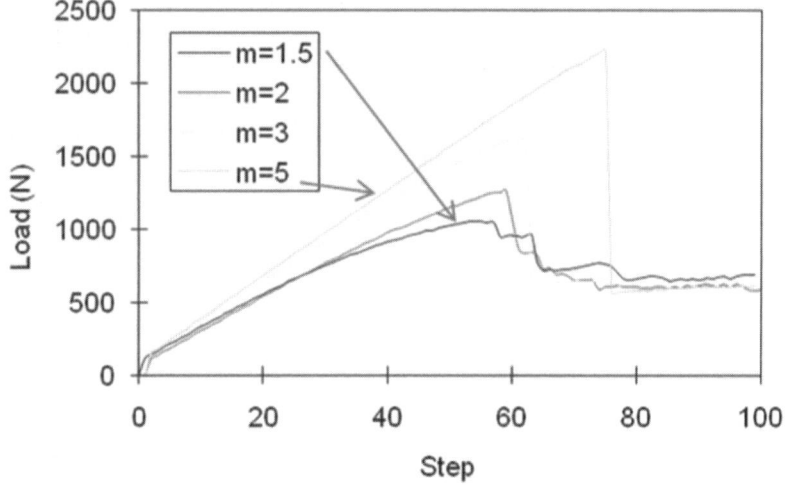

Figure 12.9 Numerically obtained load–displacement (loading step) curves for the four specimens with different homogeneity indices, *m*.

200 Rock failure mechanisms

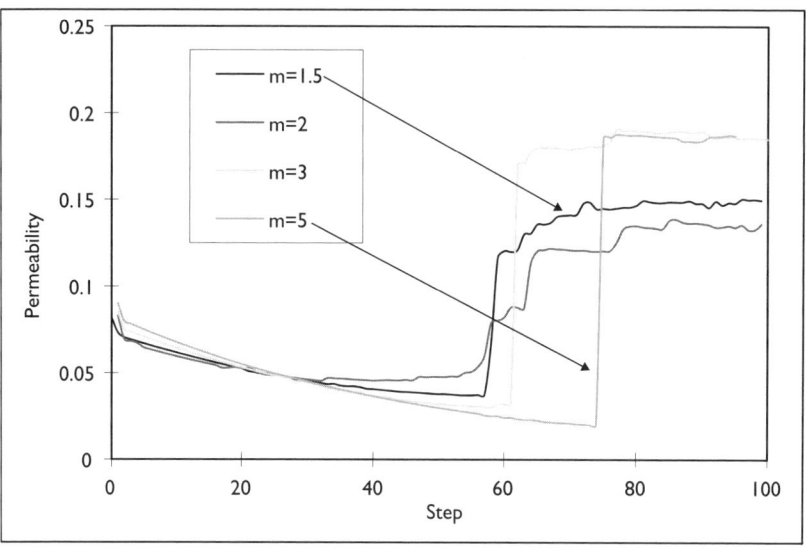

Figure 12.10 Permeability–displacement (loading step) curves for the four specimens with different homogeneity indices, m.

12.4 COMPARISON WITH THE ROCK DEGRADATION MODELLING BY YUAN AND HARRISON (2005)

The diagrams and results shown in the previous sections of this Chapter can be compared with another numerical simulation model, also constructed for the purpose of exploring the influence of rock structure breakdown on water flow (Yuan and Harrison, 2005). They state that, "… a hydro-mechanical local degradation approach is developed and used to investigate progressive damage and associated flow behaviour in heterogeneous rock. The approach is based on the establishment of an elemental scale hydro-mechanical constitutive model that incorporates simple idealised representations of degradation of strength and stiffness, confining stress-dependent dilatancy and deformation-dependent permeability. Through the use of such an idealised elemental behaviour, together with hydro-mechanical heterogeneity at the elemental scale, a finite difference numerical formulation allows both the mechanical and hydraulic behaviour of heterogeneous rocks to be simulated. Two independent statistical distributions are used to represent mechanical and hydraulic heterogeneity, and an uncoupled hydro-mechanical algorithm is used to investigate local variations of permeability induced by mechanical damage, and how these influence macroscopic flow patterns." Theirs is a hybrid continuum-discontinuum approach. They also explain that, "… the mechanical behaviour of rock evolves from elastic-brittle to elastic-brittle-ductile with increasing confining pressure. Using this observation in conjunction with idealised piece-wise linear differential stress–axial strain curves it has been shown that degradation in both strength and stiffness can be unified through a degradation index …." They define

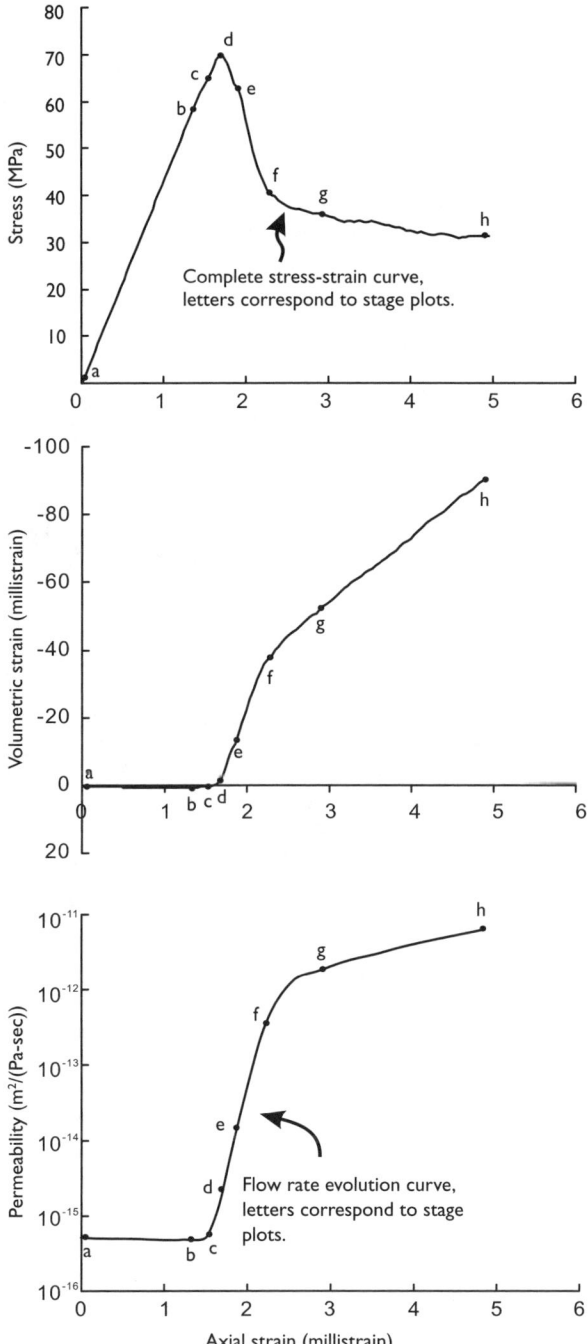

Figure 12.11 Complete axial stress–axial strain curve, volumetric strain–axial strain curve and permeability evolution curve for simulated uniaxial compression (from Yuan and Harrison, 2005).

and implement a dilatancy index in their model and its relation with the confining pressure. They also use a Weibull distribution for the elemental heterogeneity and their elemental strength failure is governed by the Hoek–Brown criterion. Finally, their discontinuum model operates within the FLAC continuum model. Thus, their model is based on heterogeneity, stiffness and strength degradation, dilatancy and deformation-dependent permeability. For further details of the approach, see Yuan and Harrison (2005).

Figures 12.11 to 12.13 show the use of the Yuan and Harrison model to demonstrate the development of the complete stress–strain curve, the degradation of the rock

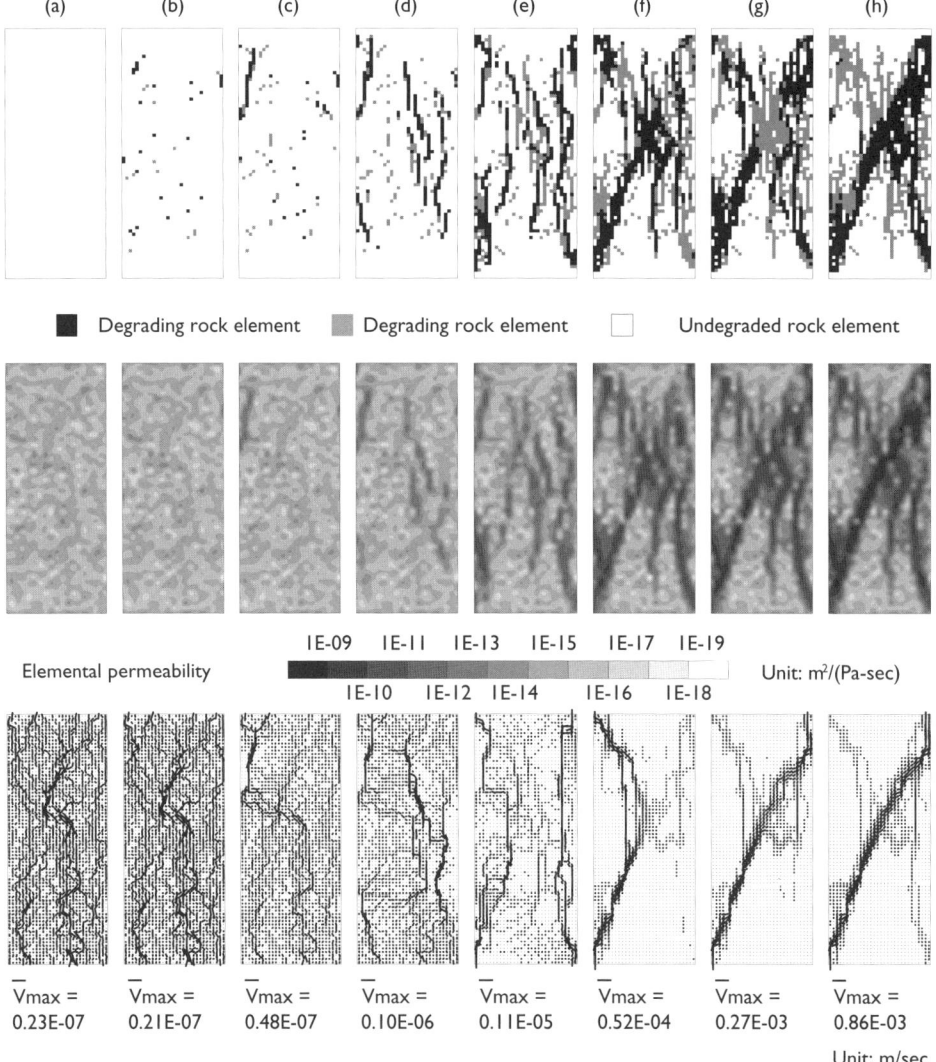

Figure 12.12 Fracture patterns, permeability plots, and flow vectors for each labelled loading stage shown in Figure 12.11 (from Yuan and Harrison, 2005).

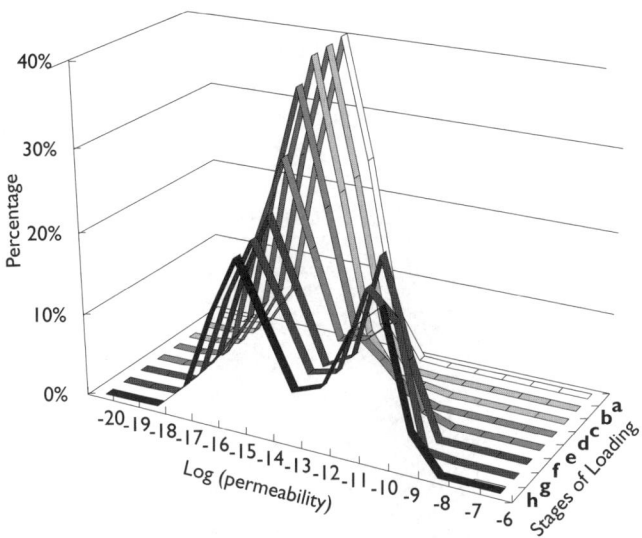

Figure 12.13 Histograms showing evolution of elemental permeability under uniaxial compression. The letters on the right-hand horizontal axis correspond to the loading stages in Figures 12.11 and 12.12, (from Yuan and Harrison, 2005).

and the associated flow, and the evolution of the elemental permeability. The partial change of elemental flow to fracture flow can be clearly seen in the Figure 12.13 histograms where the original uni-modal elemental permeability variation alters from Stage e to Stage f just after the peak of the complete stress–strain curve (see these points in Figure 12.11) as the permeability changes to a mixture of elemental and fracture flow. Yuan and Harrison (2005) also studied the effects of triaxial compression under confining pressures from zero to 80 MPa and the effects of different degrees of heterogeneity.

The results of the rock failure process analysis model shown in Section 12.3 and the Yuan and Harrison degradation model shown in Section 12.4 show similar trends in revealing the emergent properties resulting from the progressive micro-structural breakdown of rock. The models have somewhat different numerical approaches and so each has their advantages in highlighting different aspects of the process. The important point is that they both illustrate the same trends, thus verifying the modelling conclusions.

12.5 FLUID FLOW IN INITIALLY INTACT ROCK CONTAINING BLOCK INHOMOGENEITIES

In this section, the problem of fluid flow in rocks containing both small and larger scale inhomogeneities is discussed. The modelling is conducted on a numerical rectangular specimen that simulates rock aggregate, grains, grain boundaries, weak zones and water. The specimen sketch is shown in Figure 12.14. The mechanical parameters,

204 Rock failure mechanisms

such as Young's modulus, strength, and permeability coefficient, etc., of each element are all randomly assigned following Weibull's distribution with homogeneity index, $m = 2.5$. The larger grains have high but uncorrelated modulus and strength. The specimen is subjected to both hydraulic and mechanical loading. Hydraulic pressure loading, 1 MPa, is applied to the bottom of the specimen while keeping the top open

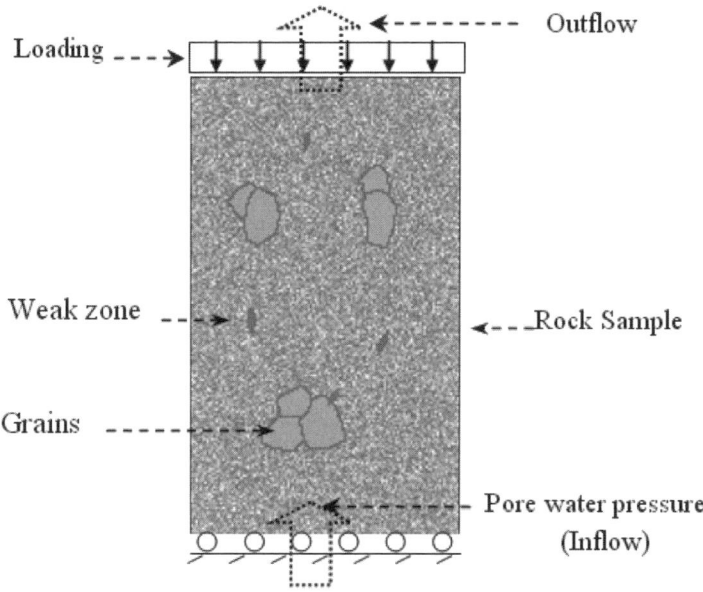

Figure 12.14 Specimen configuration with grains introduced in addition to the background inhomogeneity.

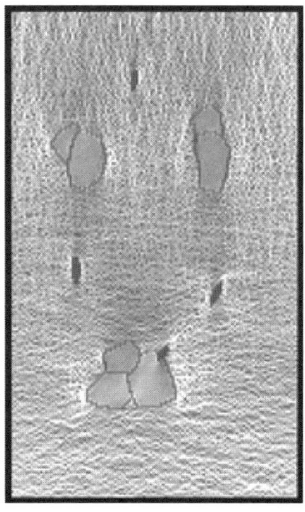

Figure 12.15 Shear stress field in the rock specimen shown in Figure 12.14.

to the atmosphere and setting zero volume fluid flow for the right and left boundaries. Fluid is allowed to flow into the specimen from the bottom boundary under the 1.0 MPa fluid pressure, and to flow out from the top boundary. The specimen is compressed in the vertical direction under displacement control to investigate the fracturing and fluid behaviour.

The main parameters involved in the modelling are as follows: homogeneity index (m) = 2.5; Young's modulus (E) = 50 GPa; internal friction angle (ϕ) = 30°; compressive strength (f_c) = 120 MPa; ratio of compressive and tensile strengths (f_c/f_t) = 10; Poisson's ratio (v) = 0.25; residual intensity coefficient $\lambda'(f_{cr}/f_c = f_{tr}/f_t)$ = 0.1, i.e. the ratio of the residual compressive and tensile strengths to their initial values is 0.1; coefficient of pore-water pressure (a) = 0.5; permeability coefficient (K) = 0.001 mD; mutation coefficient of permeability (ζ) = 20; coupling coefficient (β) = 0.05.

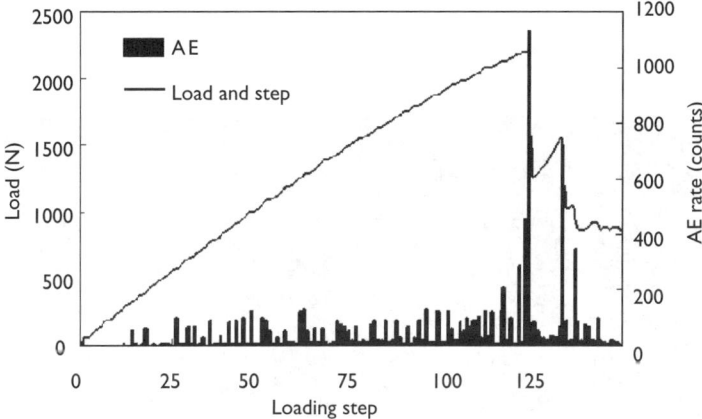

Figure 12.16 Relations between load, loading step and the associated acoustic emission.

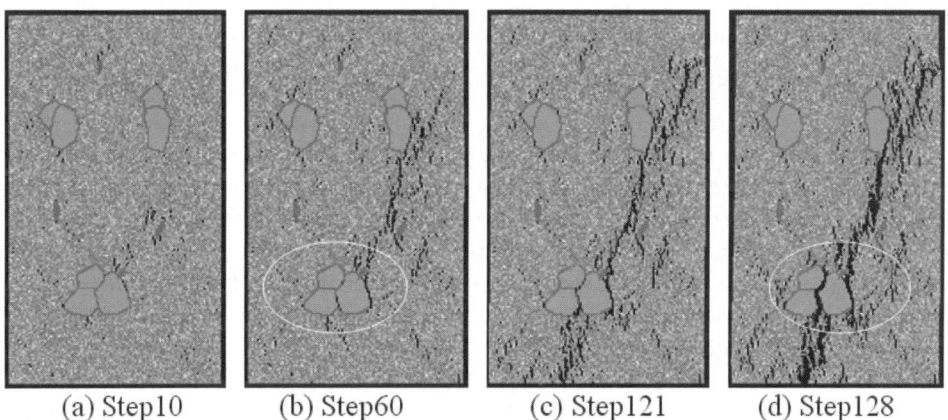

Figure 12.17 Plots of the failure process of the rock sample subjected external loading and hydraulic pressure.

206 Rock failure mechanisms

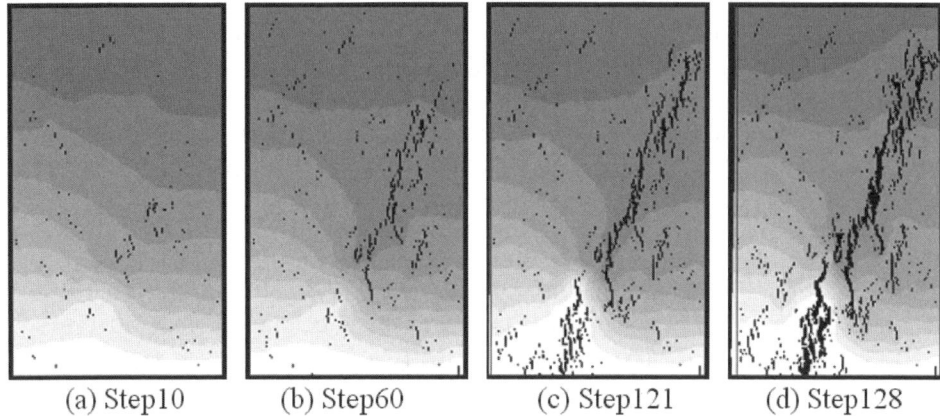

Figure 12.18 Pore pressure gradient in the rock specimen (lighter colour is higher pressure).

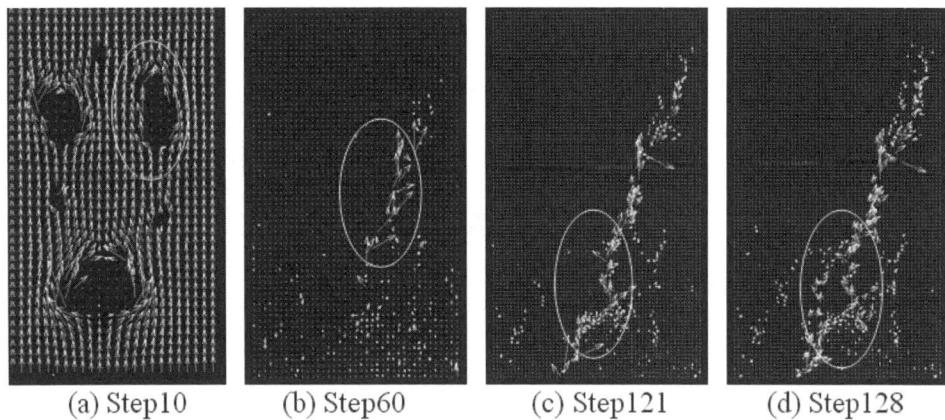

Figure 12.19 Fluid migration in the rock sample (the arrows indicating the velocity vectors).

As shown in Figure 12.15, the stress field in the rock specimen first increases along the boundary of the grains and weak zones. The bright areas in the immediate vicinity of the grains and weak zones are the zones of highest stresses. Figure 12.16 shows the relations between the load, loading step and the associated acoustic emissions (AE events) of the specimen. During the initial deformation or linearly elastic phase, small counts of AE events occurred.

Figure 12.17 shows the initiation, propagation and coalescence of fractures at different loading steps. It can be seen that the onset of failure in the specimen is first indicated by the formation of a large number of isolated fractures (Figure 12.17a). With further loading, a macro-fracture develops within the central part of the specimen, but it by-passes the grains to propagate toward the bottom of the specimen

Rock failure and water flow 207

(Figure 12.17b). Due to the weaker properties of the cementation interface among the grains, another macro-fracture propagates through the grains to converge with the upper macro-fracture (Figure 12.17c-d).

The water pressure gradient and flow rate arrows are shown in Figures 12.18 and 12.19. With the propagation of fractures, the water pressure gradient moves ahead (Figures 12.18c-d). The fluid by-passes the grains and weak zones (Figure 12.19a). Once the macro-fracture comes into being, as before, the fluid preferentially enters the fractures (Figure 12.19b). Since the macro-fracture is much larger than the micro-fractures in the matrix, the rate of infiltration into the macro-fracture is greater than it is into the micro-fractures. The first main channel for fluid flow by-passes the grains along their boundaries (Figure 12.19c). As another macro-fracture occurs, the second main channel for fluid flow forms (Figure 12.19d).

CHAPTER 13

Rock failure induced by thermal stress

13.1 INTRODUCTION

Thermal stresses play an important part in the failure of rock when it is subjected to significant low or high temperatures. Thermal stresses can be induced in two kinds of situation, i.e. thermal mismatch and thermal gradient. In thermal mismatch, when uniform temperature changes occur and due to differences in the coefficients of thermal expansion in different parts of the rock, non-uniform thermal dilatation or contraction occurs. In the thermal gradient case, there are differences in temperature at various locations caused by the gradient. For example, solar radiation causes the temperature of an exposed rock surface to rise but, because rock is a poor heat conductor, a thermal gradient occurs. The most severe condition occurs when a sunny, windless day follows a few days of cold weather (Figure 13.1). A similar case arises if there are underground heat sources from geological causes or engineered sources, (Figure 13.2). Both the thermal mismatch and the non-uniform temperature distribution can occur together.

The breakdown of rocks due to thermal stress has been recognised by engineers for some time and research has indicated that marbles as building stones show complex weathering phenomena. One of the most significant examples, already touched on in Chapter 9, is the anisotropic thermal expansion coefficient of calcite and the consequential bowing of Carrara marble slabs (Williams, 2009). This subject has been extensively explored in a European Union study (SP Institute, 2005). One of the Carrara marble quarries in Italy is shown in Figure 13.3a and the bowing of Carrara marble slabs on Finlandia Hall in Helsinki, Finland, in Figure 13.3b.

For the two kinds of thermal stress, i.e. thermal mismatch stress (TMS) and thermal gradient stress (TGS), the effect of TMS on rock failure is more remarkable, and much research about TMS has been reported. In this book, we are emphasising the influence of rock heterogeneity on rock failure and, as Kingery (1955) has pointed out, with a change in temperature, "no stresses arise providing that the body is homogeneous, isotropic and unrestrained". The reality is that rocks are generally not homogeneous, isotropic and/or free to expand in all directions.

Figure 13.1 Natural spalling of granite in the Gobi desert due to hot days and cold nights.

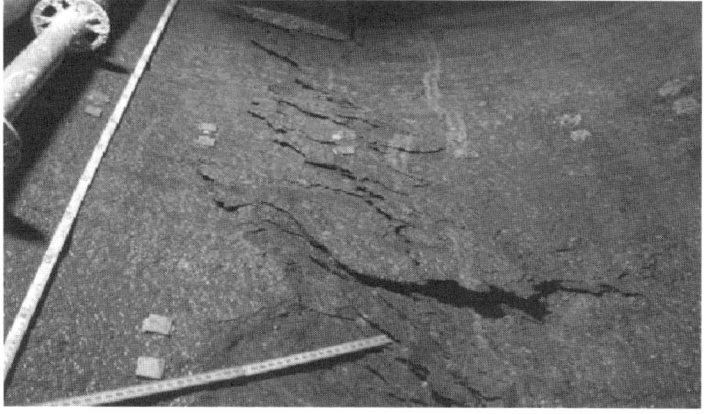

Figure 13.2 Artificial spalling of granite induced by excavation and rock heating in the Äspö Hard Rock Laboratory, Sweden (Andersson, 2007).

13.2 THERMALLY-INDUCED ROCK FAILURE

Imagine that a rock sample is composed of a number of small cubical elements of equal size which fit together to form the given continuous body. If the temperature of the sample is subjected to uniform temperature change, and its boundary surfaces are unrestrained, then each element will expand an equal amount uniformly in all directions. The elements are thus still equal-sized cubes and still fit together to form a continuous body. The important conclusion for this homogeneous body is that no stress will be induced even if the temperature change is large. If, however, when heterogeneity is taken into account so that each element may expand unequally in all

Figure 13.3a Carrara marble quarry in Italy.

Figure 13.3b Portion of Finlandia Hall in Helsinki, Finland, showing the convex bowing of the Carrara marble cladding slabs which were originally flat (see also Fig. 9.13 in Chapter 9).

directions, thermal mismatch then occurs. To evaluate the influence of heterogeneity on thermal stress and hence rock failure, four samples with different homogeneity indices, $m = 1.1, 1.5, 3, 10$, are subjected to uniform temperature increase with unrestrained boundary conditions, as shown in Figure 13.4. The model with a scale size of 100 mm × 100 mm is divided into 100 × 100, i.e. 10,000 elements. The parameters for the model are as follows: homogeneity index, $m = 1.1, 1.5, 3, 10$;

212 Rock failure mechanisms

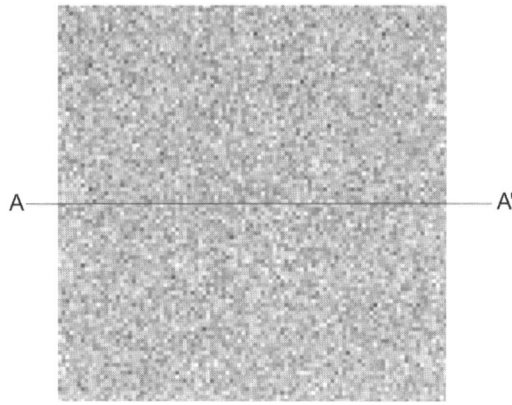

Figure 13.4 Numerical model—initial state before differential thermal expansion.

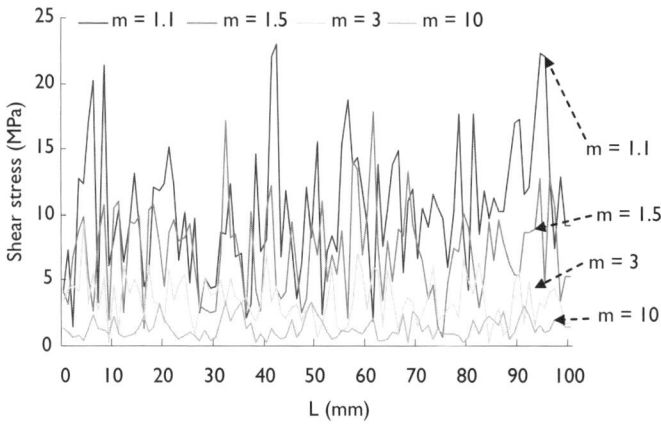

Figure 13.5 Shear stresses with different homogeneity indices.

mean elastic modulus, $E_0 = 50$ GPa; Poisson's ratio, $v = 0.2$; coeff. therm. exp. $= 12 \times 10^{-6}/°C$; temperature increment, $\Delta T = 100°C$.

Figure 13.5 shows the shear stress along the line A-A' indicated in Figure 13.4. The range of the shear stress changes with the homogeneity index, m: the smaller the homogeneity index is, the larger are the shear stresses induced. The heterogeneity-induced shear stress can be large enough to cause the intact rock elements to break down. If the model is subjected to a temperature decrease, tensile stresses will be induced and rock failure will be easier.

13.3 THERMAL CRACKING OF A DISC–RING MODEL

In order to investigate the influences of the thermal mismatch stress on the failure process and final failure patterns without any external loading, a disc–ring model is simulated, with different coefficients of thermal expansion and subjected to a uniform temperature change. The disc–ring model represents a stiff inclusion in a rock matrix. The radii of the two concentric circles are a and b (see Figure 13.6). The disc and ring are labelled 1 and 2, respectively.

The theoretical thermal stresses for homogeneous elastic materials in this geometrical configuration are known. With uniform temperature change, ΔT, and free boundary conditions, the radial and tangential stresses in the disc 1 and ring 2 regions are:

$$\sigma_{1,\rho} = \varepsilon \Delta T \tag{13.1}$$

$$\sigma_{1,\theta} = \varepsilon \Delta T \tag{13.2}$$

$$\sigma_{2,\rho} = \frac{a^2(b^2 - r^2)}{r^2(b^2 - a^2)} \varepsilon \Delta T \tag{13.3}$$

$$\sigma_{2,\theta} = -\frac{a^2(b^2 + r^2)}{r^2(b^2 - a^2)} \varepsilon \Delta T \tag{13.4}$$

where r is the distance from the centre, the subscripts 1 and 2 denote the disc and ring respectively, and the subscripts ρ and θ denote the radial and the tangential components, respectively. ε is given by:

$$\varepsilon = \frac{\alpha_1 - \alpha_2}{p} \tag{13.5}$$

and p is given by

$$p = \frac{1 - \nu_1}{E_1} + \frac{b^2 + a^2 + \nu_2(b^2 - a^2)}{E_2(b^2 - a^2)} \tag{13.6}$$

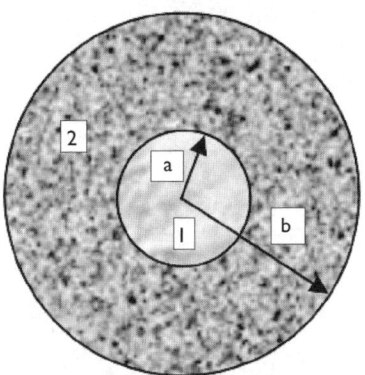

Figure 13.6 The disc–ring model.

The simulation for the disc–ring model uses the properties as listed below.

	Inner	Outer
Homogeneity index (m):	300	300
Mean elastic modulus (E_0 in GPa)	40	20
Poisson's ratio (v)	0.3	0.2
Coeff. Therm. Exp. ($\times 10^{-6}$ in °C)	15	10
Temperature increment ΔT (in °C)	100	100
Disc and ring radii (mm)	12	25

Thus, both the disc and ring have the same heterogeneity and are subjected to the same temperature increase, but the disc as an inclusion within the rock matrix is stiffer than the matrix and has a higher coefficient of thermal expansion.

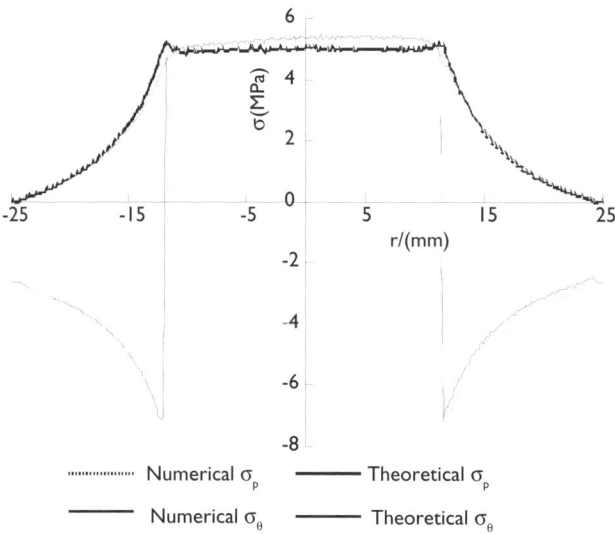

Figure 13.7 Stress distribution in the disc–ring geometry (stiff inclusion and matrix) at $\Delta T = 100°C$.

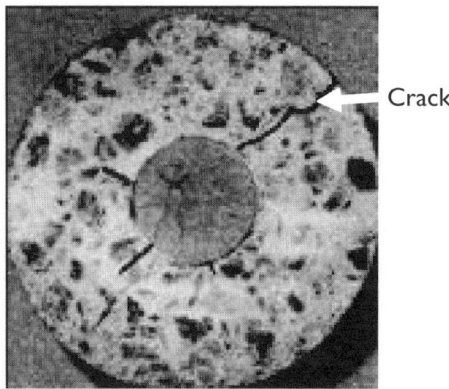

Figure 13.8 Failure pattern of the disc–ring model obtained by experiment.

Rock failure induced by thermal stress 215

Substituting the disc and ring material properties into Equations (13.1) to (13.6) and into the simulation model, we obtain both solutions, as shown in Figure 13.7. Equation (13.1) and Equation (13.2) indicate that the disc is in a hydrostatic pressure state, and Equation (13.3) indicates that the ring has a compressive stress in the radial direction and Equation (13.4) shows that the tangential stress is tensile. The theoretical and numerical solutions agree well, as evidenced by Figure 13.7. So we expect that a radial crack can occur in the disc, as is confirmed by the experiment shown in Figure 13.8.

13.4 THERMAL CRACKING IN MODELS CONTAINING IRREGULARLY SHAPED INCLUSIONS

Figure 13.9 shows models containing irregularly shaped inclusions in a matrix subjected to temperature increase. In this model, and the one shown in Figure 13.10, the coefficient of thermal expansion of the matrix material is greater than that of the inclusions.

In Figure 13.10, the inclusions and matrix are subjected to both an increase (upper images) and decrease (lower images). During temperature *increase*, the matrix expands more rapidly than the inclusions: the matrix will be subjected to compression and tensile stress will be generated in the inclusions. Larger tensile stress exists at the inclusion–matrix interface and, if the interfaces are weak, separation may occur and so-called θ-cracks will form. This type of failure process can be seen in Figure 13.10 (a)–(e). Cracks have occurred by void nucleation and growth within the matrix and

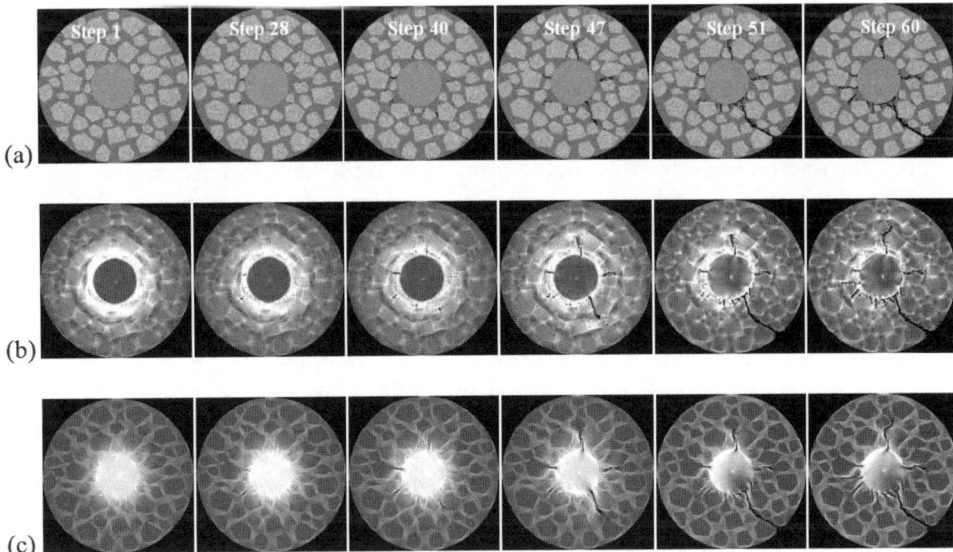

Figure 13.9 Plots of the model failure process for a uniform temperature increase. The gray levels represent the magnitudes of a) elastic modulus, b) shear stress and c) maximum principal stress—the lighter the gray, the higher the value. The black points/lines represent failed elements.

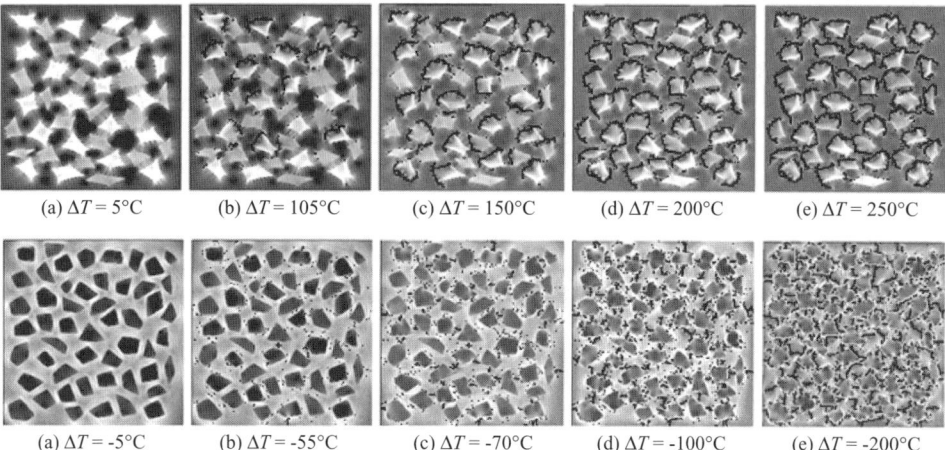

Figure 13.10 The failure process for the irregular inclusions model with a uniform temperature change (upper images for uniform temperature increase; lower images for uniform temperature decrease). The gray levels represent the magnitude of the minimum principal stress—the lighter the gray, the higher the stress. The black points represent failure elements. Cracks can be seen where the failure elements have accumulated.

de-cohesion at the interfaces between the inclusions and the matrix. No inclusion fracture is observed. Tensile stress concentrations around the inclusions and cracks are prone to occur at the interface between these stiffer inclusions and matrix, and there are radial cracks as well.

Heterogeneity is also one of the reasons for these types of cracks. At first, there are some discrete weak elemental failures in the matrix and stress redistribution will occur around the failed elements; the failed elements then induce more elemental failure around them and micro-cracks develop. When such micro-cracks emerge in the model, the micro-crack tip stress fields contribute to further crack expansion.

When the model is subjected to uniform temperature *decrease*, because the coefficient of thermal expansion of the matrix is greater than that of inclusions, the matrix shrinks more rapidly than that of the inclusions. Tensile stresses will be generated in the radial direction around each inclusion, and the inclusions themselves will be subject to compression. These stresses may lead to overall failure of the matrix or, at least, they will encourage the growth of existing radially oriented micro-cracks. It can be seen from the images in the lower part of Figure 13.10, that matrix failure between two neighbouring inclusions is along the shortest path. The radial cracks occurred by void nucleation and growth within the matrix and de-cohesion at the interface between the inclusions and the matrix. There are some θ-cracks as well due to the heterogeneity of rock.

The difference in the damage mechanisms for temperature increase and temperature decrease shows that there is a strong relation between the rock microstructural components, their properties and geometry and the type of temperature variation.

One of the many rock engineering applications in which temperature and heat flow is particularly important is in the design of an underground radioactive waste

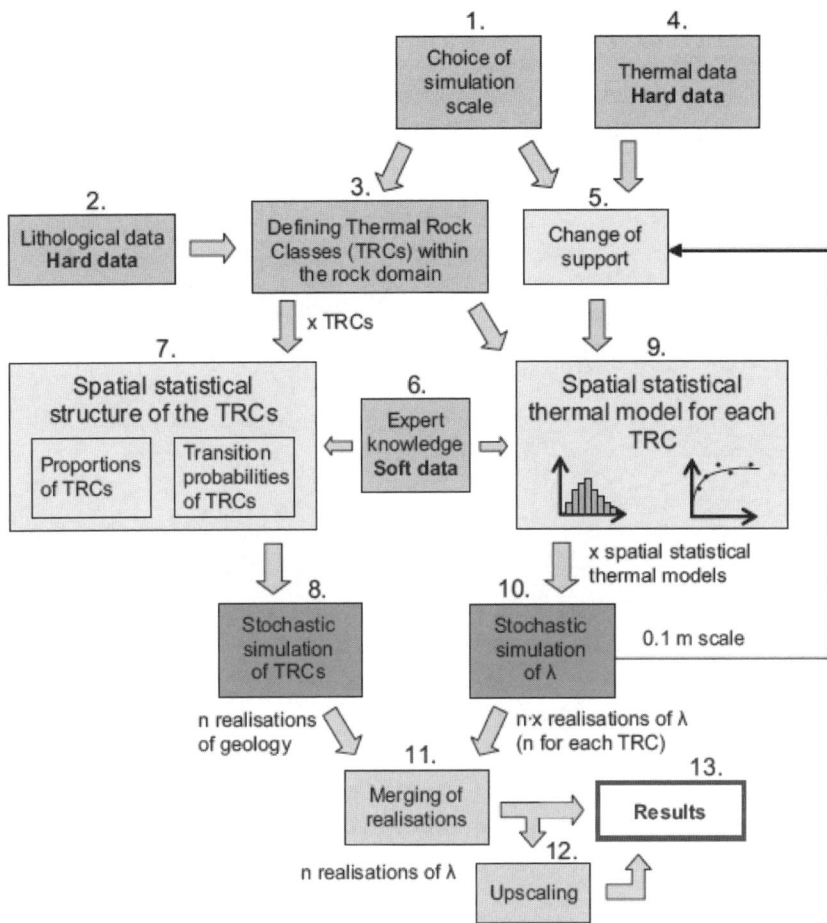

Figure 13.11 Flowchart illustrating an approach for thermal conductivity modelling of a heterogeneous rock domain (from Sundberg et al., 2008).

repository for the disposal of high-level waste. Individual copper and steel cylinders containing heat generating waste (and there can be thousands of these) are placed in small shafts bored from repository tunnels at a depth of around 500 m. In order to control the rise in temperature throughout the repository as a result of the heat generating waste, one must know the thermal properties of the host rock—and understand the effects of inhomogeneity, anisotropy and fractures on the heat flow.

The flowchart in Figure 13.11 is from work developed during the Swedish programme (Sundberg et al., 2008) and illustrates a sophisticated method of thermal conductivity modelling which includes stochastic simulations based on rock property measurements at one of the potential repository sites.

Chapter 14

Slope failure in rock masses

14.1 INTRODUCTION

The estimation of the stability of natural and engineered rock slopes is an important issue for geotechnical engineers and there has been considerable research in the development of the analysis methods. If it is anticipated that slope failure will be along pre-existing fractures (Figures 14.1 and 14.2), then plane, wedge and toppling failures can be analysed using stereographic projection methods (Figure 14.3) for hard rocks. If a new failure surface is created, then the potential for failure can be analysed by soil mechanics methods based on assumptions regarding the inclination and location of the inter-slice forces.

In strong rock masses, slope failure usually occurs along pre-existing natural fractures, as illustrated in Figures 14.1 and 14.2. However, it is not easy to characterise *in situ* fractures and so there is considerable benefit in being able to use numerical methods to capture the idiosyncrasies of the natural rock mass. Numerical methods permit the treatment of rock slope stability problems involving complexities relating to geometry, rock inhomogeneity, anisotropy and non-linear behaviour. Moreover, the distinct element method and discontinuous deformation analysis methods can be used to simulate large block movements in complex geological media consisting of many blocks which can themselves break up during the simulation process without any external intervention. Also, in cases where the candidate failure surface of a rock slope is not completely controlled by discontinuities, i.e. the failure of intact rock is also involved, conventional slope stability assessment methods are inadequate. Most importantly, the critical failure surface can be found automatically with numerical simulations.

Nevertheless, most of the current numerical methods do not take into account the heterogeneity of rock masses at macroscopic levels with complicated geological conditions. During fracturing, the heterogeneity plays an important role in determining the fracture paths and fracture patterns in the rock mass. Most unstable rock slopes undergo some degree of progressive failure development and extensive internal disruption of the slope mass. As a consequence, the factors governing initiation and eventual failure are complex and are not easily included in simple static analyses.

The modelling of progressive failure in slopes is somewhat more challenging as the model needs to incorporate both strength degradation and, ideally, brittle fracture propagation and its temporal evolution. In this Chapter, we illustrate slope failure

220 Rock failure mechanisms

Figure 14.1 Small wedge failure caused by the adverse conjunction of two joints, Loch Lomond, Scotland.

Figure 14.2 Large wedge failure at the Tectonic Bore Mine, Western Australia caused by two major fractures creating the wedge which slipped into the open-pit mine. Left: the wedge beginning to form; right: the wedge accelerating and creating dust which escapes along other pre-existing fractures.

through simulations and show the structural change and development of the rock mass, including capture of the interior structure of the intact rock, layered rock and jointed rock.

14.2 STRENGTH REDUCTION RULE AND DETERMINATION OF SAFETY FACTOR

As an alternative to the traditional approaches, the principle of strength reduction is incorporated into the constitutive simulation model. Referring to the shear strength reduction technique (Matsui and San, 1992), this is applied to each element. The strength f_0 is linearly degraded according to the following Equation:

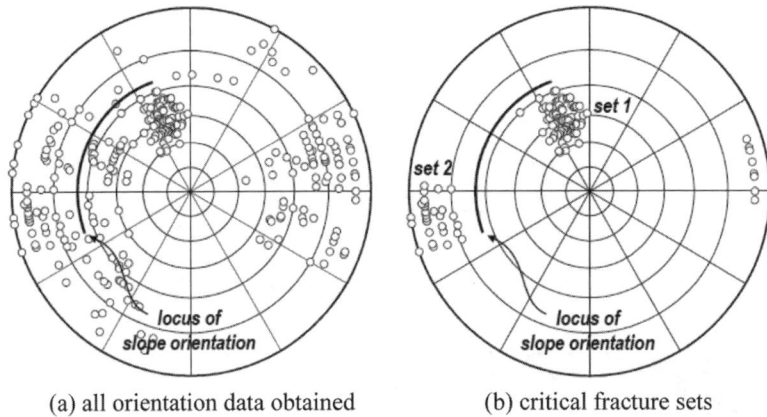

(a) all orientation data obtained   (b) critical fracture sets

Figure 14.3 Fracture orientation data and slope stability stereographic analysis for Rubha Mor rock slope, Loch Lomond, Scotland (from J.P. Harrison, Imperial College, UK).

$$f_0^{trial} = \frac{f_0}{F_s^{trial}} \tag{14.1}$$

where F_s^{trial} is the trial safety factor and f_0^{trial} is the trial strength of an element. The trial strength, f_0^{trial}, is employed in the simulation to investigate the strength of the rock medium when considering slope failure and the simulation is run with the trial strength f_0^{trial} until the critical slope failure surface is determined.

In defining slope failure, the maximum acoustic emission (AE) event rate is employed as the criterion for slope failure. Slope failure is accompanied by a significant increase in the nodal displacement within the elements; accordingly, there is a significant increase in the number of damaged elements. The AE counts are found by the number of damaged elements and the energy releases are calculated from the released strain energies. When the AE rate reaches the maximum value, a macro-failure surface forms and slope failure occurs, and the corresponding F_s^{trial} is the safety factor F_s of the slope.

Firstly, homogeneous slopes are simulated with slope angles β ranging from 30° to 50°. Figure 14.4 shows the simulation model which is 70 m × 30 m and discretised into 210 × 90 (18,900) mesh elements. The elemental parameters include the Young's modulus, E, uniaxial compressive strength, f_c, and the Poisson's ratio, v. These parameters are the equivalent parameters of the rock mass. The input parameters, the Young's modulus (E_0), compressive strength (f_0), the Poisson's ratio (v) and unit weight (γ) of the material are 80 MPa, 0.15 MPa, 0.4 and 23 kN/m³, respectively. Self-weight acts on the elements. The trial safety factor (F_s^{trial}) is gradually increased until the critical failure surface is established.

Figure 14.5 shows the critical failure surface (the brighter zone) and displacement vectors of the slope for $\beta = 45°$. The critical failure surface gives the boundary of the potential sliding plane. The displacement in the sliding plane is significantly larger

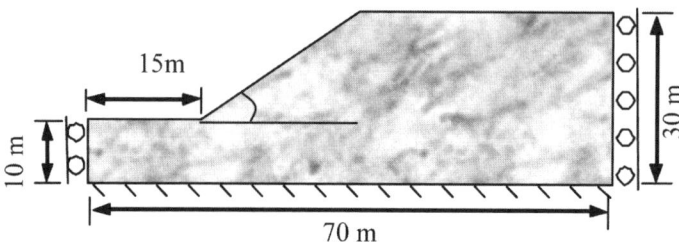

Figure 14.4 The simulation model.

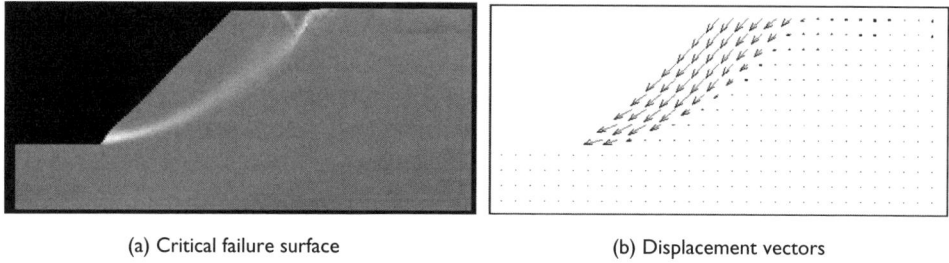

(a) Critical failure surface (b) Displacement vectors

Figure 14.5 Numerical results for the stability of a soil slope.

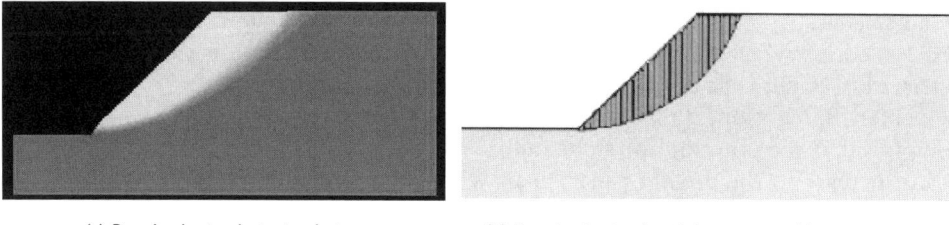

(a) Result obtained via simulation (b) Result obtained with limiting equilibrium solution

Figure 14.6 Failure pattern comparison from the numerical simulation and limiting equilibrium solution.

and the maximum displacement reaches 100 mm when the critical failure surface is completely formed. Figure 14.5a presents the failure pattern of the sliding plane in the slope. This agrees well with the result by traditional limiting equilibrium analysis (Figure 14.6b).

The number of damaged elements and associated energy release related to acoustic emission (AE) accompanying the strength reduction are recorded and shown in Figure14.7. It can be seen that many trial safety factors were attempted, ranging from 1 to 1.1. When $F_s^{trial} = 1.08$, there is a sudden rise in the AE rate, which indicates that the critical failure surface forms. Thus the safety factor of the slope is 1.08. For different slope angles, corresponding safety factors are calculated with the numerical code, as shown in Figure14.8; this Figure also gives the comparison between the numerical code results and traditional limiting equilibrium analysis.

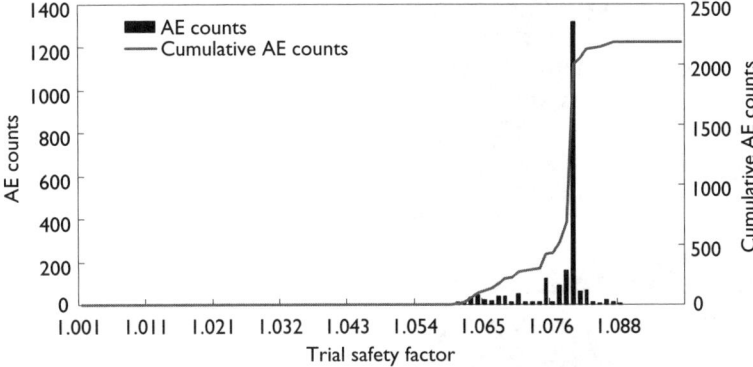

Figure 14.7 AE counts with trial safety factor.

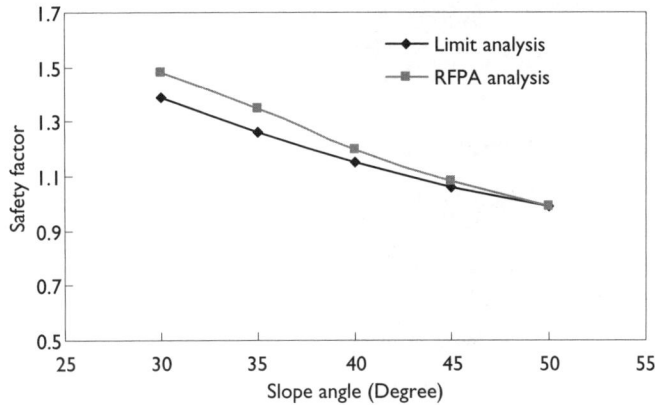

Figure 14.8 Safety factors with different slope angles for the numerical simulation and limit equilibrium analysis.

14.3 A SLOPE IN A LAYERED ROCK MASS

Figure 14.9 illustrates the development of slope failure for a model containing alternating soft and hard rock layers. The failure process and associated displacement are shown in Figure 14.9. In this case study, shear failure first occurred near the slope foot and tensile failure is initiated in the body of the slope. As pointed out by Griffiths and Kidger (1995), both the shear and tensile failure are all triggered at the weakest elements because the strength profile of the rock material is randomly distributed with specified mean and variance. As a result, a major crack is initiated from the central part. With the propagation of the crack, a critical failure plane is gradually formed. Finally, the rock slope collapses, the major failure mode of the rock slope being a combination of shear and tensile fracturing. However, detailed study of the fracture mode from the simulation shows that the slope favours the development of brittle tensile fracture propagation sub-parallel to the maximum principal stresses. Shear

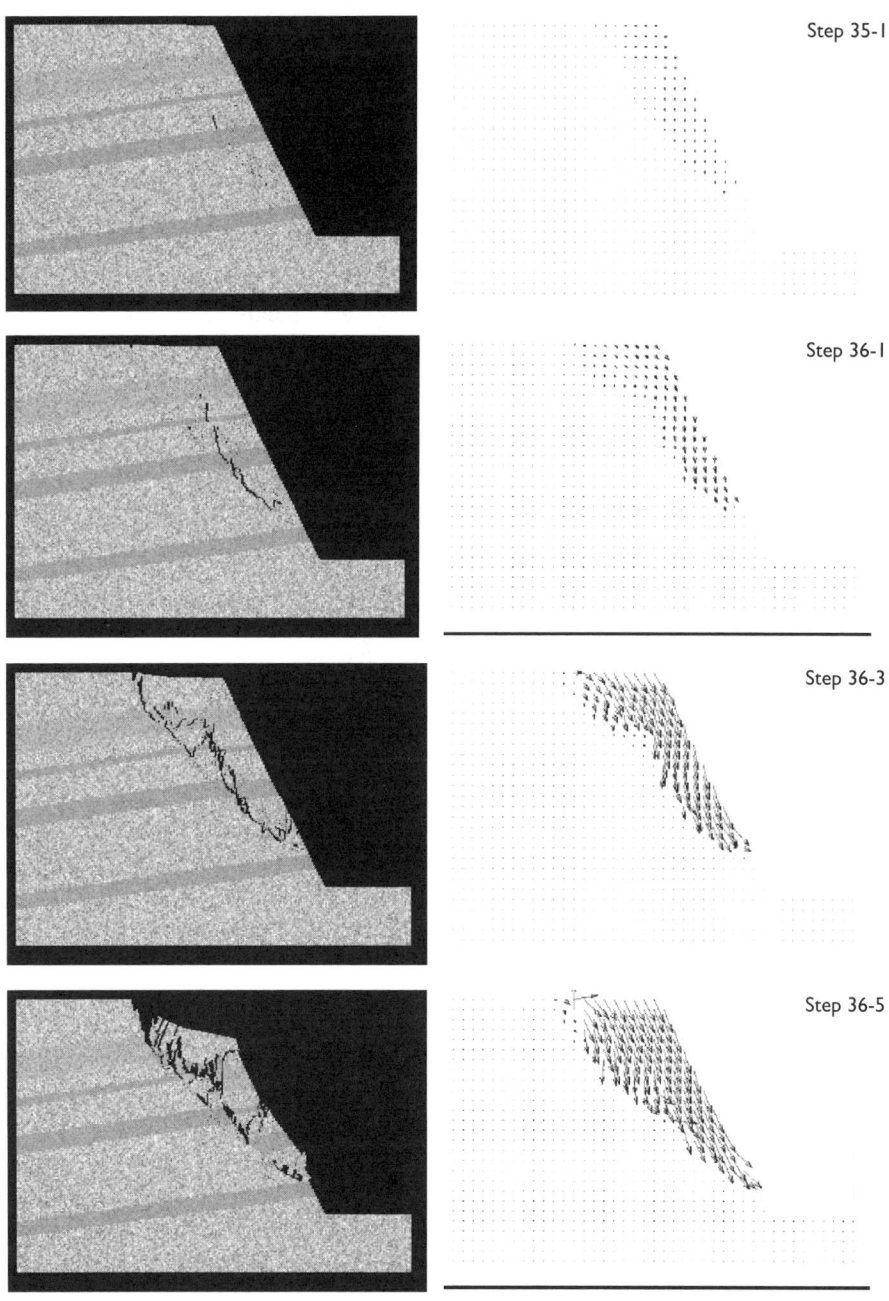

(a) Shading indicates Young's moduli and the layers. (b) Displacement vectors.

Figure 14.9 Simulation results for the stability analysis of a slope in a layered rock mass.

along the failure surface, therefore, only becomes a factor after the failure surface is nearly fully developed and mobilisation becomes possible. As stated by Eberhardt *et al.* (2004), the failure surface only becomes a 'shear' surface once tensile fracturing has progressed to the point where significant cohesion loss has occurred along the path of coalescing fractures and larger displacements become possible through kinematic feasibility and slide mass mobilisation.

This case example indicates, as have previous rock failure examples in this book, that the heterogeneity of the model has a remarkable influence in determining the fracture paths and the final failure patterns. The modelling considering model heterogeneity reproduces the dimensions and irregular stepped nature of the realistic failure surface well. The modelling results show that the development of slope failure facilitates the understanding of where the failure originates, how it starts and develops. It also shows where breakage of intact rock is involved during the development of the failure and how each of the destabilised blocks of rock moves. As stated by Wang *et al.* (2003), with this kind of knowledge, a rock engineer can undertake further study on the stability control strategy so that it is more relevant and effective and hence the rock slope can be designed on a more coherent basis.

14.4 A SLOPE IN A JOINTED ROCK MASS

Joints are the most common type of geological discontinuities in rock masses. They usually occur in sets—which are more or less parallel and can be regularly spaced. There are usually several sets with different orientations so that the rock mass is divided into blocks. Thus, as we have emphasised, a rock mass is not a mathematical continuum: instead it is divided into a number of blocks by joints along which sliding or opening can take place.

Similar to the PFC model (Wang *et al.*, 2003), joint planes can be defined individually or as a set of planes. Such sets can also be combined to create a multiple-intersecting, jointed or blocky system. Each joint set may have multiple joint planes that are separated from each other by a specified spacing. In this Section, a rock slope, 70 m wide and 50 m high, on a rock mass consisting of two joint sets is studied. A total of 25 joints were created in the rock mass model. Their orientations are illustrated in Figure 14.10. The joint set dip angles are 30° and 75°. The mechanical properties of the intact rock and the joints are assigned to each element individually. For instance, assigning a different strength, stiffness and Poisson's ratio to elements along a joint plane can be used to simulate a wide range of mechanical properties of a joint. The Young's modulus (E_0), compressive strength (f_0), the Poisson's ratio (v) and unit weight (γ) of the intact rock material are 20 GPa, 50 MPa, 0.2, and 25 kN/m³ respectively and those for the joints (E_0, f_0, v, γ) are 500 MPa, 5 MPa, 0.3 and 17 kN/m³ respectively. Gravity loads were applied to the elements and the trial safety factor (F_s^{trial}) gradually increased until the critical failure surface formed.

The failure processes of the jointed slope are shown in Figure 14.10. The bright area at the toe of the slope is the zone with the highest stresses. Under these conditions, failure initiates at the toe of the rock slope, i.e., where the stresses are highest, and propagates upwards through the slope along an existing

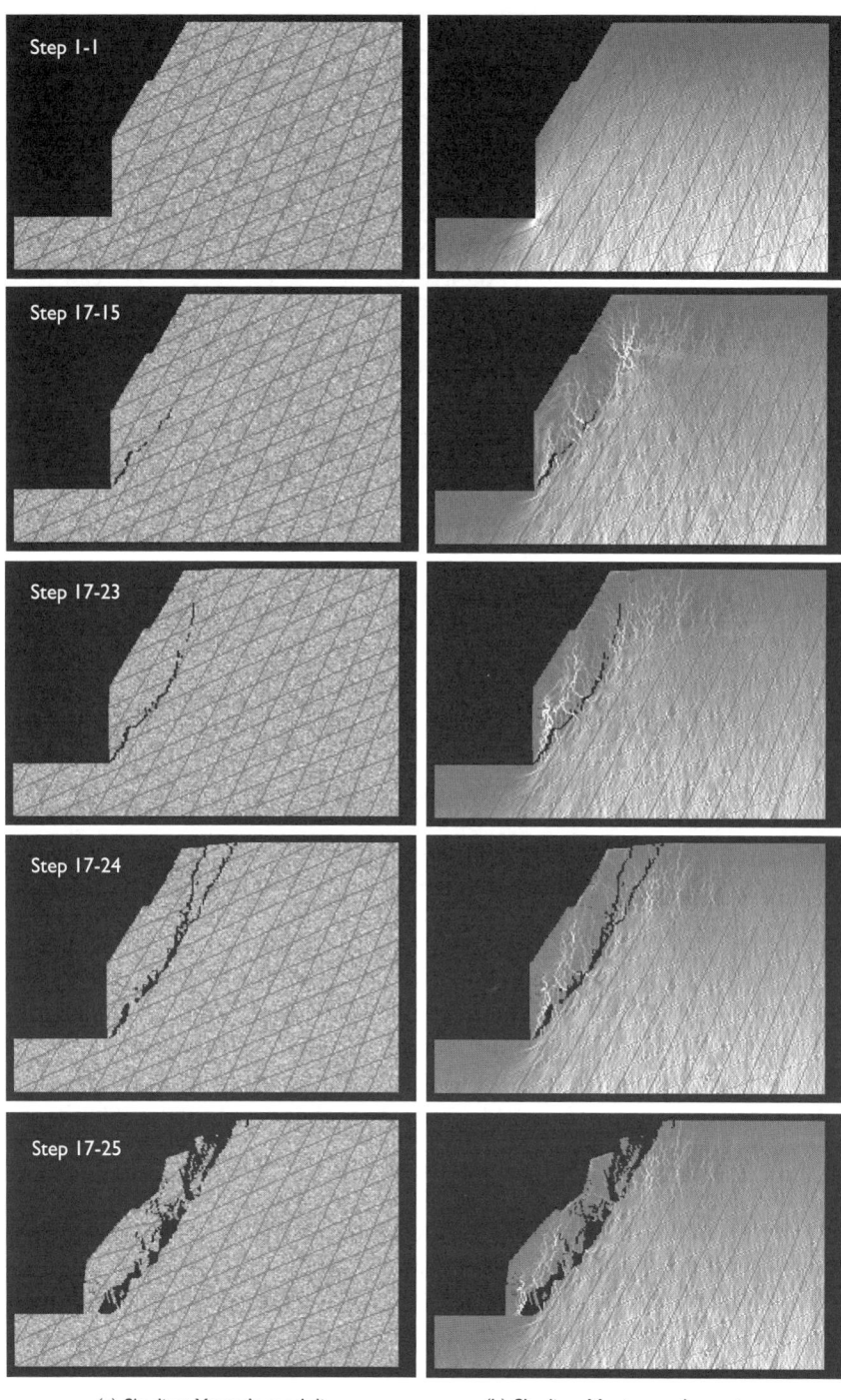

(a) Shading: Young's moduli   (b) Shading: Maximum shear stress

Figure 14.10 Failure mode of a rock slope with two sets of joints.

joint with dip angle of 75°. However, with the stress concentration, the fracture abruptly deviates from the existing joint and breaks through two rock blocks before reaching another existing joint. Then the fracture keeps on propagating along the second existing joint until it reaches the slope top surface. Following the collapse of the rock mass during the first fracture event, the model results show the development of a second major fracture that begins to initiate along a joint and coalesces with the first major fracture. The processes together form the critical failure surface.

The process is largely driven by the initial formation of the brittle tensile fractures, eventually leading to shear failure and mobilisation once the rock mass coherence becomes significantly degraded. It can be seen that the final failure mode is a combination of major sliding and minor toppling with an irregular failure surface. This final profile of the failure surface is largely influenced by the interaction of the prevailing discontinuity and fractures generated through tension and shear. Again, as observed in the last Section for the intact rock model, shear only becomes a factor after enough tensile fracture damage has been incurred to allow mobilisation. The greater part of the critical failure surface develops along the joints without cutting through blocks. As can be seen in Figure 14.10, as soon as the critical failure surface forms, the rock blocks above the critical failure surface are destabilised and suddenly move towards the slope toe. This is a typical slope failure pattern. The final safety factor F_s of the slope in this modelling exercise is 1.52.

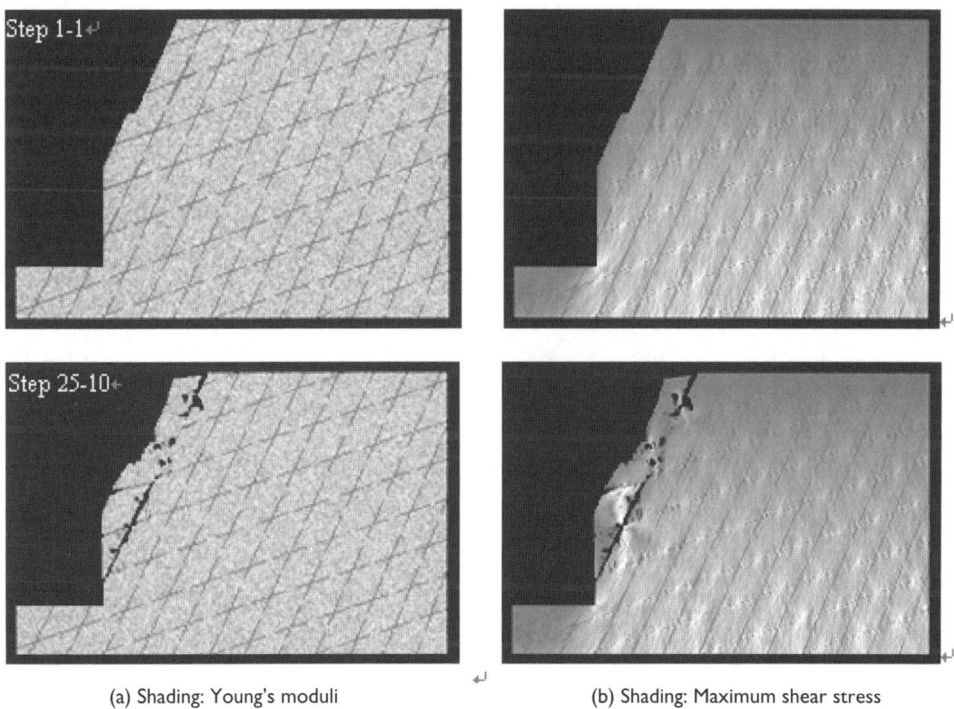

(a) Shading: Young's moduli   (b) Shading: Maximum shear stress

Figure 14.11 Slope failure mode for a rock mass with 80% joint persistence.

14.5 A SLOPE IN A JOINTED ROCK MASS WITH DIFFERING JOINT PERSISTENCE

In rock slope stability analyses, the failure surface is often assumed to be structurally controlled and pre-defined as a continuous plane or series of interconnected planes. However, in real rock slopes, as emphasised by Terzaghi (1962), most rock masses contain discontinuous joints varying in persistence, such that both their stress-conditioned shearing resistance and the cohesion of intact rock bridges between discontinuous joints combine to resist shear failure.

Figure 14.11 illustrates this rock bridge effect. When the joint persistence is reduced by 20% (i.e., so it is 80% of the original full persistence), the slope becomes relatively stable (Figure 14.11), as one would expect. The final safety factor F_s of the slope is 2.0. The behaviour of rock slopes with different joint persistence clearly indicates that, given similar intact rock material properties, slope stability, failure scale and failure pattern are controlled by the properties and quantity of the joints. As can be seen from the numerical simulation, the key to progressive failure in this rock slope is that the process is predominantly driven by the propagation of fractures through the intact rock along the line between existing discontinuities.

Chapter 15

The fracture process when cutting inhomogeneous rocks

15.1 INTRODUCTION

Until quite recently, the detailed mechanisms of rock cutting have not been well understood. Although numerous experimental and theoretical investigations have been carried out in the past decades right up to the present time (e.g. Copur, 2010), fracture evolution in real materials during mechanical cutting is rarely observed directly in standard laboratory tests. Also, in these tests, only the outward appearance of the failure phenomena can be observed because of the rock's opacity. In terms of interpreting and characterising the rock failure in this context, Mishnaevsky (1998) has discussed damage and fracture of heterogeneous materials in terms of rock fragmentation and damage evolution using theories of information, fractals, fuzzy sets and the theory of complex systems, but empirical and idealised analytical models cannot predict the internal fracture evolution fully due to their simplified assumptions. The failure process in detail is, however, traceable through numerical modelling and so we will discuss and present examples of rock cutting simulations to reveal the detailed rock breakdown process in these circumstances.

Several numerical methods have been developed for the purpose. Bit penetration into rock was studied by Wang and Lehnhoff (1976) using a finite element method (FEM). A numerical model with FEM was also used by Zeuch et al. (1983) to model the cutting problem. A DIANA code based on FEM was used by Korinets et al. (1996) to simulate rock breakage with a spherical/flat-bottomed indenter and disc roller cutter. In the work of Tan et al. (1996), a modified expansion cavity model is implemented in a boundary element method (BEM) code, together with a fracture model, to simulate the formation of fractures under the action of single and multiple indenters. The use of other codes, such as FRANC, UDEC, FLAC and NUMA, has also been reported for research and consulting work on rock indentation and cutting problems. However, most of the previous studies have not related their work to rock heterogeneity.

15.2 MODELLING ROCK CUTTING AND THE FAILURE MECHANISM

Rock cutting can be simplified to a two-dimensional plane strain problem, as shown in Figure 15.1. The steel cutter is modelled as a quasi-isotropic homogeneous material with a high strength compared with the rock; whereas the rock is modelled

230 Rock failure mechanisms

Figure 15.1 Model for the numerical simulation of inhomogeneous rock cutting.

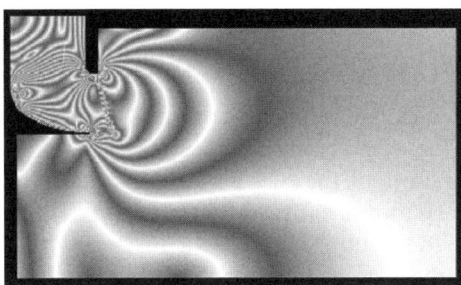
a. Homogeneous rock

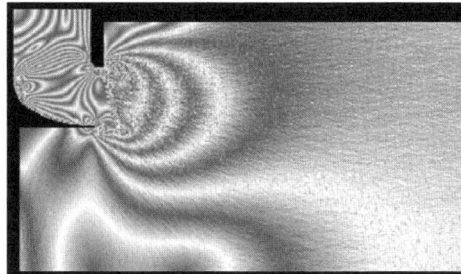
b. Heterogeneous rock

Figure 15.2 Quasi-photoelastic stress fringe pattern in cutter and rock (contours of shear stress).

as a heterogeneous material. An external displacement in the horizontal direction (0.005 mm/step) is applied to the steel cutter from the left side, and the upper side is kept at the same horizontal level without any movement in the vertical direction. The whole model is divided into $250 \times 125 = 31{,}250$ mesoscopic finite elements.

15.2.1 Quasi-photoelastic fringe pattern

The process of rock fragmentation under mechanical loading generally includes the following stages: the build-up of a stress field, the formation of a crack system, the formation of a crushed zone, surface chipping, crater formation and formation of sub-surface cracks. In the initial stage of cutting, the rock is deformed as an elastic body. Therefore, before considering the failure analysis, the quasi-photoelastic stress fringe patterns induced in the cutter and in both homogeneous and heterogeneous rocks are simulated, as shown in Figure 15.2.

It is apparent that the stress fields in the rock consist of three zones: the interaction zone, the confining pressure zone and the zone outside the confining pressure zone. The interaction zone is caused by a mismatch between the elastic moduli of the rock and cutter. The rock in the confining pressure zone is generally in a state of rather high triaxial compression. In the post-failure analyses, it is found that the confining pressure zone defines, to a great extent, the shape and size of the so-called crushed zone. Comparing Figure 15.2(a) with Figure 15.2(b), it is seen that, because of the heterogeneity, the stress trajectories in the confining pressure zone become coarser, which makes the boundary of the crushed zone become vague, as is shown in the post-failure analyses. The stress contours in the cutter, which are caused by the reaction from the rock, are also clearly shown in the fringe pattern. The high stress concentration in the corners often causes tools to break or wear. The stress distribution coupled with the heterogeneity will determine the sites of crack initiation and the directions of crack growth in the following failure analysis.

15.2.2 Fracture pattern

Figure 15.3 shows the development of the fracture process and the progressive evolution of the stress fields as the cutting displacement increases. When the cutter is

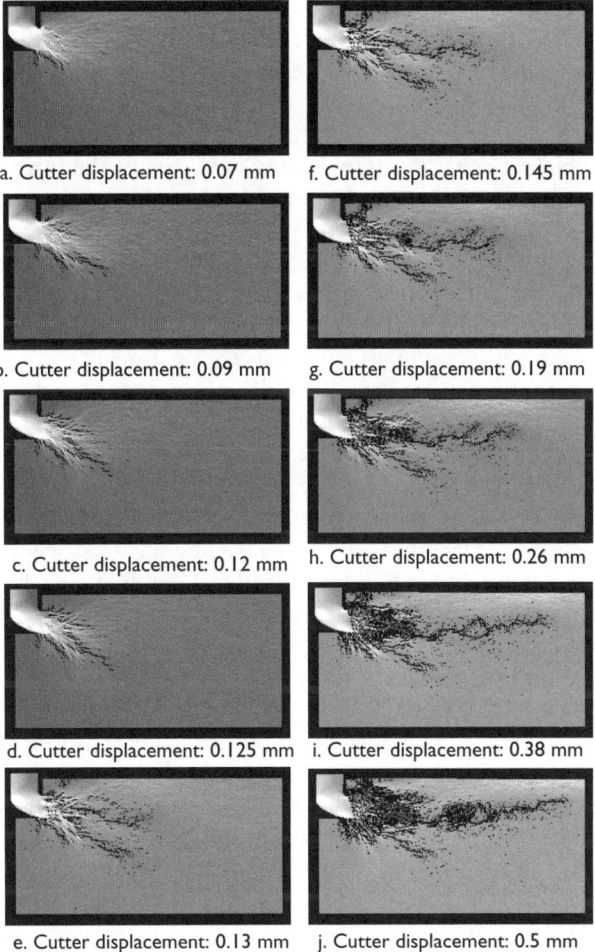

Figure 15.3 The fracturing process in cutting (maximum principal stress distribution).

applied to the rock, a high stress concentration is induced in the rock near the cutter, and the rock distant from the cutter is almost unaffected. As shown in Figure 15.3(a), cracks initiate first at the lower edge of the cutter and propagate, dipping into the rock at approximately 45° in the direction of the cutter displacement. The initiation mechanism of the cracks is tensile failure. This is because the tensile strength of brittle rock material is less than the compressive strength.

Ahead of the cutter, the rock is highly stressed and it is in a triaxial state due to the high confining pressure. Therefore, the rock in this zone does not fail at the initial stage of cutting. As the cutter displacement increases, tensile cracks initiate at the upper edges of the cutter and, at the same time, the cracks initiated from the lower edge of the cutter propagate downwards, as shown in Figure 15.3(b). With continuous cutter displacement, compressive failure occurs ahead of the confining pressure zone, as shown in Figures 15.3(c) and (d). With the tensile failure and compressive failure releasing the confinement for the triaxial compressive zone, the rock in the initial confining pressure zone is compressed into failure and the crushed zone gradually comes into being with a dimension similar to that of the cutter–rock interface, as shown in Figure 15.3(e). From the results, it is found that the crushed zone is the zone with a high density of micro-cracks, and it is not necessary to consider the crushed zone as some separate body which interacts with the tool and the rock.

The crushed zone has an important influence on the succeeding development of the crack system, which changes the transferring direction of the force applied by the cutter. As shown in Figure 15.3(a)-(d), before the formation of the crushed zone, all of the cracks show a tendency to dip down into the rock. Associated with the crushed zone, there must be a volumetric expansion and a tensile stress field, which result in tensile crack propagation and the formation of chipping cracks. Therefore, after the formation of the crushed zone, two main categories of cracks, chipping cracks and sub-surface cracks, come into being, as shown in Figure 15.3(e).

As the displacement increases, one of the chipping cracks rapidly intersects with the free surface and small pieces of rock are chipped, as shown in Figure 15.3(f). At the same time, the other main chipping crack propagates in a curvilinear path, but approximately parallel to the free surface of the rock. The sub-surface cracks propagate in a curvilinear path to dip into the rock at a certain angle in the direction of the cutter. It is thought that the curvilinear path is caused by the heterogeneity of the rock. The mechanism of the chipping crack is tensile failure mainly induced by tensile stress associated with the crushed zone. With the continuous increase in the loading displacement, the main chipping crack, which is driven by the tensile stress, propagates and bifurcates as shown in Figure 15.3(g).

As the displacement increases further, the bifurcated chipping cracks propagate and interact, as shown in Figure 15.3(h). At the same time, more and more cracks form around the crushed zone, which makes the crushed zone enlarge and its boundary become vaguer. With a continuous increase in the loading displacement, the bifurcated chipping cracks coalesce and form a single major chipping crack again, as shown in Figure 15.3(i). As the chipping cracks propagate and intersect with the free surface, the major chip is expected to come into being as shown in Figure 15.3(j). The cracks which propagate to dip into the rock form the sub-surface cracks after the formation of the major chip.

According to these numerical model results, when cutting heterogeneous brittle material, the process of chip formation is caused by a complicated stress state, with the elements mainly failing in a tensile mode or a mixed shear and tensile mode and sub-fracturing occurring in a dominantly tensile mode. The sub-surface fractures are actually the bifurcation of some major tensile fracture. Therefore, in cutting brittle materials, as differentiated from cutting plastic materials, the mechanism of chip removal is not shear, but tensile or mixed-mode crack growth. Similar phenomena were also noted by Mishnaevsky (1995).

15.2.3 The chipping process

Figure 15.4 shows the simulated chipping process when cutting heterogeneous brittle materials. In this simulation, there are mainly two major chips. When the loading displacement is applied to the rock, tensile cracks initiate first at the upper and lower

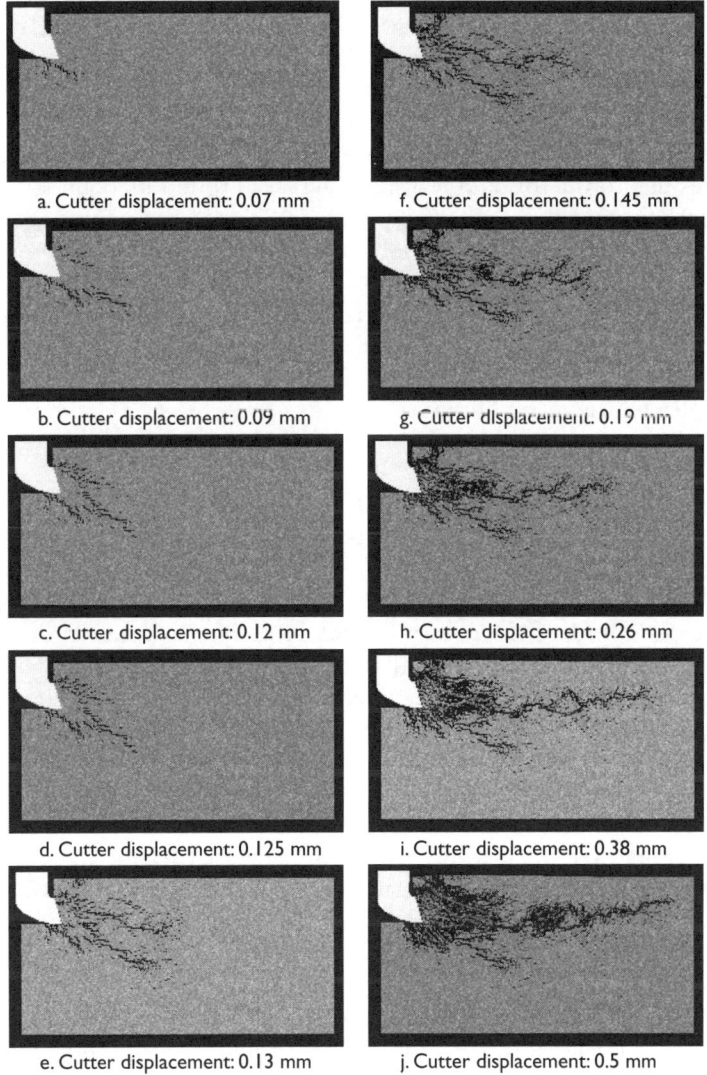

Figure 15.4 Modelling the chipping process in rock cutting (Young's modulus distribution).

edges of the cutter, as shown in Figures 15.4(a) and (b), the loading displacement then being 0.09 mm. As the loading displacement increases from 0.09 to 0.125 mm, as shown in Figures 15.4(c) and (d), both the tensile cracks propagate at approximately 45° to the direction of the cutter displacement. When the cutter moves to 0.13 mm, with the formation of the crushed zone, the cracks initiate from the upper edge of the cutter and then bifurcate. One of the bifurcated cracks propagates upward to form a minor chipping crack and the other bifurcated cracks coalesce with the micro-cracked zone, as shown in Figure 15.4(e). At the same time, the cracks initiated from the lower edge of the cutter coalesce with the micro-cracked zone. Associated with the micro-cracked zone, tensile stress fields are induced because of the expansion of the micro-cracked zone. A major chipping crack initiated from the micro-cracked zone is driven to propagate by tensile stress from the tensile stress fields.

When the cutter displacement increases to 0.145 mm, the minor chipping crack bifurcated from the upper tensile crack intersects with the free surface and small pieces of rock are chipped, as shown in Figure 15.4(f). The major chipping crack initiated from the micro-cracked zone propagates in a curvilinear path, but approximately parallel to the upper free surface of the rock, and is expected to form a flat chip. With the increase in the cutter displacement from 0.145 to 0.38 mm, the crushed zone enlarges due to the formation of micro-cracks around its boundary. The major chipping crack propagates and exhibits a complicated mechanical behaviour. As shown in Figures 15.4(g)-(i), the major chipping crack propagates in a curvilinear path. During the propagation process, the bifurcation and coalescence phenomena occur frequently. It is found that tensile failure is the main mechanism of chipping crack propagation. When the cutter displacement increases to 0.5 mm; the major chip is expected to come into being as shown in Figure 15.4(j). The pieces of chipped rock have a complicated geometrical shape.

15.3 THE LOAD–DISPLACEMENT RESPONSE WHEN CUTTING INHOMOGENEOUS ROCK

The load–displacement response, as shown in Figure 15.5(a), is actually determined by the fractures ahead of the cutter. On the basis of the assumption that there is consistency between the damage and the seismicity associated with the fracture process, the acoustic emission (AE) phenomenon during the fracture process is also observed, as shown in Figure 15.5(b). From the two curves, we know that the load–displacement response is therefore closely related to the fractures in the rock induced by the cutter. This is made clear when we compare Figure 15.4(a) and (b) with Figure 15.5.(a) and (b).

The increasing load at the very beginning is caused by the quasi-elastic deformation of the rock surface, as shown in Figure 15.5a marked by 'a'. At this stage, the load–displacement relation is quasi-linear and, correspondingly, there is almost no acoustic emission in the rock, except for that caused by a very small amount of damage occurring. This quasi-elastic deformation was then followed by the deformation of the crushed zone and cracked zones. At this stage, the movement of the rock fragments in the crushed zone was restrained by the surrounding intact rock. In order to break the rock, continuous loading was required. This would have caused

The fracture process when cutting inhomogeneous rocks   235

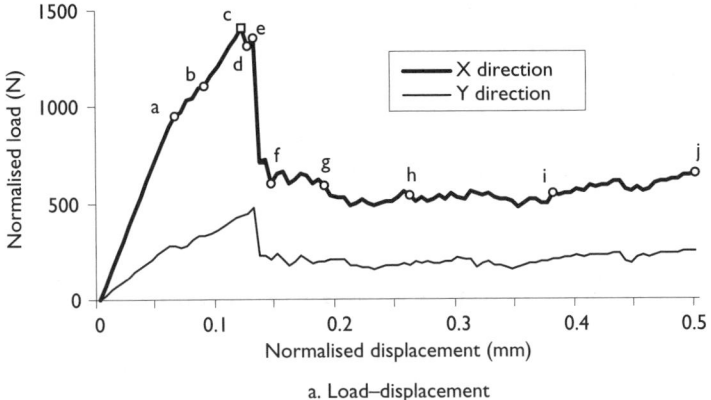

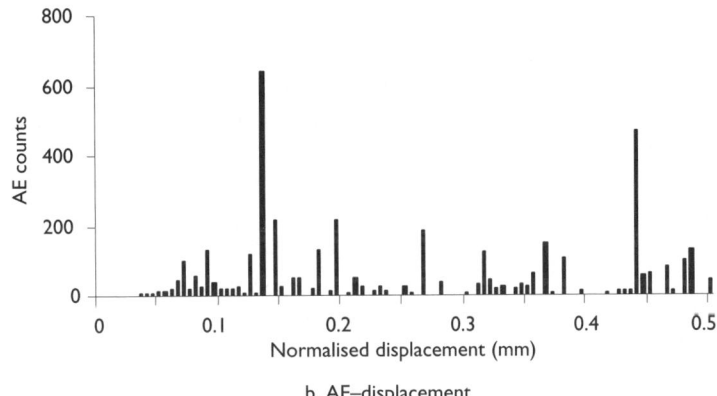

Figure 15.5 Load–displacement curve and associated AE (letters corresponding to the stages in Figures 15.3 and 15.4).

intense comminution of the rock fragments if the confinement had been sufficiently high. However, in the present case the moving tendency of the rock fragments in the crushed and cracked zones pushed the surrounding rock upwards, due to a smaller confinement on the upper side, and resulted in chipping cracks.

During the process of rock fracturing, the cutter gradually penetrated into the rock. When the chipping cracks reached the rock surface, the surrounding rock, which had offered the upper confinement at the earlier stage, together with part of the rock underneath the cutter, moved away forming the chips. The movement of the cutter met less resistance in this case. The force acting on the cutter was thus reduced to a rather low value, as shown in Figure 15.5a, and the cutter penetrated the rock to a greater depth with a decreasing load. After that the rock fragments could be removed with a very small load.

It is clear from the curve of the AE counts (Figure 15.5b) that this load drop is related to a sudden increase in the AE counts (or the number of failed elements). Generally, every large increase in the AE counts results in a significant drop. The load–displacement curve for the cutting process obtained from the simulation shows that the simulated chipping process is similar to the chipping process found through experimental observation of brittle cutting. It is also somewhat similar to the results obtained from indentation experiments carried out in the laboratory. In addition, from the above-described simulation, it is easy to understand that the cutting depth is mainly dependent on the load magnitude, the macro-mechanical properties of the rock, and the heterogeneity of the rock. However, it is also influenced by the cutter geometry, which characterises the load distribution.

15.4 THE CRUSHED ZONE DURING ROCK CUTTING

According to the above simulated results and the review by Mishnaevsky (1995), a zone of highly fractured and inelastically deformed rock, the crushed zone, is created near the cutting tool. The crushed zone has an important influence on the chipping process and energy utilisation in cutting. Some researchers consider the crushed zone as a separate body which interacts with the tool and the intact rock. In the simulations presented here, the crushed zone is in fact the zone with a high density of micro-cracks and comes into being because the rock in the initial confined stress zone is compressed into failure as the tensile cracks at the two edges of the cutter and the compressive failure in front of the initially confined stress zone release the confining stress. Therefore, the mechanism of the crushed zone is mainly compressive failure.

In some experiments, due to very high confining pressure, the rock in the confined zone may fail in a ductile cataclastic mode with the stresses satisfying the ductile face of the double elliptic strength criterion. However, in the present simulation, the stresses of few elements satisfy this criterion. Perhaps this is because the depth of the cut is not great enough or there is no confinement on the rock. Further research work is needed to clarify this phenomenon. Previous research in mechanical rock fragmentation indicates that about 70% to 85% of the energy is consumed by the formation of the crushed zone. Lindqvist (1982) has shown that the energy loss associated with friction, micro-cracking and so on consumes 89% of the transmitted energy in disc cutting, whereas the useful energy is only about 2–3%. From the load–displacement curve shown in Figure 15.5b, the work performed by the cutter can be calculated to be 0.320 J, which is a small amount of energy.

In fact, as evidenced above, little energy is required to cause rock failure. In Harrison and Hudson (2000), one of the worked examples includes a cylindrical rock specimen (100 mm long and 50 mm in diameter) subjected to uniaxial compression throughout the complete stress–strain curve. It only requires 39 J for this complete rock failure process—which is the energy required to power a 100 W light bulb for just 0.4 seconds. In the case of tunnel boring machines, it appears that considerable energy is required because of the large amount of energy used by the machine in moving forward, but only a small fraction of this energy is used for actually cutting the rock: the rest is lost in friction, stress waves induced in the rock mass, etc.

Chapter 16

Rock failure around tunnels in jointed rock

16.1 INTRODUCTION

The two main ways in which rock can fail around tunnel, cavern and mine openings are by a) rock blocks falling or sliding into the excavation or b) the rock failing due to the concentrated stresses around the excavation periphery. The former case tends to occur at shallow depths where there is usually a greater jointing frequency; the latter case tends to occur at deeper depths where the *in situ* stresses are higher. Hoek *et al.* (1995) describe the process of rock failure and the associated support requirements in underground rock excavations.

Ortlepp (1997) has compiled a series of illustrative examples of stress-induced rock failure in the deep South African gold mines, one of which is illustrated in Figure 16.1. In this case, the rock failure is stress-induced but there is also the influence of pre-existing weakness planes in the rock mass.

Joints usually occur in sets within which the joints are more or less parallel and can be regularly spaced; also, there are often several sets occurring in different directions so that the rock mass is broken up into a blocky structure. Because of the low shear and tensile strengths of these discontinuities, as well as the looseness of a rock mass due to the unloading by excavation, rock masses can slide along these structural planes or detach, flex and break. The zone around the excavation which has been disturbed or damaged by the process of excavation (this process creating space for rock blocks to move and creating stress concentrations) is known as the Excavation Disturbed/Damaged Zone, the EDZ, and has been extensively studied in recent years (e.g. Bäckström, 2008; Jonsson *et al.*, 2009; Hudson *et al.*, 2009).

In order to understand the issues involved in the process of designing support for a tunnel in a jointed rock mass, it is necessary to examine how a jointed rock mass surrounding a tunnel deforms and how this deformation can induce the associated failures. A fundamental understanding is particularly required of the fracture initiation and growth processes. A general way to investigate the deformation and failure characteristics is to conduct a model test and numerical analysis. For a rock mass blocky structure cut by joints, Goodman and Shi (1985) suggested block theory to analyse the stability of rock blocks around underground openings. Chan and Goodman (1987) calculated the mean number and volume of removable blocks using the Baecher disc model (Baecher *et al.*, 1977). Hoerger and Young (1990) applied the Baecher disc model with a bivariate normal distribution of joint orientation to the analysis of block occurrence around a tunnel. Song *et al.* (2001) used a three dimensional statistical

238 Rock failure mechanisms

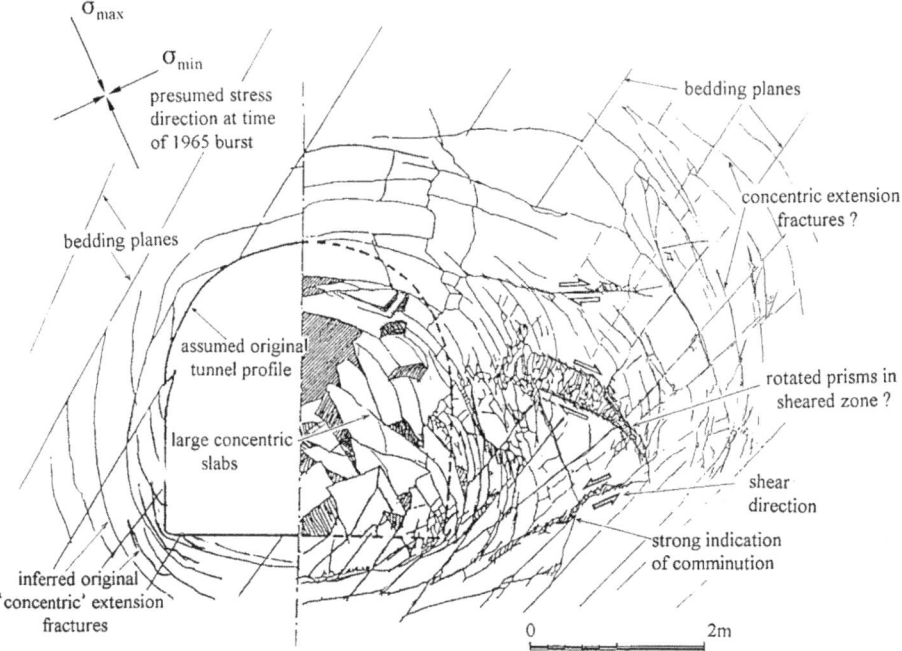

Figure 16.1 Schematic illustration of stress-induced rock failure around a deep level South African mine tunnel (from Ortlepp, 1997).

joint modelling technique to analyse the stability of rock blocks around a tunnel and reached the conclusion that the removable blocks occurred more frequently as joint persistence, degree of scatter in joint set orientation and volumetric frequency of the joints increased.

Although the influences of rock joints and the stress state on the stability of underground structures have been studied, both analytically and experimentally, the understanding of the failure mechanism of underground excavations under complex geological conditions is still far from being complete and satisfactory. The main objective of the illustrative study presented here is to investigate the influence of different dip angles of layered joints and the effect of the lateral stress coefficient on the stability of a tunnel excavated in a jointed rock mass.

16.2 PROGRESSIVE FAILURE AROUND A TUNNEL IN A JOINTED ROCK MASS

In the following simulations, five numerical models are considered (see Figure 16.2). For comparison, there are no joints in the first model; in the other four models, there are layered joints at dip angles of 0°, 30°, 45°, and 60°. The cross-section of each model has an arch profile in the crown and vertical sidewalls with dimensions 5 m wide and 4 m high. The distance between joints is 4 m. In order to show the

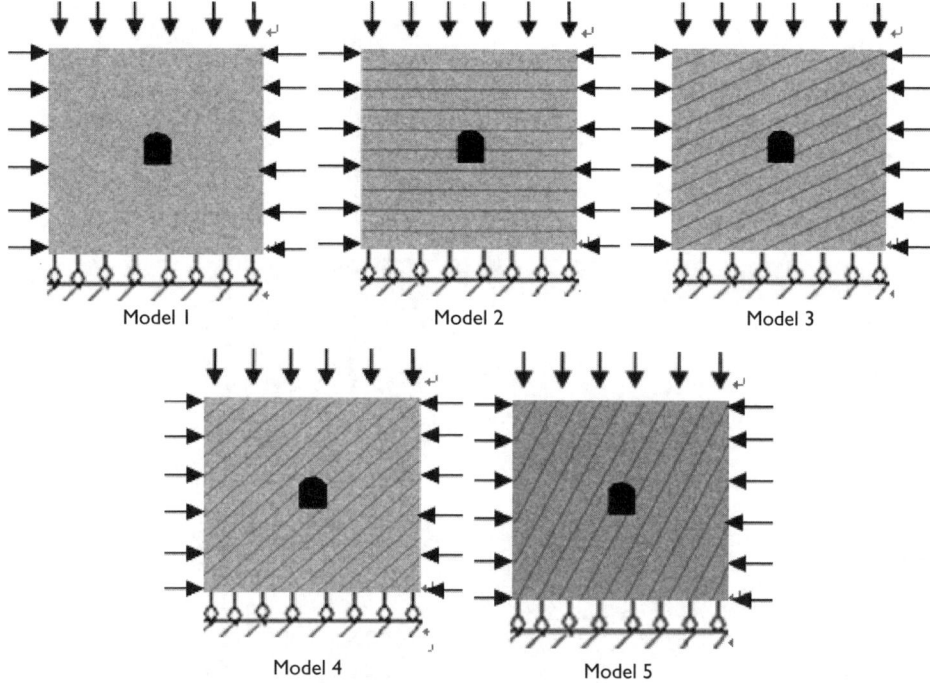

Figure 16.2 Models with different dip angles of the joints (these models consist of 40,000 elements; the horizontal stress is 1 MPa and the vertical stress is 5 MPa).

initiation, propagation and coalescence process clearly, the effect of gravity has not been included. These five models have a common dimension of 40 × 40 m. For a better understanding of the failure mechanism of the tunnel in this jointed rock mass, the strength of the rock mass is reduced by 1% at each step until the tunnel fails.

The parameters for the heterogeneity (m), elastic modulus, strength, Poisson's ratio, internal friction angle: rock mass 3, 1.5 GPa, 9 MPa, 0.3, 32°; joints 3, 100 MPa, 2.16 MPa, 0.4, 23°.

16.2.1 Effect of dip angles on the stability of the tunnel

The failure process observed in the numerical models after excavation and the induced shear stress redistribution are presented in Figure 16.3. The notation at the upper-left corner indicates the strength reduction step, e.g. 41–1 means that this is at the 1st iterative step of the 41st reduction step. From Figure 16.3, we can see that dip angle has a distinct effect on the failure mode of the tunnel. It is clearly seen that the distributions of failure zones around the underground cavern are strongly dependent on the joint dip values.

For model 1 (there are no joints), after excavation, cracks initiate at the roof and floor of the tunnel. With the strength reduction, more cracks appear at the two sidewalls and extend in the direction of the maximum principal stress until failure of

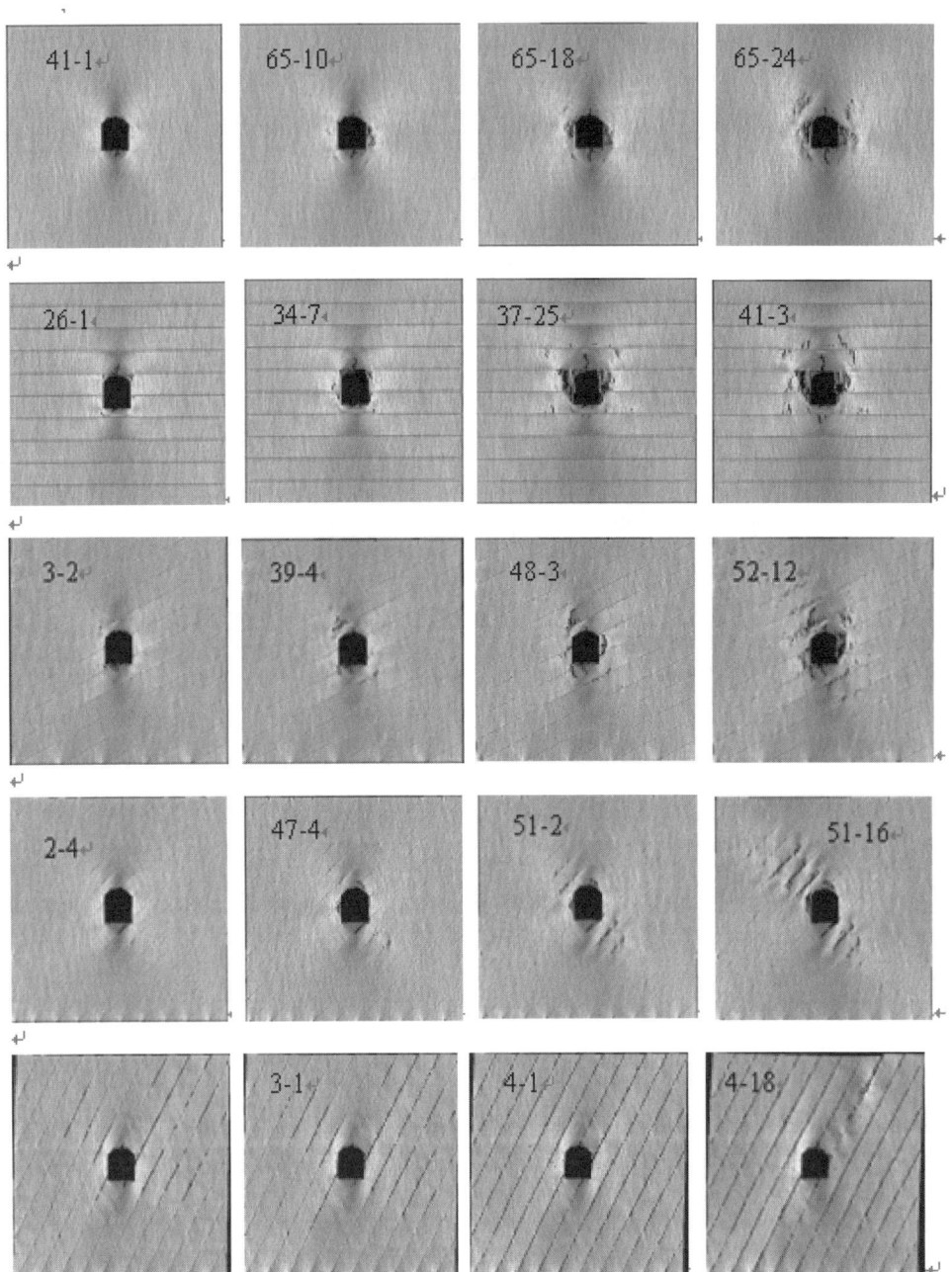

Figure 16.3 Failure modes of a tunnel in a layered rock mass after excavation and the induced stress redistribution. For the model types, see Figure 16.2.

the whole tunnel occurs. For an intact rock mass, the failure pattern of the tunnel is mainly controlled by the lateral stress coefficient, which has been investigated by Zhu *et al.* (2005). But in a jointed rock mass, the failure mode will be affected by joints, which can be seen clearly from Figure 16.3. For model 2 (the dip angle of the joints is 0°), cracks first occur at the roof of the tunnel after excavation and extend upwards. The layered rock mass above the crown performs more like a beam. With the strength reduction, a crack extends upward and the 'beam' breaks into the tunnel. Meanwhile, cracks at both sidewalls extend from the bottom of the sidewall and eventually the rock mass located at the two sidewalls spalls into the tunnel. The final shape of the tunnel is in the form of a bowl.

For models 3 and 4 (the dip angles of the joints are 30° and 45° respectively), after excavation, joints near to the left shoulder of the tunnel first begin to detach from the surrounding rock mass and, with strength reduction, this tendency becomes clearer. Here we should notice that the detaching tendency is perpendicular to the joint set, which corresponds closely with engineering observations. So it will be more efficient if rock bolts are fixed in the normal direction to the layered joints in this region. For model 5 (the dip angle of the joints is 60°), the cohesive strength of the joints cannot resist the sliding force, thus causing the failure of the whole tunnel.

This confirms the intuitive idea that the existence of joints changes the failure mode of the tunnel. For horizontally layered joints, the failure mode is the breakage of the 'rock beam' and the spalling and crushing of sidewalls. For joints at dip angles of 30° and 45°, the failure mode is the sliding-in of the sidewall and the detaching, flexing and breaking of layered rock mass near the shoulder of the tunnel. For joints at a large dip angle, the failure mode is the sliding of the rock mass along the interface of joints and rock mass.

Figure 16.4 shows the major principal stress vectors after excavation. From this Figure, we can see that, for an intact rock mass, the stress distribution around the

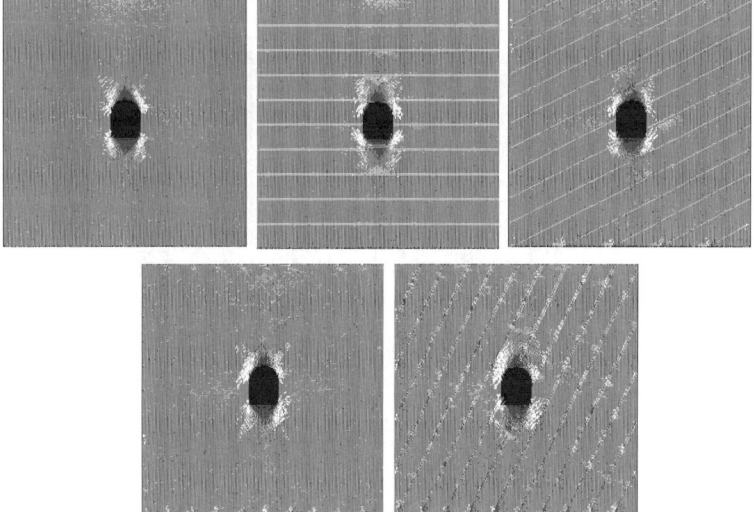

Figure 16.4 Maximum principal stress vector diagram around an excavated tunnel in a layered rock mass.

242 Rock failure mechanisms

tunnel is symmetrical but, for a jointed rock mass and with an increase in the dip angle, the asymmetry of stress around the tunnel increases gradually, causing the asymmetry of the failure mode.

In order to investigate the effect of layered joints on the displacement and stress at the perimeter of the tunnel, four key demonstration points are chosen (see Figure 16.5). Figure 16.6 illustrates the displacement of the four key points in models 1–5.

From Figure 16.6, we find that the existence of the joints increases the displacement around the tunnel, especially the displacements at the two sidewalls. For example, for model 1, the displacements of key points 2 and 3 on the left and right sidewalls are small; but, for models 4 and 5, the displacements of two sidewalls increase sharply. The larger the dip angle is, the larger is the displacement on the sidewall. In fact, the displacements of points 4, 1 and 3 in model 5 are 6.1, 0.74 and 0.66 times those in model 1.

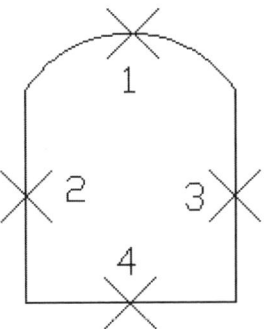

Figure 16.5 Key points on the tunnel perimeter.

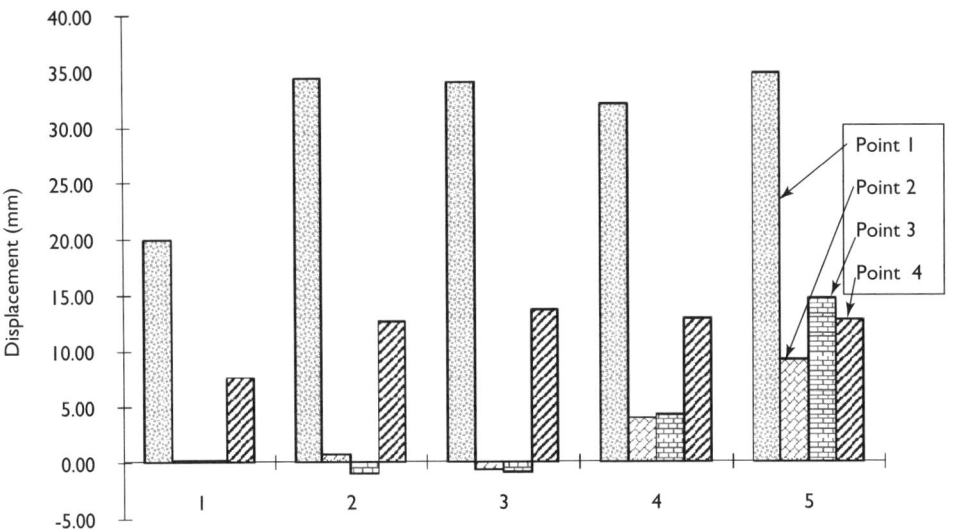

Figure 16.6 Displacements of the key points in the different models. The horizontal ordinate is the model number.

16.2.2 The effect of the lateral stress on the mode of tunnel failure

In order to investigate the tunnel failure mode under different loading conditions, a lateral stress of 0.5, 0.7, 1.0, 1.2 and 1.5 MPa is applied in model 3, and the vertical stress (1.0 MPa) and parameters of the rock mass and joints remain the same. Figure 16.7 shows the failure process for each model. In this Figure, the failure patterns at four typical reduction steps are shown. We can see that the lateral stress magnitude has a distinct effect on the failure mode of the tunnel. For a lateral: vertical stress coefficient of $k = 0.5$ and 0.7, cracks mainly concentrate at the two sidewalls where the rock mass is cut by joints. At a lateral stress coefficient of 0.7, the detachment of joints is only confined to the regions near the left shoulder of the tunnel, and no remote detachment is formed. The failure pattern of the tunnel is spalling of the two sidewalls and detaching, flexing and breaking of the layered rock mass.

For the lateral stress coefficient of 1.0, cracks initiate at the bottom of the left sidewall and the joint location in the right sidewall. Then, with strength reduction, cracks begin to form in the roof and the bottom of the right sidewall and propagate in the direction of the major principal stress and coalesce with the joints. For lateral stress coefficients of 1.2 and 1.5, the failure patterns around the tunnel change: fractures only occur in the roof and floor of the tunnel, and there are no fracture zones at both sidewalls. The fractured zones in the floor of the tunnel are mainly caused by shear stress. The shape of the fractured zones in the roof and floor take on the shape of two triangles.

From these simulation results, it is evident that the lateral:vertical stress coefficient, k, has a significant effect on the failure mode of the tunnel. For a lateral:vertical stress coefficient value of $k < 1$, the fracture zones mainly concentrate at the two sidewalls and at the upper-left part of the tunnel, where the rock mass cut by joints begins to detach, flex and break into the tunnel. For a lateral stress coefficient value of $k = 1$, the fracture zones occur at the two sidewalls, roof and floor of tunnel. But, for a lateral stress coefficient of $k > 1$, the fracture zones only occur in the roof and floor of the tunnel. So, although joints and other structural planes will have an effect on the failure mode of the tunnel, a high horizontal tectonic stress will have a dominating effect on the failure of the tunnel.

Goodman (1976) suggested a graphical method to estimate the failure zones around a tunnel in these circumstances (Figure 16.8): draw the normal line OO' to the joint planes, and two lines OA and OB with an angle ϕ_j (the joint friction angle) to the normal; then draw tangent lines parallel to OA and OB to the tunnel periphery. Zones where sliding can occur are then identified as indicated in Figure 16.8.

This method is easy to use, although it does not take the effect of the lateral stress coefficient into account. From the simulation results, we can see that, for a low lateral stress coefficient (e.g. $k = 0.5$), the failure zones around the tunnel are similar to those indentified by the graphical method (see Figures 16.8 and 16.9). But, for high lateral stress coefficients (e.g. $k = 1.2$ and 1.5), the failure zones change due to the effect of the lateral stress.

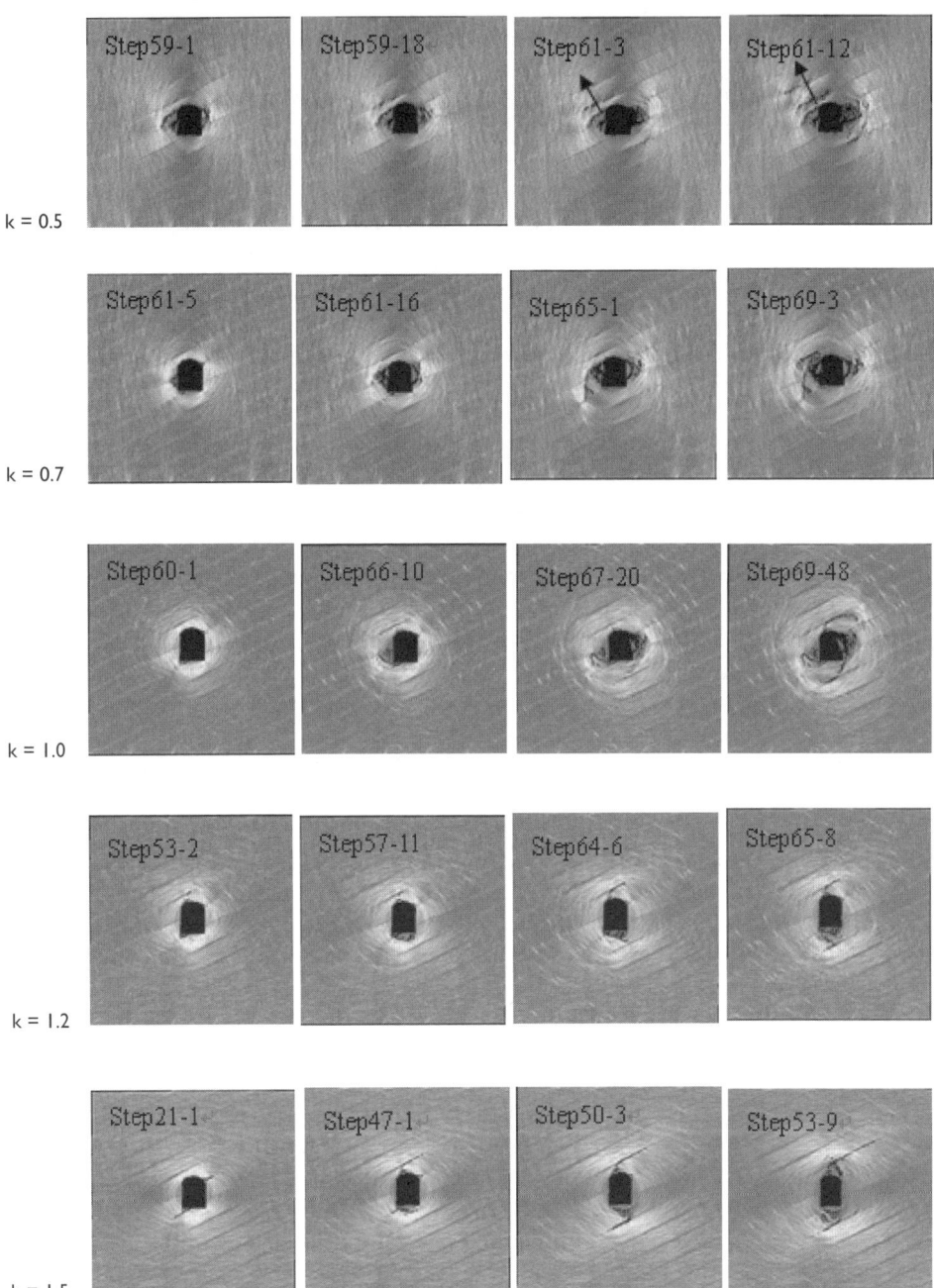

Figure 16.7 The failure process of a rock mass with inclined joints (dip angle is 30°) under different lateral stress coefficients. The vertical load is 1.0 MPa and the lateral stress coefficients are 0.5, 0.7, 1.0, 1.2, and 1.5.

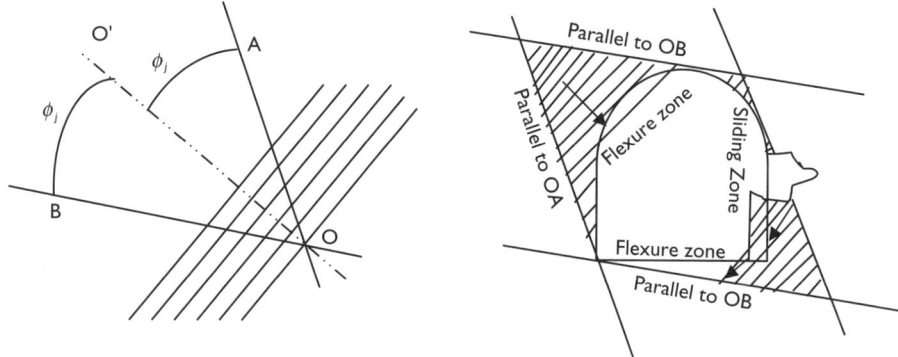

Figure 16.8 Graphical method to distinguish the sliding zones and bending zones (from Goodman, 1976).

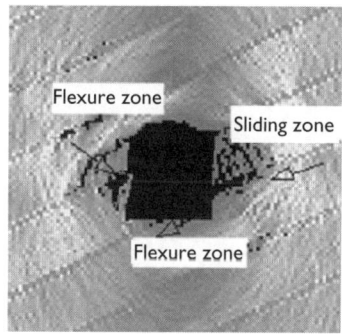

Figure 16.9 Failure zone when the lateral stress coefficient is 0.5.

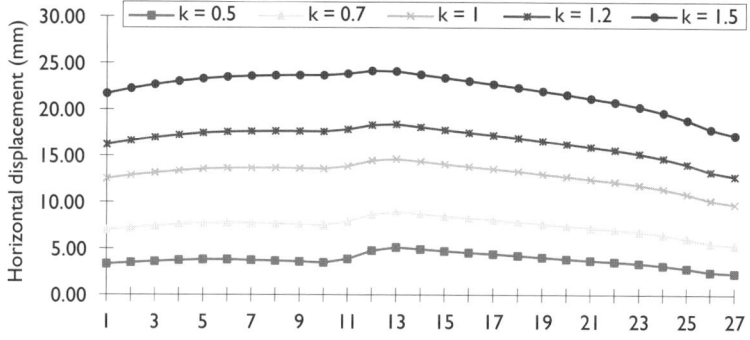

Figure 16.10 Horizontal displacement of the left sidewall after excavation. The horizontal co-ordinates indicate evenly spaced points on the two sidewalls, point 1 being at the top and point 27 at the bottom.

246 Rock failure mechanisms

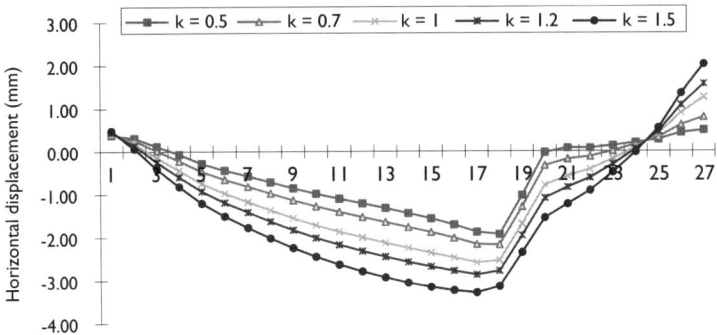

Figure 16.11 Horizontal displacements of the right sidewall after excavation. The horizontal co-ordinates indicate evenly spaced points on the two sidewalls, point 1 being at the top and point 27 at the bottom.

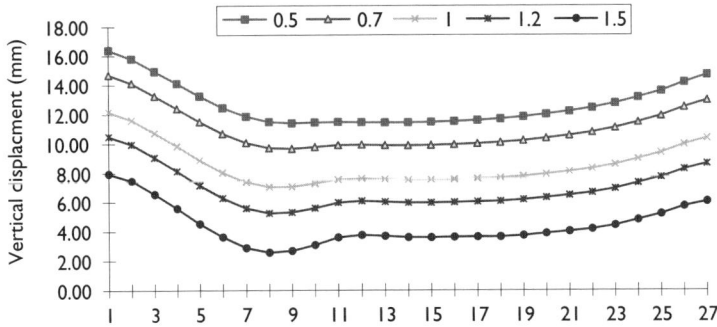

Figure 16.12 Vertical displacement of the tunnel floor after excavation. The horizontal co-ordinates indicate evenly spaced points on the tunnel floor.

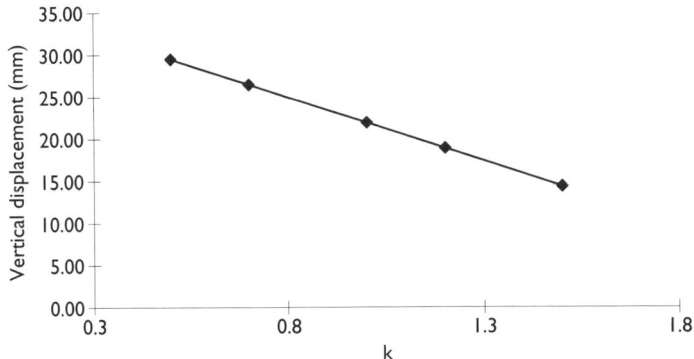

Figure 16.13 Vertical displacement of the crown-midpoint after excavation as a function of the lateral stress.

16.2.3 Displacements at the tunnel periphery

Figures 16.10 and 16.11 show the horizontal displacements at the left and right sidewalls. The horizontal co-ordinates indicate evenly spaced points on the two sidewalls. Point 1 is at the top of the sidewall and point 27 is at the bottom of the sidewall. With increasing lateral confinement, the displacements of the two sidewalls are all increased, especially in the central part of the sidewall. For the right sidewall, the upper part of the sidewall slides into the tunnel along the interface of the joints.

The vertical displacement of the floor and the mid-point of the roof are shown in Figures 16.12 and 16.13. The displacements of the floor and the mid-point of the roof decrease with increasing k. Thus, the increasing of the lateral stress will make the displacements at the roof and floor of the tunnel decrease, but the displacement of the two sidewalls increase.

It is a paradigm shift in rock tunnel design that we can now conduct such detailed numerical modelling to support the design and also to be able to incorporate progressively more details of the rock mass characteristics into the model. Moreover, the capability of such numerical modelling is sure to increase in the years ahead.

Chapter 17

Rock failure induced by longwall coal mining

17.1 INTRODUCTION

There are two main techniques in coal mining: in the room and pillar method, the coal is extracted in rooms, usually in a regular geometrical pattern and sufficient coal is left in pillars to support the roof; in longwall coal mining, the coal is removed in a long strip, the longwall, which advances forward in a direction perpendicular to the strip, leaving the superimcumbent strata to collapse, or cave, behind it. The way in which this collapse of the overlying strata occurs is not straightforward—because of the vagaries in the overburden rock structure, the complexity of the stress redistribution, and the deformation and non-linear characteristics due to failure. Consequently, research into this rock failure process is one of the most active and challenging research activities.

The natural stress distribution in the rock strata is disturbed by the longwall excavation and high stress zones are created in the adjacent coal and at the ends of the longwall because of the redistribution of the pre-existing stresses in the rock mass. When the longwall face has advanced sufficiently far, the immediate roof collapses behind the face at a certain distance depending on geological conditions (Yavuz, 2004). Note that this is a case where continuous failure of the rock is a desirable outcome; if the roof does not cave at a reasonably close distance behind the longwall face, its eventual collapse will be more energetic and can cause air blast damage. Failure of the roof continues until the roof and floor are in contact (Figure 17.1).

Although there are many numerical methods that can be applied to calculate the stress distribution and deformation in such a mining situation, to date, few attempts have been made at a comprehensive understanding of the mining-induced fracturing and collapse. Moreover, *in situ* monitoring of mining-induced fractures and collapse, as well as the stress and deformation in the strata and caved zone, are difficult due to the inaccessibility of the waste area. Therefore, various assumptions have been put forward by investigators in the past. Most of these assumptions have been based on indirect methods of prediction, rather than measurement *in situ* or through direct modelling with numerical methods that can capture the failure process during excavation.

In physical modelling, the most commonly used model simulating strata movement has been through representing the mining extraction and overlying rocks by using a similar combination of materials, such as sand/plaster/water, suitably scaled in strength and size so as to allow accurate simulation of the strata movement following coal extraction. Here numerical simulation is used to study the behaviour of mining excavation induced-failure and collapse of the mining structures.

250 Rock failure mechanisms

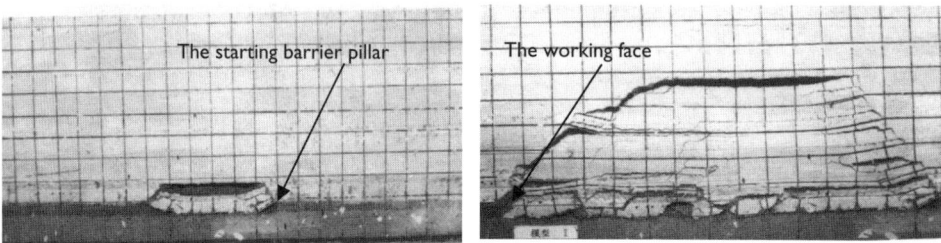

Figure 17.1 Physical model test illustrating the strata collapse behind an advancing longwall face (from Yu, 1998).

17.2 ILLUSTRATIONS OF LONGWALL MINING SIMULATIONS

To make a detailed study of the strata movement, fracture formation and development, numerical modelling studies have been conducted, as shown in Figure 17.2. The grayscale reflects the rock mass stiffness, the lighter the grayscale the stiffer the rock, and the model has layers with different mechanical properties. The model represents a mining zone 300 m long and 100 m high. A coal seam with a thickness of 6 m is located between the simulated strata, which lies 84 metres below the surface. The simulated model is partitioned into 30,000 elements and the mining procedure was simulated by cutting the coal seam from left to right at intervals of 5 m per step.

The model employs gravity loading to generate caving, fracturing and strata movement above the mining extraction. Figure 17.3 illustrates the general character of the strata movement above a caved longwall extraction, the situation being modelled corresponding to a relatively shallow mining setting, with strong overburden. Well-defined fracture lines occur at the edges of the extraction area. Fractures in the roof strata were numerically traced as mining advanced. Bed separations are prominent features exhibited in the model, and indicate the potential for further fracturing. The central region of the collapsed roof strata shows the beds have lowered in an almost intact condition near to the extraction horizon, although significant separations exist at higher horizons. The fracture lines above the edges of the longwall extraction indicate that such regions can represent flow paths for water and gas from upper horizons, thus gaining access to the working horizon. The major separation above the longwall could act as a temporary reservoir for water percolating from adjacent or overlying strata, and result in later release through interconnecting fractures to the working longwall (Whittaker and Reddish, 1989).

Large bed separations appear to be a significant feature during the early stages of longwall face extraction. Additionally, large blocks cave in up to a certain distance from the start position and thereafter cave more regularly in the form of controlled cantilever failures. The fracture initiation and development process in roof strata has been traced and it is found that there can be a periodic fracture pattern with the advancement of the mining face.

Immediate roof weighting: at the beginning of the simulation, a set-up entry of 5 m is cut along the face at the starting point of the panel. Thereafter, as the coal

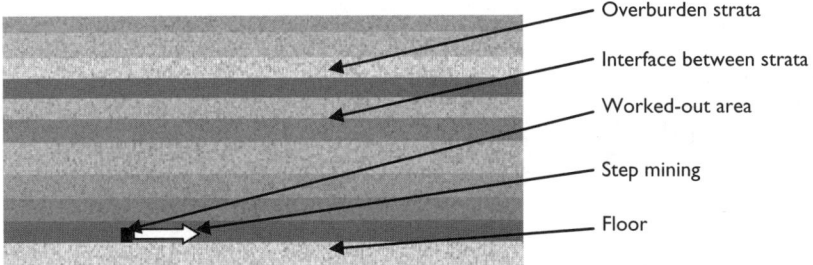

Figure 17.2 Numerical simulation model of mining and overburden strata (shading represents the strata stiffnesses). The longwall face is in a direction into the page and it advances to the right of the page.

winning operation proceeds, the goaf (the area behind the working face) containing the intact roof will be left open for a distance of around 10–20 m from the starting barrier pillar edge, depending upon the mechanical properties of the roof rock and the stress induced by the excavation. The immediate roof, which is affected by the swelling deformation caused by excavation, disintegrates from the composite main roof layer at the layer middle, and starts to bend and sag as the width of the open goaf widens up. As shown in Figure 17.3(A), cracks develop first in tension stress concentrated areas. In Figure 17.3(B), the immediate roof caves in after the sag of the rock beam above reaches its limiting value. The caving height of the immediate roof is about 2 m. Except for the case of a thick, stronger rock layer, this immediate first composite roof layer becomes detached from the face, as shown in Figure 17.3(B).

However, in mining practice, if the first composite layer does not naturally fail and cave in, then blasting must be carried out to induce its failure and hence avert the occurrence of an air blast. It must always be presumed during the selection of the capacity of powered roof supports that the bottom-most layer/composite layer will be detached during the coal mining process and hence its total dead load must be taken into account.

Main roof primary weighting: As the width of the open area behind the longwall face widens, the first composite main roof will detach from the upper composite roof layer at the middle, and start to bend and sag, as shown in Figure 17.3(B). Until that time, the main roof rock behaves like a fixed end beam. Cracks are always present inside the rock beams. More cracks are then generated and widen up during the deflection process, which may cut the layer after a certain amount of deflection and ultimately the central portion of the beam may fail. As the span increases beyond the limit, curvilinear shear cracks will develop in the void created by disintegration of the immediate roof from the first main roof composite layer. These cracks will then completely cut the first main roof composite layer and fixed end beam forming a smaller simple fixed beam.

When the face progresses to about 50 m, as shown in Figure 17.3(C), this simple ended beam loses support, falls and is broken into various sizes—depending on the height of the face, the amount of shear cracks, number of beds in the composite layers, physico-mechanical properties of the rock, etc. In general, weak to moderately stronger rock beams always break beyond the 'weighting zone' and hence no roof

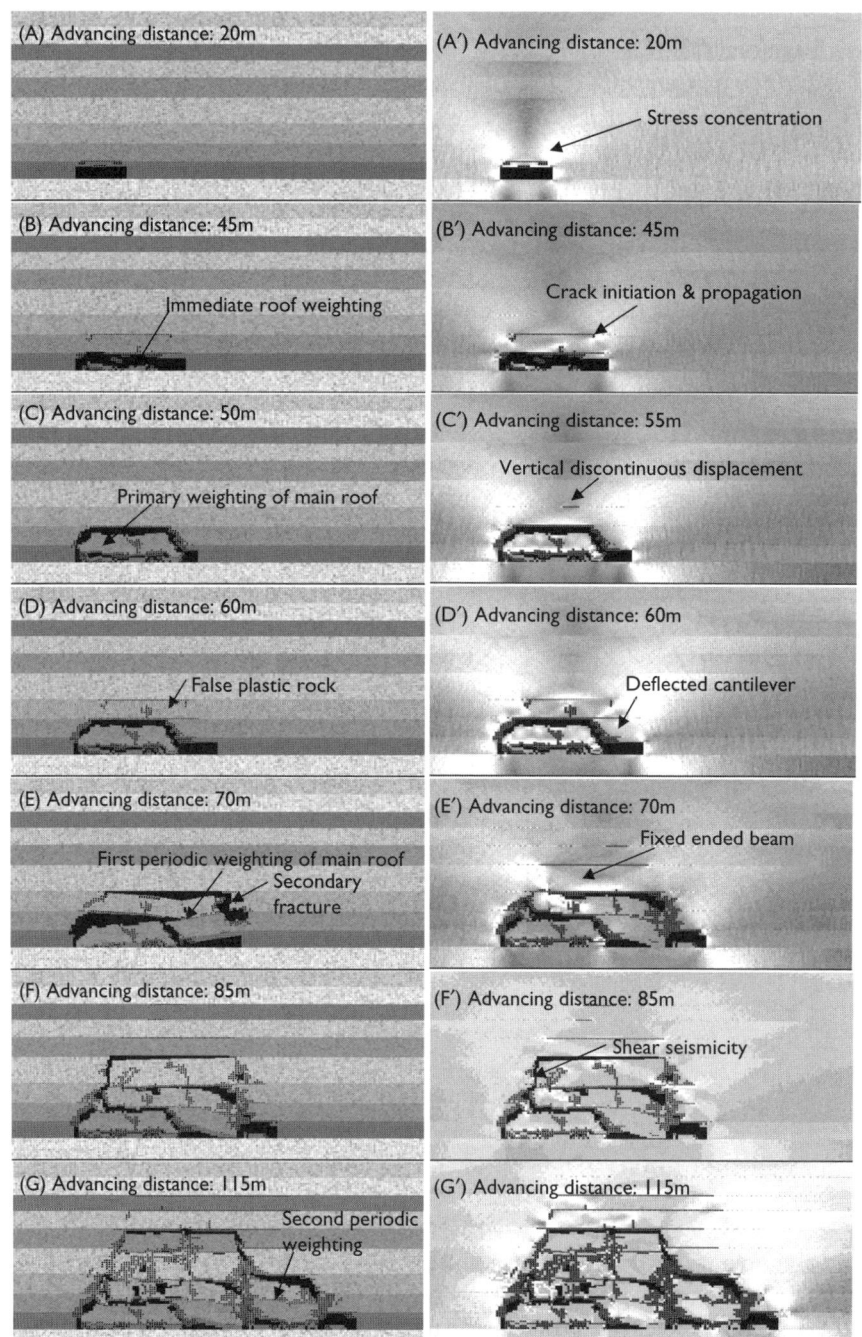

Figure 17.3 Overburden strata failure induced by longwall mining (left: elastic modulus distribution; right: shear stress distribution). The longwall face is in a direction into the page and it advances to the right of the page.

weighting is felt by the face supports. However, since the length of the cantilever after the first failure of the lowest main roof beam near to the face is longer, as shown in Figure 17.3(C), a higher weighting compared to the periodic weighting may be felt in the face, providing that detachment occurs inside the 'weighting zone'. This somewhat higher load should be taken into consideration and should be included in the safety factor during calculation of the capacity of the powered roof support.

Main roof periodic weighting: After the failure of the simple ended beam, the deflected cantilever of the first main roof composite beam over the powered supports will shorten in size gradually by the cutting of the shear cracks generated into it and will ultimately either overhang behind the powered supports or cut just behind the powered supports leaving no overhang, as shown in Figure 17.3(C). The length of the overhang depends on the mechanical properties of the rock, the thickness of the roof rock layer/composite layer, the type and nature of roof rock, the rupturing thrust by the powered supports, etc. With the work face advancing about 60 m, vertical discontinuum displacement and fracture phenomena occur in the upper roof, as shown in Figure 17.3(D). Then the upper main roof fails, deflects and at last caves in as shown in Figure 17.3(E). Because the upper roof is relatively thick, shear is the main failure mode, as shown in Figure 17.3(F).

The upper roof completes the first periodic weighting when the working face advances about 85 m. In this case, the roof is strong and the generated shear cracks are not of a perfect curvilinear shape, and also the upper portion does not bend towards the goaf but is inclined towards the face. Then, during caving, the simple ended beam may offer a slight load over the powered supports through the intact cantilever of the same rock layer/composite whose one end is fixed with the face as shown in Figure 17.3(F). The secondary periodic weighting occurs when the working face advances about 115 m. The upper roof continues its periodic movement with the working face further advancing, as shown in Figure 17.3(G). After the caving of the third layer, the fourth composite layer will fail in a similar manner, as has happened in the case of the first and second composite layers. In this way, layer after layer of superincumbent strata will be disturbed.

However, in the absence of thick stronger strata, the so-called main and periodic weighting may never be felt at the face. In that case, the failure of strata from the top of the support follows a chronological sequence. Hence the main and periodic weighting terminology is limited to site-specific conditions, i.e. it is valid for the case of thick and stronger strata being present within the caving height.

After the immediate roof caves in, tensile cracks are propagated into the first and also in the second composite layers in a non-systematic manner but in a curvilinear shape with the upper portion bending towards the goaf, as shown in Figure 17.3(B'). The curvilinear shape of the crack is caused by the heterogeneity of the rock beam. The location of the maximum axial tensile stress always lies in the lower surface at the mid-point and the upper surface at the two ends of the rock beam. However, as the extraction operation proceeds, the location of the maximum axial tensile stress moves forward continuously and the damage area switches forwards. When damage accumulates to an extreme extent with the step excavation proceeding, a crack will develop in the maximum damage area, which departs from the mid-point of the rock beam and is close to the working face as shown in Figure 17.3(C'). Then the asymmetrical failure shape comes into being, its shape being caused by step excavation.

When the work face advances about 70 m, a new phenomenon, 'secondary fracture', occurs in a quarter part of the upper surface of the second main roof close to the working face. The secondary fracture can be explained by damage theory. As the extraction operation proceeds, the main roof close to the starting point of the panel is affected by tensile stress caused by each step excavation and is damaged severely. Therefore, a crack first develops in the upper surface close to the starting point of the rock beam. However, in the upper surface of the rock beam, close to the working face, the maximum tensile stress area continuously moves forward and so does the maximum damage area with the stepped excavation working face proceeding. When the damage of the rock beam reaches the strength limit, the crack develops again in the upper surface of the rock beam, as shown in Figure 17.3(E'). The main roof repeats its failure process with the working face further advancing, as shown in Figure 17.3 from (F') to (G').

From a physical sense, there is no doubt that the thickness of the roof strata plays an important role in controlling the length of the roof's periodical collapse. Generally speaking, if the immediate roof comprises thick strata, it overhangs a large area

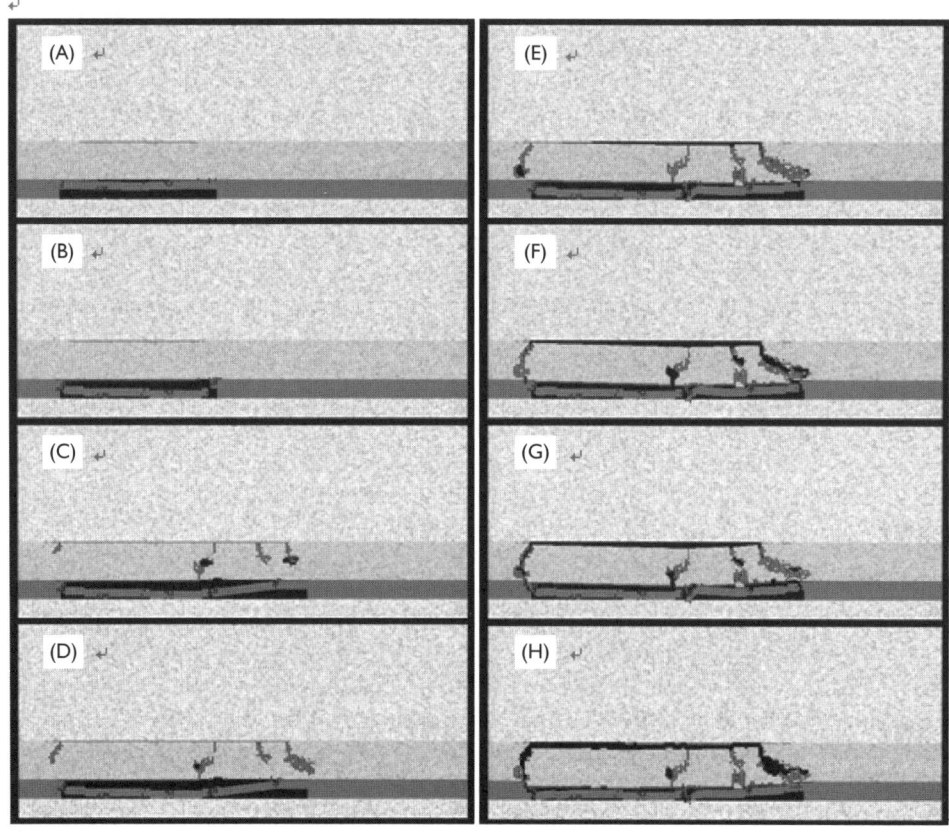

Figure 17.4 Numerically simulated result of overburden strata failure induced by mining (with a thicker immediate roof).

Rock failure induced by longwall coal mining 255

for a longer period. Also, a good roof consisting of stronger strata always overhangs a considerable distance. As shown in Figure 17.4, in the case of a thicker immediate roof strata, this distance becomes longer, which indicates that the collapse distance depends upon the thickness of the immediate roof layer and the mechanical properties of the roof rocks (Figure 17.4).

Both Figures 17.3 and 17.4 demonstrate the observed experimental phenomena of strata separation. Parting planes are generally weak zones of fractured beds of macroscopically identifiable fissile surfaces in the strata constituting the roof. The stronger strata layers are deflected and a void is always noticed in between the deflected strata layer and the top intact strata layer. Comparing Figures 17.3 and 17.4, it can be seen that the weaker strata cave in quickly with or without any distinct deflection; whereas, the stronger strata deflect first before their shear failure.

17.3 THE DALIUTA COAL MINE IN CHINA

Figure 17.5 shows a practical example from the Daliuta coal mine of Shenhua Corporation. The lower 2^{-2} coal stratum is mined by the longwall excavation method after the upper 1^{-2} coal stratum has been mined by room and pillar methods. In order to evaluate the influence of the existence of the upper worked-out area on the security of the lower coal stratum, the overburden rock strata failure and movement process are simulated. Based on the simulated results, the mining stress distribution and its appearance, the earth surface horizontal movement and vertical subsidence are analysed with the work face advancing.

Strata formations and thicknesses, etc., are shown in Figure 17.5, and were established from the study of a number of borehole lithologs. The mechanical parameters of the rock were obtained from the collected rock cores; these were tested by the standard test procedure for uniaxial compressive strength, tensile strength (by Brazilian test) and shear strength (by punch shear test as well as by triaxial testing). The mechanical

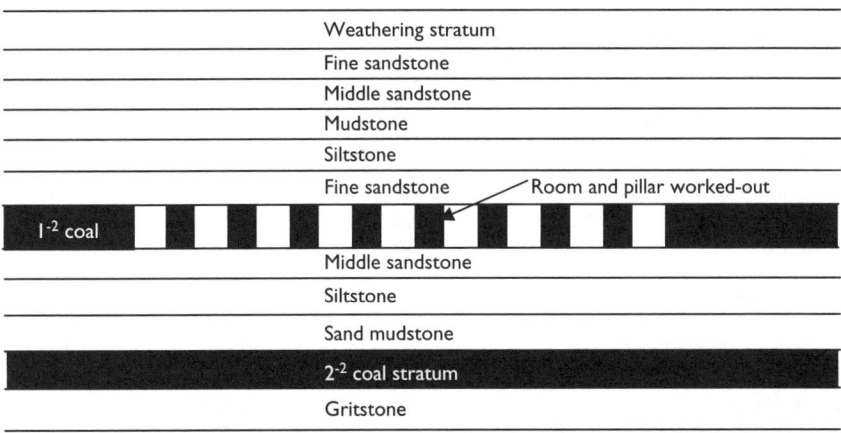

Figure 17.5 Schematic diagram of the overburden strata layout and lithology (Daliuta coal mine, Shenhua Corporation in China).

256 Rock failure mechanisms

parameters of the overburden rock strata, as shown in Table 17.1, were obtained from the field study and statistical analysis. Composite coal is mined by the longwall mining method along the coal deposit direction, with the excavated height being 4 m. The upper coal is mined by the room-and-pillar mining method, with a maximum excavation height of 7.5 m, with permanent coal pillars remaining. The distance between the two coal strata is 33 m. The advance distance is 3 m per day.

17.3.1 The strata failure process

Again, the numerical modelling shows the regular progression of the fractures which appear related to the predicted tensile strain during extraction. With the working face advancing after slotting, a crack first develops in the lower surface at the mid-point and upper surface at the two heads of the rock beam, as shown in Figure 17.6(A). When the span increases beyond the limit, the crack will propagate and eventually lead to the caving in of the immediate roof shown in Figure 17.6(B). Later, the immediate roof detaches from the face when it is exposed to the worked-out area, as shown in Figure 17.6(C).

When the extraction operation proceeds to about 51 m, the primary weighting of the main roof occurs, as shown in Figure 17.6(D). Thereafter, the first and second periodic weighting of the main roof occurs, and the height of caving is close to the upper remaining permanent coal pillar, as shown in Figure 17.6(E, F, G). From the simulated results, we know that before the caving height approaches the permanent coal pillars, the weighting of the roof stratum shows clear periodic characteristics as the working face advances. The simulated results are consistent with the results of equivalent material model tests and *in situ* monitoring.

When the caving height is close to the upper remaining permanent coal pillars, because of the specific rock strata structure, the weighting of the main roof loses its periodic characteristic. Since the rock stratum below the pillar is very strong and sufficiently thick, it offers sufficient resistance against itself breaking, as shown in Figure 17.6(H). The thick rock stratum is termed the key stratum. With the longwall working face advancing, tensile cracks are generated and propagated into the first,

Table 17.1 Physical and mechanical parameters of the overburden rock strata.

No.	Composite strata lithology	Elastic modulus (GPa)	Compressive strength (MPa)	Weight ($10^{-5} N/mm^3$)	Thickness (m)
1	Weathered stratum				7
2	Fine sandstone	8	50	2.56	8
3	Middle sandstone	10	60	2.65	8
4	Mudstone	2	36	2.38	7
5	Siltstone	4	45	2.42	6
6	Fine sandstone	8	50	2.56	10
7	Coal	1	33	1.4	8
8	Middle sandstone	8	60	2.56	12
9	Siltstone	6	40	2.5	10
10	Sand mudstone	4	20	2.42	9
11	Mudstone	1.5	5	2.38	2
12	Coal	1	33	1.4	4
13	Gritstone	10	100	2.5	10

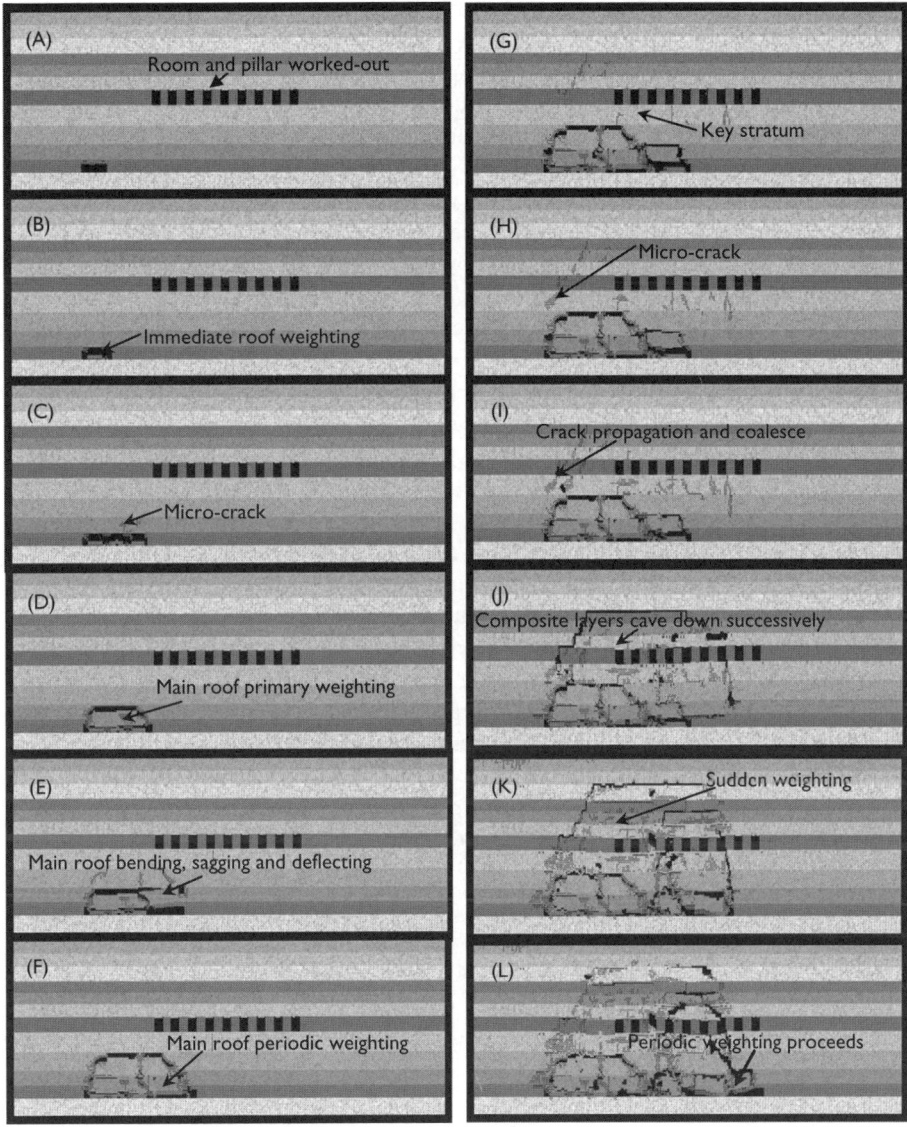

Figure 17.6 Numerical simulation of the overburden failure process when a multi-coal stratum is mined.

second or third composite layer under the key stratum in a non-systematic manner, but in a curvilinear shape with the key stratum bending towards the goaf, as shown in Figure 17.6(G, H, I). The layers next to the key stratum do not play any significant role in offering load over the powered supports.

The key stratum is vital to the whole rock mass stability. Once the key stratum caves in, all the composite layers above the key stratum will cave down successively one after another within a short time span for a considerable height, as shown in

258 Rock failure mechanisms

Figure 17.6(J). In this model, the caving height is about 43 m. From Figure 17.6(J), we observe that a considerable length of the immediate roof is compressed into failure before the lower coal stratum is mined. A sudden weighting may be felt in the face, as shown in Figure 17.6(K). The phenomenon is caused by the specific overburden rock strata structure, which belongs to 'secondary weighting' accompanied by a large energy release. An earthquake may be induced because of the large rock movement. The 'secondary weighting' has a great influence on the lower longwall working face and threatens its security Therefore, in the practical excavation process, extra support measures should be added. Thereafter, the overburden rock strata caves in periodically with the work face advancing, as shown in Figure 17.6(L).

A further feature shown by Figure 17.6 is the ability of a strong overburden at the surface to cause bridging across the lower caved beds overlying the room and pillar mining and the longwall extraction. The capacity for bridging of the strata in this manner is dependent upon the strength and general competence of the rocks near to the surface, in addition to the width of the extracted region. Figure 17.6 shows tensile fractures at the surface.

17.3.2 Pillar stresses

Figure 17.7 shows the stress distribution in the left boundary of the coal pillar with the various advancing spans of the working face: $L = 18$ m, 48 m, 70 m and 102 m. The x-axis represents the length from the left boundary of the model to the starting barrier pillar. The y-axis represents the stress passed on to the coal by the overburden rock strata. With the extraction operation proceeding, the stress in the starting barrier pillar increases because of the stress concentration and the force passes onto the coal stratum by the upper roof rock beams. Before the stress reaches the limit strength of the rock beam, the whole coal stratum from the boundary of the model to the starting barrier pillar remains in an elastic compressive state and the distribution of

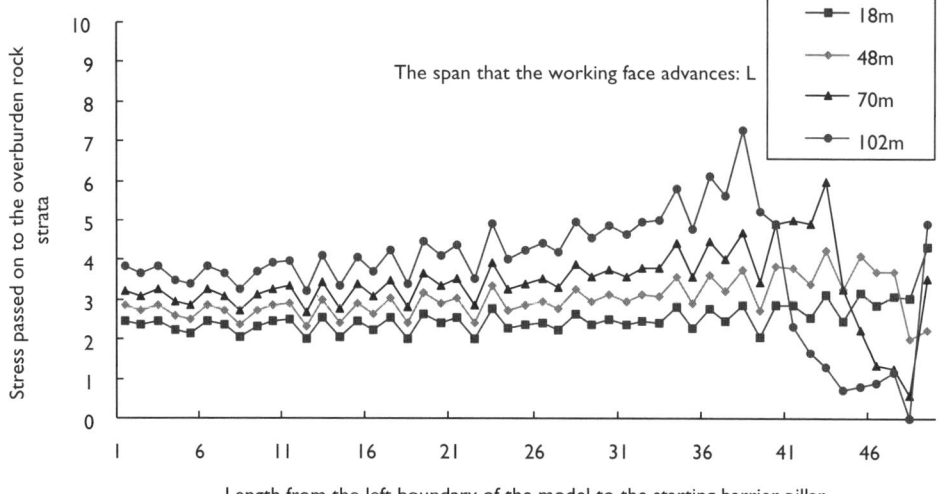

Figure 17.7 Support stress within the pillars of the boundary coal pillar.

support stress follows a decreasing curve, whose peak lies in the coal wall, as shown in Figure 17.7 ($L = 18$ m).

With the working face further advancing, the stress in the coal stratum close to the starting barrier pillar will reach the strength limit and enter the plastic state. Thereafter, the support pressure of the coal stratum close to the starting barrier pillar will decrease and the peak point of the support stress distribution curve will move to the inner portion of the coal stratum, as shown in Figure 17.7 (for $L = 48$ m). As the extraction operation proceeds, the peak point of the support stress distribution curve will further depart from the starting barrier pillar, shown in Figure 17.7 ($L = 70$ m). After the coal stratum enters the plastic state, the distribution of support stress in the coal stratum close to the starting barrier pillar will be separated into two parts, i.e. the 'inner stress field', which is decided by the fractured rock beam's weight between the fracture line and the starting barrier, and the 'outer stress field', which is decided by the whole overburden rock strata weight outside the fracture line, as shown in Figure 17.7 ($L = 102$ m). After the two stress fields come into being, a stress peak develops in the reverse direction against the boundary line.

Figure 17.8 indicates the support stress in the coal stratum before and after the sudden caving of the main roof. The x-axis is the length of the simulated model from the left boundary to the right boundary. The y-axis is the support stress in the coal stratum. Because of the existence of the key stratum, which will offer sufficient resistance against itself breaking, the weight of the upper strata after failure will act on the key stratum in the form of a direct load. Thereafter, the load will pass on to the coal stratum and the support by the overburden rock strata. Therefore, higher and higher stress will be concentrated on the working face as shown in Figure 17.8(A) with the span of the goaf increasing.

After the failure of the key stratum, the load on it will be gradually held by the coal stratum and support. In this case, the coal stratum will be compressed to failure before it is mined, as shown in Figure 17.8(B). When the key stratum caves in, all the composite layers in a specific area will cave down successively one after another

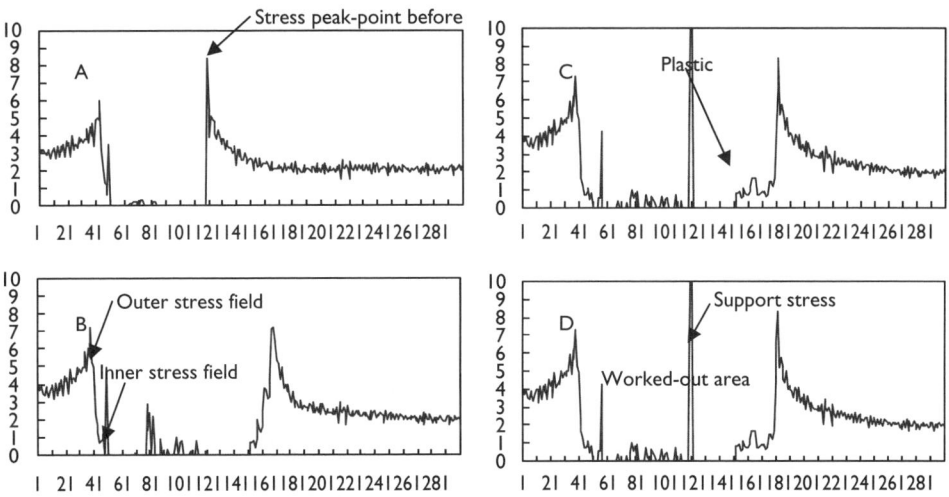

Figure 17.8 Stresses (MPa) in the coal pillar before and after the large weighting.

260 Rock failure mechanisms

within a short time span for a considerable height, depending upon the type, nature and strength of the immediate roof rock mass. At this time, the load on the key stratum acts completely on the coal stratum. A larger plastic area comes into being, as shown in Figure 17.8(C). After the sudden caving occurs, the weight of the large caving overburden rock strata will be held by the goaf. Therefore, the goaf in the worked-out area is subject to a larger stress, as shown in Figure 17.8(D).

Figure 17.9 shows the surface horizontal movement curve and Figure 17.10 shows the surface vertical subsidence curve with the different advancing spans. From field

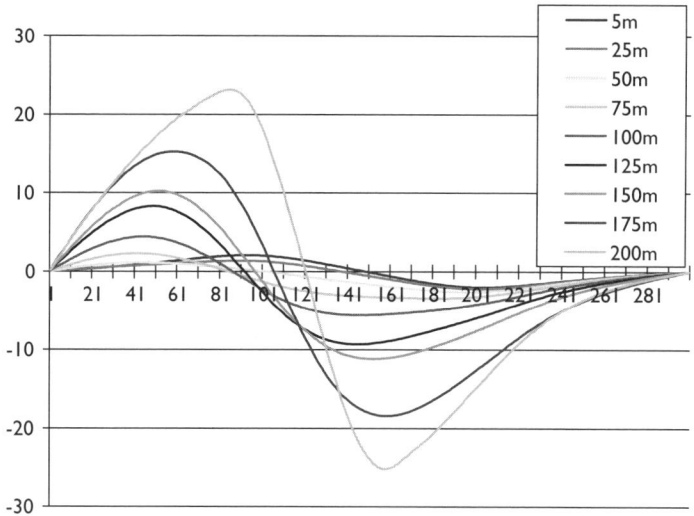

Figure 17.9 Horizontal displacements (mm) of the surface with the working face advancing. The y-axis is the subsidence in mm and the different curves represent the amount of horizontal displacement corresponding to the different advance distances of the working face.

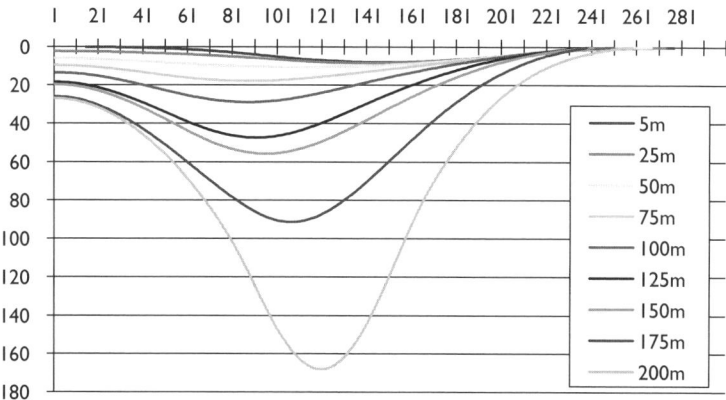

Figure 17.10 Vertical subsidence (mm) of the surface with the working face advancing. The y-axis is the subsidence in mm and the different curves represent the amount of subsidence corresponding to the different advance distances of the working face.

studies, it is confirmed that the bulk volume of the caved rock never fills the total void near the face, nor is the deflected rock layer/composite layer likely to touch the caved rock near the face. All the field observations confirmed that the ultimate bulk volume fills around 45–60% of the total caving height and the other 40–55% caving height remains void causing surface subsidence. Based on the field study and calculations, it is found that the ultimate *in situ* bulking factor of coal measure strata at a longwall face is less than 1.05 and its value decreases with the increase of caving height of the superincumbent strata. This is because the caved strata do not have much space to extend themselves. Besides this, at higher caving heights, the superincumbent strata only fractures and subsides as a larger mass in sequence. Therefore the surface subsidence is inevitable over the caved longwall workings of the coal seam(s), which are situated deeper. However, the time lag to disturb totally the superincumbent strata after finishing a longwall panel depends upon the geo-technical parameters of the roof rocks, viz. types, nature, strength, thickness and dip of the strata, geological disturbances, water condition, superincumbent strata pressure, etc.

In this model, when the lower longwall working face advances a small span, the horizontal movement and the vertical subsidence of the earth surface is mainly influenced by the worked-out room-and-pillar area. As the extraction operation proceeds, both the horizontal movement and the vertical subsidence at the earth's surface are mainly influenced by the lower longwall work face, which can be observed from the symmetrical point of the horizontal displacement curve and the vertical subsidence curve, as shown in Figures 17.9 and 17.10. It is shown by the simulated result that the horizontal movement and vertical subsidence of the surface increase continuously with the work face advancing. Though the upper coal pillars and the key stratum below them change the overburden rock structure and caving characteristics, symmetry is shown in the horizontal movement and vertical subsidence curve.

Chapter 18

Gas outbursts in coal mines

18.1 INTRODUCTION

Violent ejections of coal and gas from the working coal seam have plagued underground mining operations for over a century. These phenomena are referred to as instantaneous outbursts and have occurred in virtually all the major coal producing countries and have been the cause of major disasters in the world mining industry. The coal and gas outbursts range in size from a few tonnes to thousands of tonnes of coal with corresponding gas volumes from tens of cubic metres to hundreds of thousands of cubic metres. In fact, coal and gas outbursts can release over one million cubic feet of gas, fractured and even pulverised coal and rock (Figure 18.1). Thus, the occurrence of coal and gas outbursts in coal mines and caverns has posed great potential threat to facility operators and has challenged researchers from the rock mechanics and rock engineering community. In the last century and a half, since the first reported coal and gas outburst occurred in the Issac Colliery, Loire Coal Field, France, in 1843 (Lama and Bodziony, 1998), it is estimated that as many as 30,000 outbursts have occurred in the world coal mining industry. The majority of outbursts, more than one third of the total, have occurred in China.

These disastrous mine outbursts have resulted in loss of equipment, production time, even entire mines, and sometimes the lives of numerous miners all over the world. For instance, on October 20, 2004 at the Daping coal mine in Xinmi city, Henan province, China, 148 fatalities resulted from such an outburst. A large portion of these was due to secondary factors, such as the following gas explosion, suffocation, and poisoning. Similar disasters have befallen mines in many other countries. These occurrences have forced mining leaders and researchers to develop an understanding of the complex phenomenon, and procedures to minimise the effect of outbursts or to eliminate them completely. Some safety procedures that have been adopted do lead to reduced production rates.

For almost half a century now, considerable research attention has been paid to this complex problem (Beamish and Crosdale, 1998). The preliminary investigations concerning the coal and gas outburst mechanism through *in situ* observation, physical and theoretical studies, and numerical modelling, have been made to mitigate the coal and gas outburst hazard during the last decades. Some empirical hypotheses, criteria and analytical models have been proposed for understanding, analysis and prediction of the coal and gas outburst conditions. Kidybinski (1980) took into consideration the three components of gas content and flow, stress, and coal failure and

Figure 18.1 Coal and gas outburst induced by underground mining.

proposed the presence of three zones in the coal seam ahead of the mining operations starting at the coal face: 1) protection/degassed zone, 2) high gas pressure/active zone and 3) abutment pressure zone.

Within this model, three fundamental conditions are assumed to be met for an outburst to occur: 1) failure of the coal in compression within the active zone; 2) penetration of a hole through the protection zone; and 3) fluidised bed outflow of the products from the outburst. Gray (1980) considered two gas-initiated-failure mechanisms to exist: either tensile failure of unconfined coal or piping of sheared material. Paterson (1986) took the general view that, when gas is released from coal, there are body forces on the coal equal to the pressure gradients of the flowing gas. His models therefore were based on the fundamental assumption that an outburst is the structural failure of coal due to excess stress resulting from these body forces. A model proposed by Litwiniszyn (1985) was based on the gas existing in a condensed state within the coal. When a shock wave passes through the coal, a phase transformation occurs of the substance into a gaseous state. This sudden creation of gas causes the skeleton of the medium to be destroyed and an outburst is initiated. Support for this model is found in the following observations: 1) sometimes 'bumps' and instantaneous outbursts occur together, and some 'bumps' are regarded as initiation of instantaneous outbursts, and 2) in hand-working, especially without the noise of machinery, successive 'knocks' in the coal were often precursors to an instantaneous outburst (Hargraves, 1983). However, Paterson (1986) identified several flaws in this model, in particular cause and effect; where do the shock waves originate? Thermodynamic descriptions have also been proposed for outburst modelling (Jagiello *et al.*, 1992). Williams and Weissmann (1995) used a schematic of an outburst in frequently encountered Australian conditions to discuss gas content thresholds for outbursts. They placed emphasis on a gas pressure gradient existing ahead of the working face. However, they also believed "the most important parameter is gas desorption rate, in conjunction with the gas pressure gradient ahead of the face".

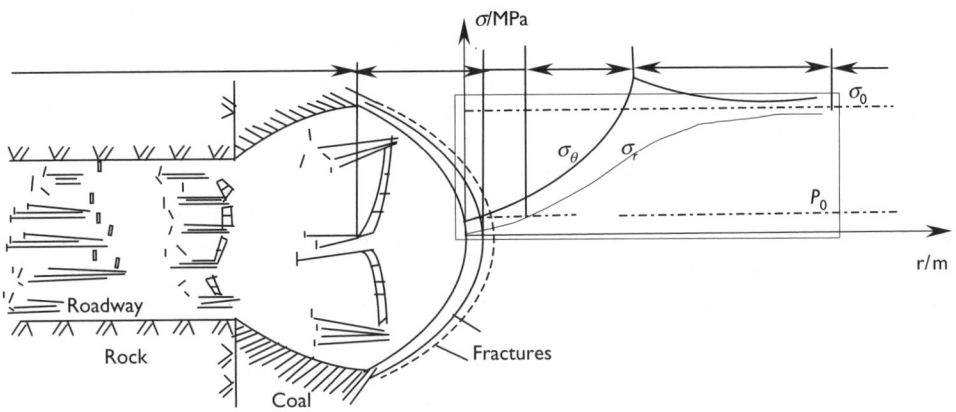

Figure 18.2 'Spherical shell losing stability' model during outbursts (after Jiang and Yu, 1998).

Jiang and Yu (1998), based on many laboratory tests, presented the 'spherical shell losing stability' model during outbursts (as shown in Figure 18.2). They believed that the outburst process consists of six phases, viz., 1) intact stress phase, 2) stress concentration phase, or abutment pressure phase, 3) coal crushed by rock stress, 4) coal split by gas pressure, 5) expulsion of coal and gas due to spherical shell losing stability, and 6) movement of coal and gas desorption.

In addition, many previous studies have considered tectonic deformation and the micro-structure of the deformed coal to be important factors influencing outburst occurrence. Farmer and Pooley (1967) found that outbursts only occur in districts subject to severe tectonic movement—hence, their association in many places with anthracite—and in association with such deformational and depositional structures as folds, faults, rolls and slips and, in particular, with rapid fluctuations in the seam thickness. Shepherd *et al.* (1981) reported on outburst occurrences in Australia, North America, Europe, and Asia, and found that probably over 90% of significant outbursts have been concentrated in the narrow strongly deformed zones along the axes of structures such as asymmetrical anticlines, the hinge zones of recumbent folds, and the intensely deformed zones of strike-slip, reverse, and normal faults. These narrow deformed zones, whether in mesoscopic or mine-scale geological structures form the loci for stress and gas concentrations.

Similar studies in China revealed that outbursts have nearly always occurred in long, narrow outburst zones along the intensely deformed zones of strike-slip, reverse or normal faults, within which coal has been physically altered into cataclastic, granular, or mylonitic micro-structures (Peng, 1990). The other occurrences are associated with bedding-plane faults and intense folds, which may produce these micro-structures in broader zones. In either case, the outburst-prone zones generally cover no more than 20% to 30% of the mine area. Some fault zones do not exhibit altered micro-structure, and these are not prone to outburst. Thus, the presence of these altered micro-structures is considered as the first essential factor for outburst occurrence, and outburst-prone districts could be predicted by studying the spatial distribution

266 Rock failure mechanisms

of altered coal and geological structures. It has also been found that outburst danger increases with the intensity of deformation and alteration of the coal micro-structure. Many studies have compared coal actually expelled from an outburst cavity to coal *in situ* with similar micro-structure, based on their physical and morphological characteristics. To date no significant difference has been found (Evans and Brown, 1973; Cao *et al.*, 2001).

Naturally, a deeper understanding of the outburst mechanism and reliable methods for the prediction of outbursts must be not only based upon long years of practical experience in mines, but also on scientific research and experimentation. Despite extensive research about the violent coal and gas outbursts which have occurred in coal mines, surprisingly little progress have been achieved in the past 150 years towards understanding or prediction. In particular, a quantitative model that describes the progressive failure process, as well as the violent outbursts process in coal mines, has not appeared. It is the aim of this Chapter to present such a model and to show how the model explains the observations associated with outbursts.

18.2 OUTBURSTS INDUCED BY CROSS-CUTTING FROM ROCK TO COAL SEAM

The numerical model shown in Figure 18.3 is designed to simulate the instantaneous outbursts which occurred in the course of cross-cutting induced by mining. In the model, the gassy soft coal seam is enclosed by an impermeable hard roof and floor rock. Moreover, a layer of thick hard rock acts as a protective screen ahead of the coal seam. The layer of hard rock is instantaneously opened by drilling and the coal seam behind the protective screen is therefore exposed in the course of this cross-cutting. The simulation model is discretised into a 150×200 mesh (30,000 elements). The gas

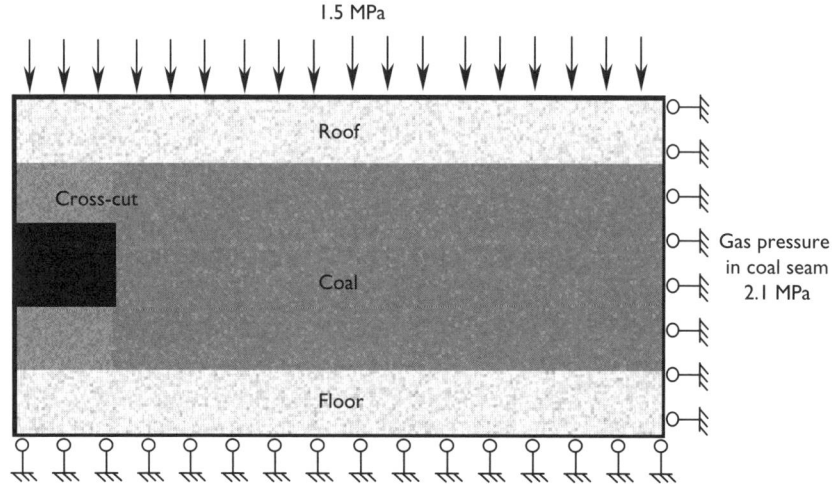

Figure 18.3 Numerical mechanical and seepage model of coal and gas outbursts induced by 'cross-cutting' penetration.

pressure saturated in the coal seam is 2.1 MPa and the Young's modulus and compressive strength of the coal are 10 GPa and 15 MPa, respectively. In addition, the Young's moduli and strengths of the rock roof and floor far exceed those of the coal seam.

The mechanical and seepage parameters in the numerical model are presented in Table 18.1.

Figures 18.4a and b show the cross-cutting induced instantaneous outbursts and the associated stress fields distributions in rock roof, floor and gassy coal seam. Comparing Figures 18.2 and 18.4, it can be seen that the numerically simulated instantaneous outbursts agree well with the 'spherical shell losing stability' model.

It can be seen from the numerically simulated results induced by cross-cutting shown in Figure 18.4 that the whole process of coal and gas outbursts can be divided into four stages:

1   *Stress concentration stage.* At the beginning of cross-cutting, the loads from the upper rock strata are mostly carried on the freshly exposed coal due to stress concentrations. The stress in the coal is not uniformly distributed at the mesoscopic scale because of the heterogeneity of the materials.
2   *Coal/rock fracture and splitting induced by rock stress.* Micro-fractures in coals are predominant under the abutment stress in this stage. The mechanical properties of coal progressively degrade due to the effect of the stress concentration and creep, as well as the three-dimensional stress state in the coal near the working face gradually transforming into a two-dimensional stress state. As a result, fracturing and splitting parallel to the free exposed face first occurs in coal near the working face. In the course of splitting, the stress in the coal near the coal face decreases and the peak stress location gradually moves away from the coalface and into the deep coal, with release of the elastic energy stored in the coal. It is noticeable that a cluster of cracks begins to form along with the transfer of the stress peak.
3   *Crack propagation driven by gas pressure.* High-pressurised gas saturated in the coal seam gushes into the 'gas way' and quickly and violently splits the fractured coal. The effect of high gas pressure eventually leads to the formation of the 'gas way' during the propagation and coalescence of clusters of cracks.
4   *Ejection of coal induced by gas pressure, i.e. outbursts.* During the process of crack propagation and coalescence induced by gas pressure, the cracks volumetrically

Table 18.1 Mechanical and seepage parameters used in the numerical model.

Mechanical and seepage parameters	Coal seam	Roof and floor	Cross-cut
Homogeneity index,	2	10	2
Mean elastic modulus, E_0 (GPa)	5	50	10
Mean compressive strength, σ_0 (MPa)	100	300	150
Internal friction angle, ϕ (°)	30	32	30
Bulk weight (kg/m³)	2.7	1.4	2.0
Ratio of UCS to UTS	20	10	10
Poisson's ratio, v	0.3	0.25	0.3
Gas permeability, λ, m²/(MPa² · d)	0.1	0.001	0.01
Coefficient of gas content, A	2	0.1	0.1
Coefficient of pore pressure	0.5	0.01	0.1

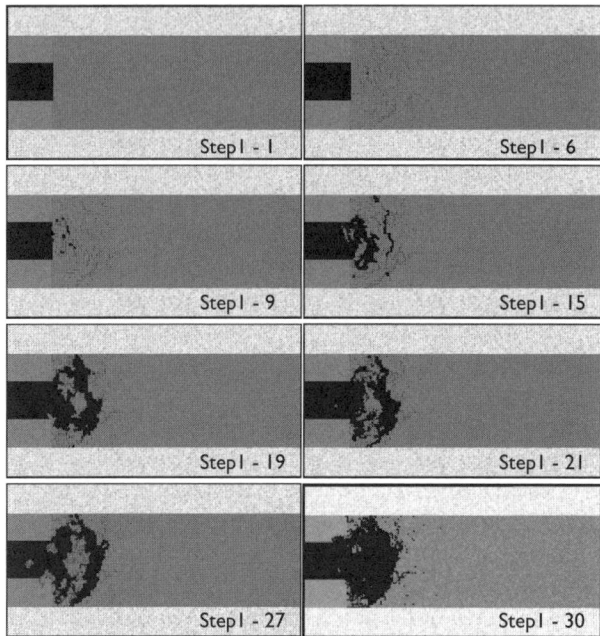

(a) Failure pattern (element stiffnesses and fracturing)

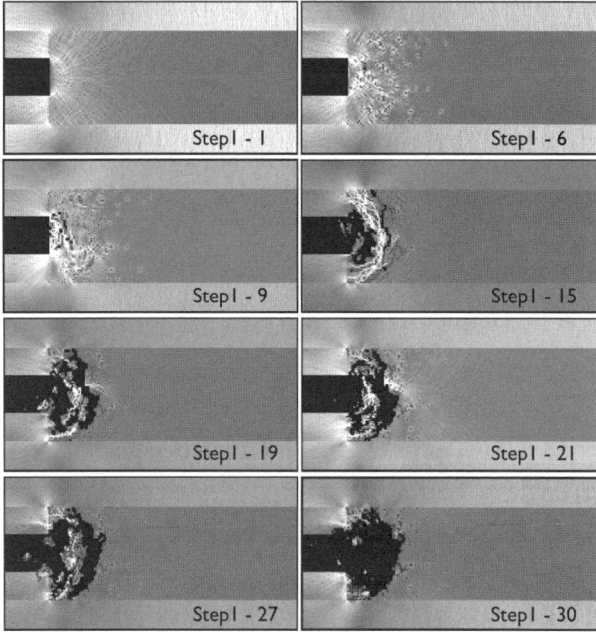

(b) Maximum shear stress

Figure 18.4 Numerically simulated coal and gas instantaneous outbursts process and shear stress distributions during instantaneous outbursts. Legend: e.g. Step 1–9 stands for the ninth calculated failure step in the first time step for modelling the outbursts process.

expand and the gas gushes into the cracks. There is a great gas pressure gradient because the gas pressure saturated in the cracks is far beyond the pressure in the freshly exposed coalface, which causes the ejection of the fractured and splitting coal, i.e. coal and gas outbursts.

It is seen from Figure 18.4 that the outburst cavity is 'pear-shaped', which agrees well with experimentally obtained results (Li, 1995).

The numerical simulated results reveal that *in situ* stress, gas pressure and the physico-mechanical properties of coal and rock are the main contributing factors affecting coal and gas outbursts. In addition, numerical simulated results not only trace the initiation, propagation and coalescence of cracks in the coal, but also present the associated evolution of the stress field in the coal seam and the roof and floor of the rock strata, i.e., the stress redistribution in the coal seam and rock roof and floor at every stage.

18.3 OUTBURST AS THE WORKING FACE APPROACHES HIGH METHANE PRESSURE IN THE COAL SEAM

As mentioned by Li (2001), a small pocket of highly deformed coal has been known to be one of the main factors which influences the locations of gas outbursts in underground mines. Figure 18.5 shows another numerical simulation with high pressure methane inclusions in the coal seam. Two soft coal inclusions containing high pressure gas are contained in the coal seam.

Usually, the inclusions in coal seams are fractured weakened zones that act as reservoirs for gas that is sealed above and below by hard coals. The zones sometimes contain large amounts of coal powder and consequently can be considered as

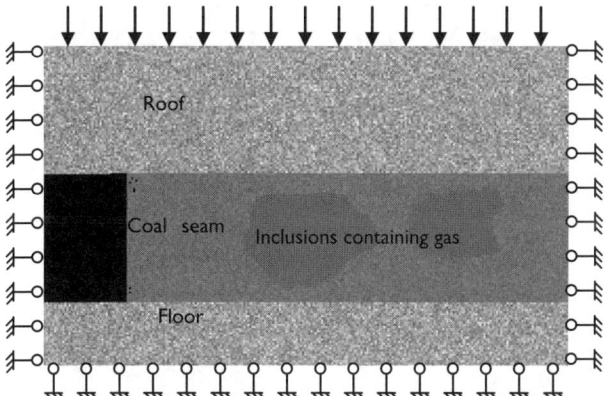

Figure 18.5 Simulation model for a coal seam with gas-containing inclusions. (The grayscale represents the relative Young's moduli for the elements, brighter elements having a higher modulus.)

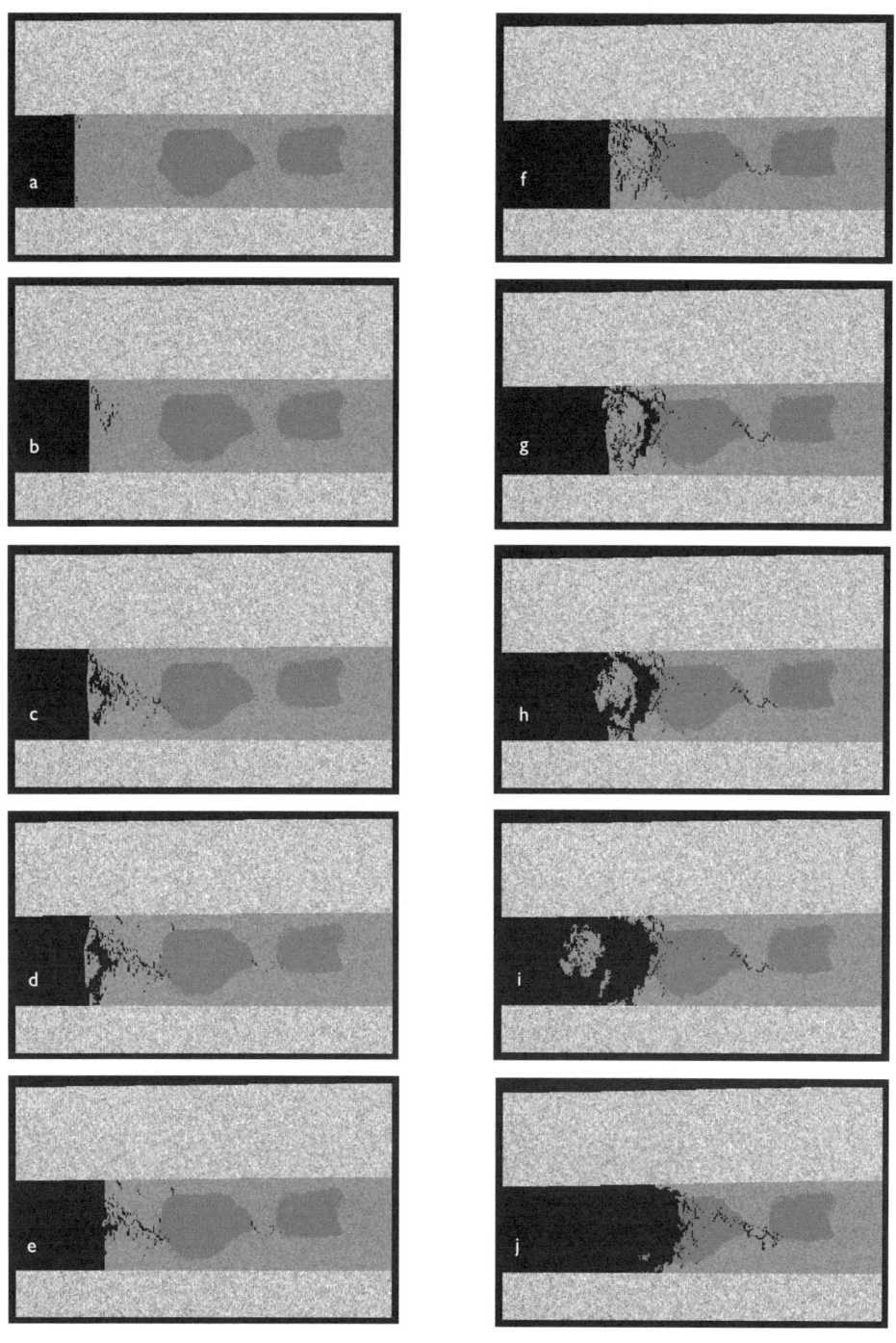

Figure 18.6 Simulated outburst when the gas-containing inclusions are approached during progressive mining (Shading indicates relative stiffnesses of the elements).

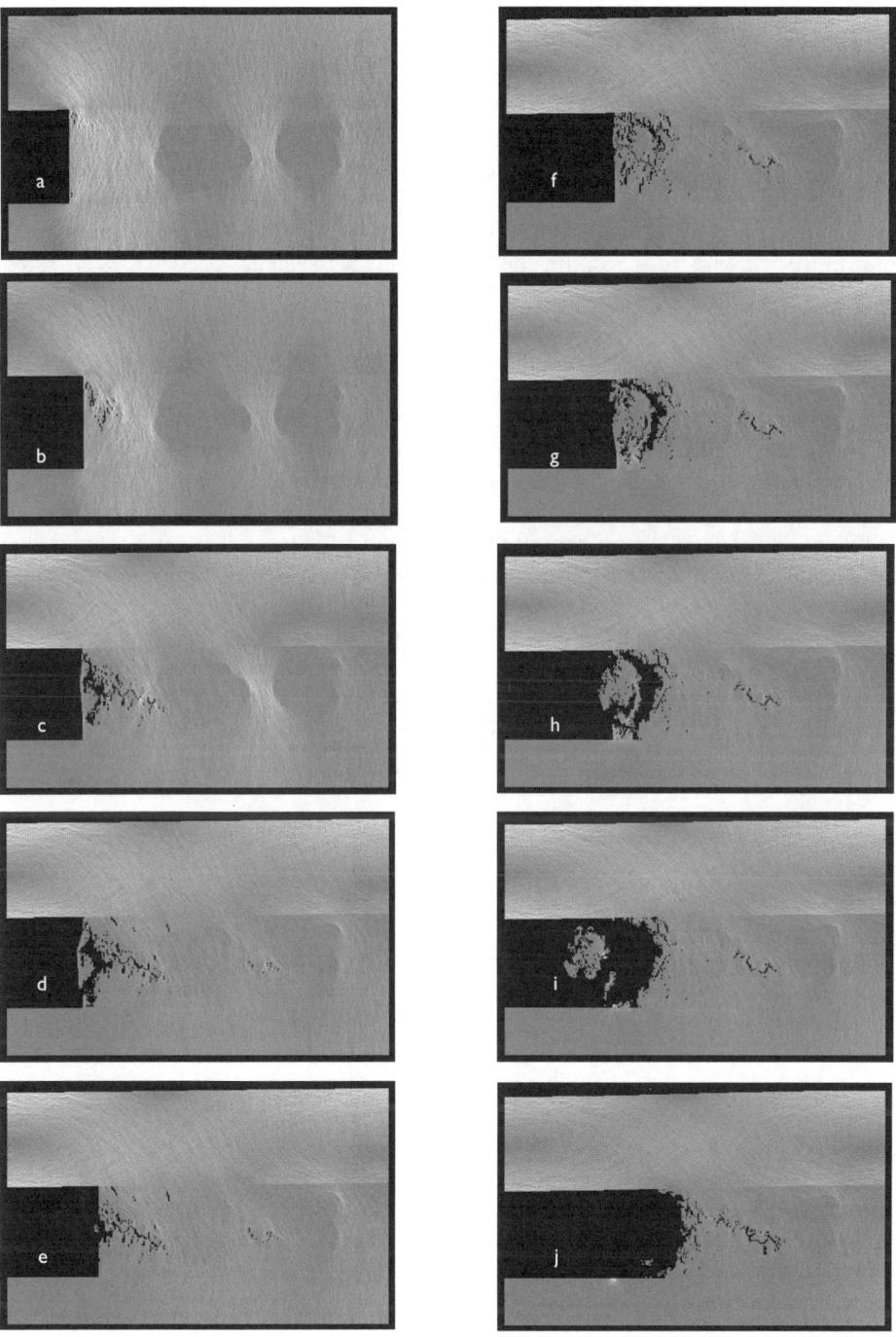

Figure 18.7 Associated evolution of the stress field (Figure 18.6) in the coal seam and the roof and floor rock strata. Shading indicates the relative shear stresses.

a potential outburst risk. In this model, the coal/rock is assumed to be impermeable, but the gas is allowed to flow instantaneously with full pressure to the initiated fracture surfaces. No gas pressure exists inside the isolated micro-fractures unless a connection is established by new micro-fractures. The input parameters for the modelling study include gas pressure, coal/rock mechanical parameters (such as Young's modulus and strength), and geometry parameters. Figure 18.6 shows the development of the numerically simulated outburst when the gas-contained inclusions are approached during the progressive mining. Figure 18.7 shows the associated evolution of the stress field in the coal seam and the roof and floor rock strata. The results indicate the importance of taking into account the interactions between gas pressure, *in situ* stress and coal/rock strength.

From the characteristics of outbursts and outburst coals, it appears that highly stressed and fractured coals are a prerequisite for outbursts. Figure 18.7 shows that mining stresses play an important role in triggering the outbursts. When the peak stress associated with a mining operation moves to the region of a high methane pressure area (inclusion) located ahead of the working face (Figure 18.7), newly broken fractures will develop with an audio emission (release of elastic energy) due to overlapping mining stress and enhanced methane pressure in the new fractures, and consequently many fractures rapidly link through fracture coalescence. Under these circumstances, the gas pressure moves through the fractures and migrates rapidly to the working face where the pressure is relatively low. Thus, whilst the basic physical process can be postulated from first principles, the advantage of the simulation is that the whole process can be studied in detail and the relative importance of the constituent parameters assessed. This then leads to a coherent strategy for evaluating various coal mine locations for their gas and coal outburst potential.

Chapter 19

Particle breakage and comminution

19.1 INTRODUCTION

Mechanical crushing, the breaking or grinding up of a material to form smaller particles, i.e. comminution, is widely used in both the aggregate-producing industry and the mineral industry. The purpose of crushing is to reduce the particle size of rock materials or to liberate valuable minerals from ores. However, although widely used, mechanical crushing is a complex process that is still not fully understood. At its most fundamental level, all industrial mechanical crushing involves the breakage of individual particles through contact with other particles or with the grinding media, or with the solid walls of the mill.

Two breakage modes have been identified in mechanical crushing: single-particle and inter-particle breakage. Single-particle breakage occurs when the distance between the chamber walls is equal to or smaller than the particle size, which is relatively simple. Inter-particle breakage occurs when a particle has contact points shared with other surrounding particles. The understanding of single particle breakage is based on the knowledge obtained through indirect tensile strength tests of particles (Chapter 3). A single particle is subjected to one of the four instant loading conditions irrespective of whether it is loaded directly by the rollers or by other mineral particles: 1) point-to-point loading, 2) plane-to-plane loading, 3) point-to-plane loading and 4) multi-point loading.

Under point-to-point loading conditions, the particle is loaded between two points. In fact, the failure mode for particles loaded between two points is rather similar to that induced in the conventional Brazilian test (Chapter 3). Under plane-to-plane loading conditions, the particle is loaded between two approximately parallel planes. The plane-to-plane loading is somewhat similar to the conventional uniaxial compression test in rock mechanics (Chapter 4). Under point-to-plane loading conditions, the particle is loaded in a mixture of point-to-point and plane-to-plane loading. Multi-point loading conditions (especially the three-point loading condition) are similar to the conventional three-point bending test used to obtain the Mode I fracture toughness in fracture mechanics.

19.2 SINGLE PARTICLE BREAKAGE

19.2.1 Breakage of single particle under diametral loading without confinement

An important step towards understanding particle crushing is to first understand the breakage of a single particle subjected to simple loading conditions. The numerical simulation of the Brazilian test shown in Chapter 3 is a good start; however, in the crushing industry, the particles are of irregular shapes as well as irregular sizes and the breakage behaviour of a single particle depends on its size, shape and the loading conditions. A better understanding of the breakage behaviour and the strength characteristics of such particles calls, therefore, for simulating particles with irregular shape, size and under different loading conditions.

A 2-D irregular shaped particle model is established for the simulation, as shown in Figure 19.1. As in the numerical Brazilian test, the model (including the particle and the loading platens) is discretised into 36,000 elements with randomly distributed mechanical properties following Weibull's distribution. All other mechanical parameters are the same as in the Brazilian test simulation.

To show the complex stress field in such an irregularly shaped particle, the fringe patterns for a homogeneous particle with the same irregular shape are given in Figure 19.2. Then, the normalised stress distributions, σ_x and σ_y, along the loading axis inside the heterogeneous particle are shown in Figure 19.3.

The influence of the material heterogeneity on the stress distribution in the particle is illustrated clearly in Figure 19.3. The first impression of these results is that the primary reason for the simulated failure mode of the irregular shaped particle is the contact condition between the loading platens and the particle. Due to the irregular shape of the particle, the contact area between the lower platen and the particle is larger than the upper one. As a result, the distributed stresses in the vicinity of

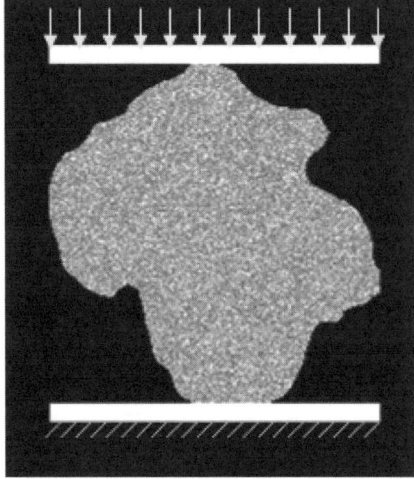

Figure 19.1 2-D irregularly shaped particle model and the loading conditions. Shading indicates the variation in the Young's moduli of the elements.

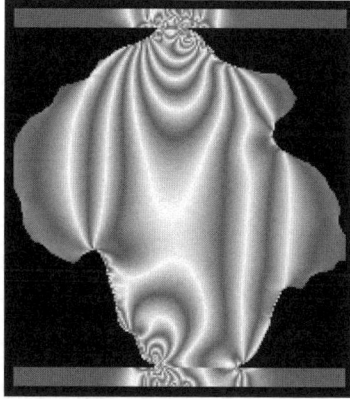

Figure 19.2 The shear stress fringe contours in a homogeneous particle subjected to diametral loading without confinement.

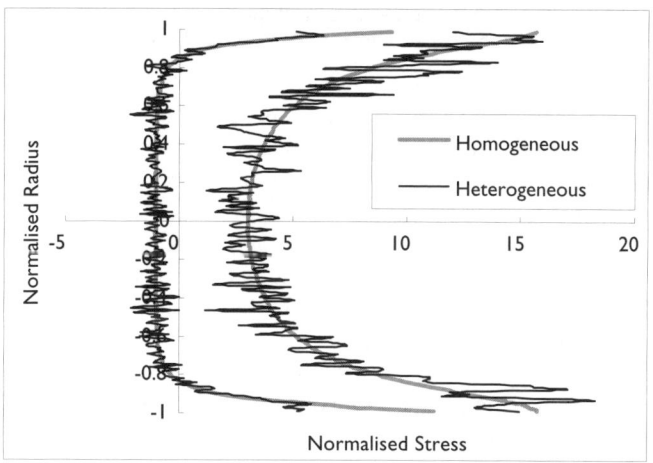

Figure 19.3 The normalised stress distributions, σ_x (compression positive, acting horizontally and to the left in the Figure) and σ_y (right), along the loading axis inside the irregularly shaped particle (both homogeneous and heterogeneous) under diametral loading without confinement.

the upper contact point are somewhat higher than those in the lower region, as can be detected in Figure 19.3. This will undoubtedly result in a somewhat different fracture behaviour compared to the Brazilian disc.

Figure 19.4 shows plots of the load and the released energy during failure versus the respective applied displacements. Figures 19.5 and 19.6 illustrate a sequence of deformed configurations corresponding to the steps shown in Figure 19.4. Initially, the particle is elastic and exhibits a stiff and stable response but in which the particle deforms essentially non-uniformly due to the different contact conditions between the platens and the irregularly shaped particle. Due to the resultant asymmetrical stress field, the crack does not originate from the centre of the particle but from a location between the upper loading

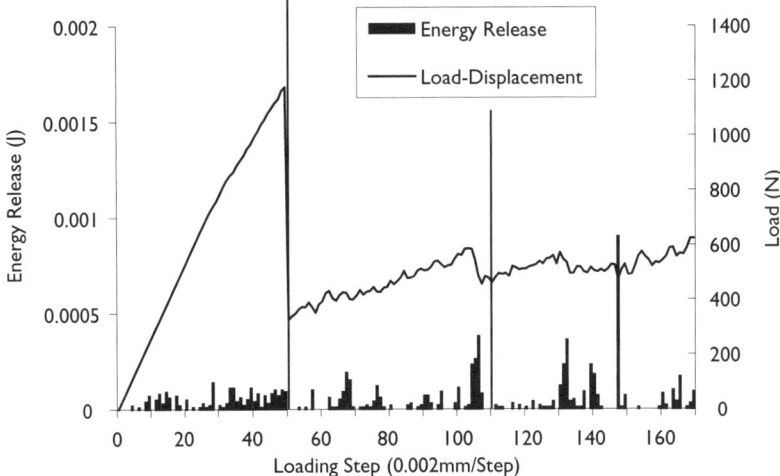

Figure 19.4 Load–displacement and energy–displacement curves for the particle subjected to diametral loading without confinement, see Figure 19.5.

point and the centre of the particle, as evident in Figures 19.5 and 19.6. It is seen that the elastic regime is terminated by the onset of instability, involving a tensile-type mode of particle splitting in which the particle shatters into several pieces (Step 51a–d in Figures 19.5 and 19.6). It is seen from Figure 19.4 that the fracture takes place with considerable violence in a brittle manner (i.e., large load drop and energy release in Step 51).

Continued loading after the maximum load is reached results in the formation of interesting failure patterns, as seen in Step 70 to Step 148 in Figures 19.5 and 19.6. We find that the zones of localised deformation with high maximum principal stresses, seen clearly in Figure 19.5, are at least partly due to the constraint effects from the loading platens. However, despite the presence of some associated zones of localised deformation and failure initiated at the contacts between platens and particle, the overall trend is towards a splitting pattern which can be considered as a tensile-type failure.

Due to the irregular particle shape, the stress concentration at the bottom contact first occurs on the left side, which results in the splitting failure first occurring between the upper loading point and the bottom left corner of the particle (Step 51d in Figure 19.5 and Figure 19.6). Then the stress concentration location moves to the bottom right corner of the particle (Step 70), which results in another splitting fracture on the right side, as shown in Steps 111–148. The same simulation is shown with the minor principal stress illustrated in Figure 19.6.

19.2.2 Breakage of single particle under diametral loading with confinement

A natural extension of the above simulation regarding the single particle breakage under axial load without confinement is the investigation of the richness and complexity of the mechanisms of particle breakage under biaxial compression. The model and the boundary conditions are shown in Figure 19.7.

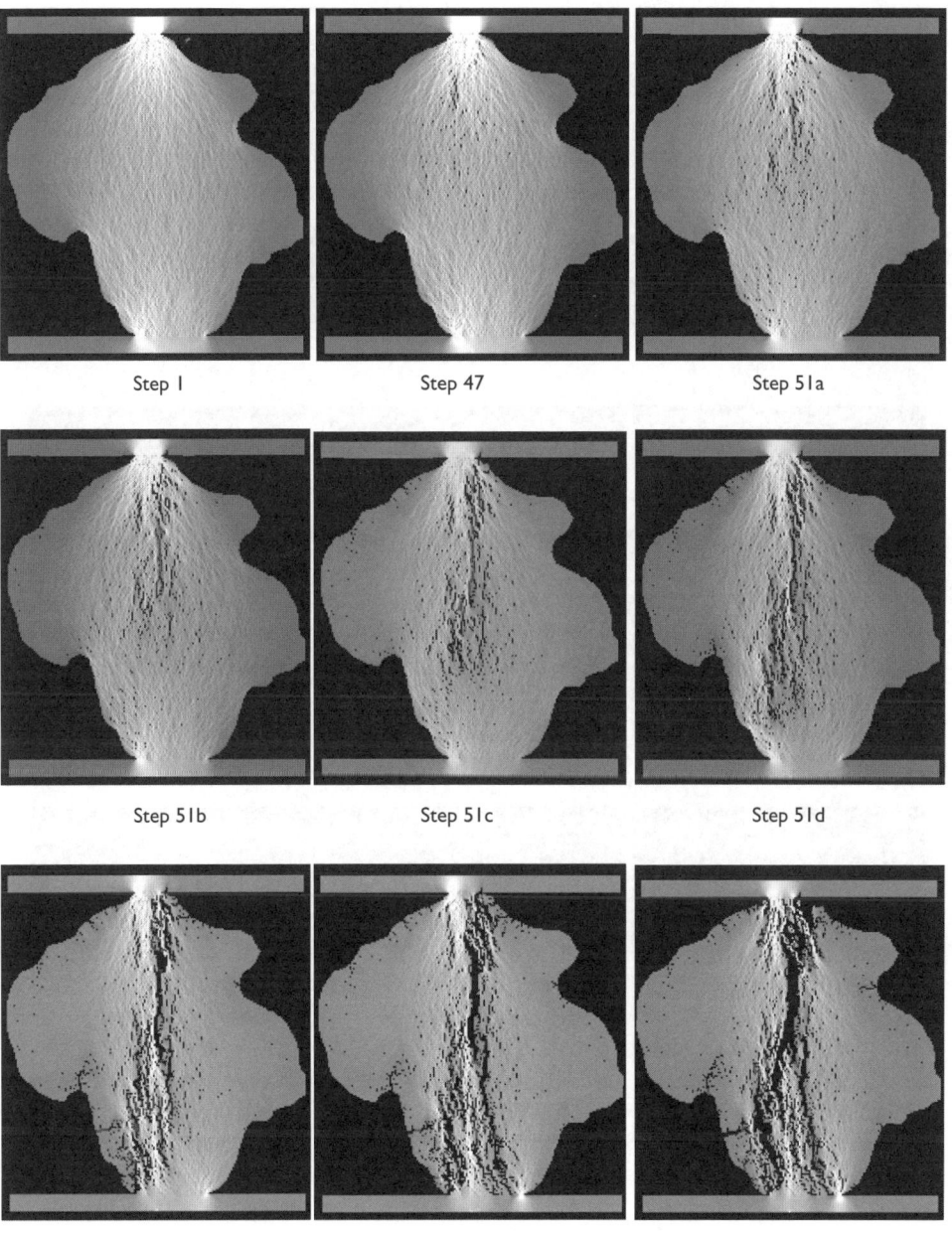

Figure 19.5 Failure mode and associated *major principal stress* field in the particle under diametral loading without confinement.

278 Rock failure mechanisms

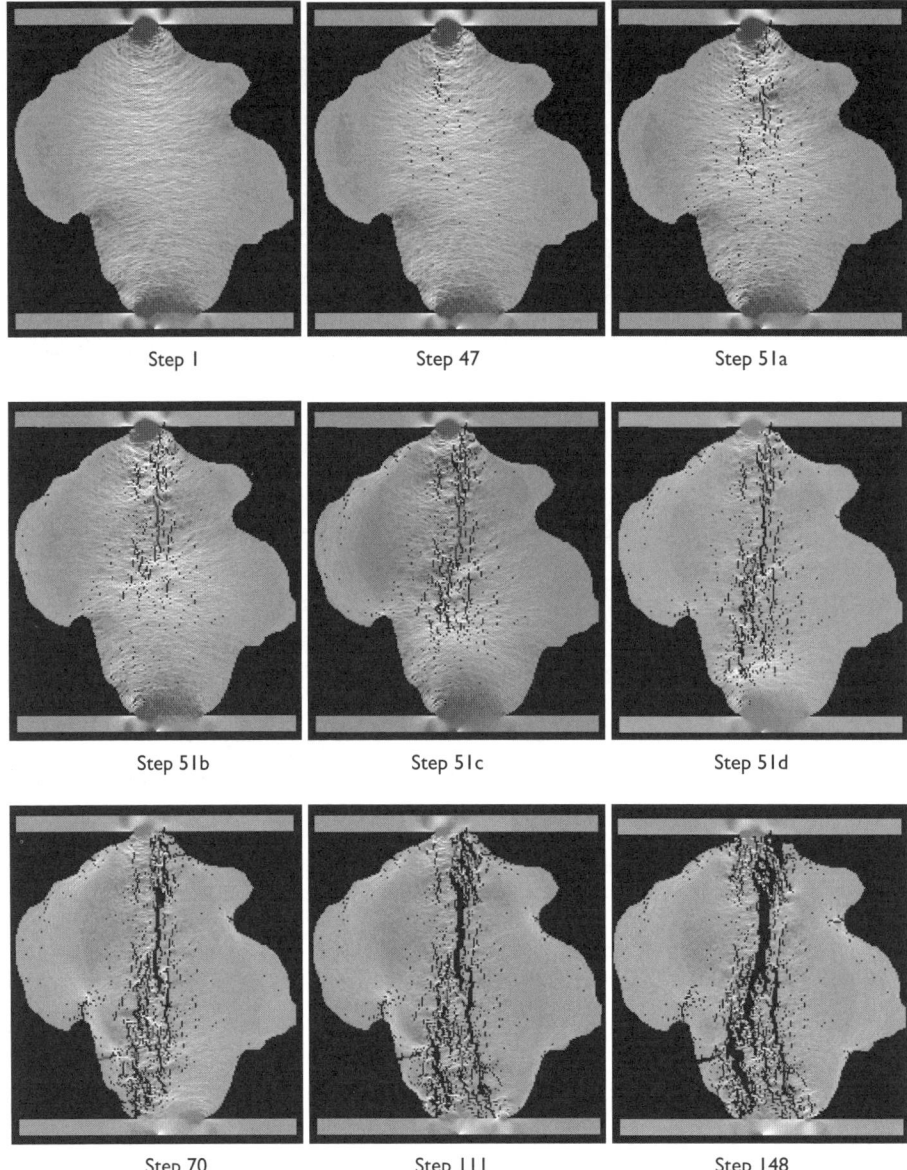

Figure 19.6 Failure mode and associated *minor principal stress* field in the particle under diametral loading without confinement.

The numerical particle sample is compressed by a relative vertical motion between two rigid platens but is confined in the horizontal direction by fixed points that are friction free in the vertical direction (Figure 19.7). The Poisson's ratio of the particle material is 0.25 and, thus, a compressive stress develops in the horizontal direction under this constraint. Figure 19.8 shows the fringe patterns in a homogeneous particle with the same irregular shape and loading condition as shown in Figure 19.7. This differs from the particle under diametral loading without confinement shown in Figure 19.2.

Figure 19.9 shows the normalised stress distributions, σ_x and σ_y, along the loading axis inside the heterogeneous particle under diametral loading with confinement. Comparing this with the previous stress distribution shown in Figure 19.3 for the particle under diametral loading without confinement, we observe similar effects: in particular, all the stresses σ_x and σ_y being in compression near the contact areas; however, the relative values of the stresses σ_x compared to σ_y are smaller than for the case of diametral loading without the confinement condition (as shown in Figure 19.3).

The confined crushing response in terms of load–displacement and energy release curves for the particle under diametral loading with confinement is shown in Figure 19.10.

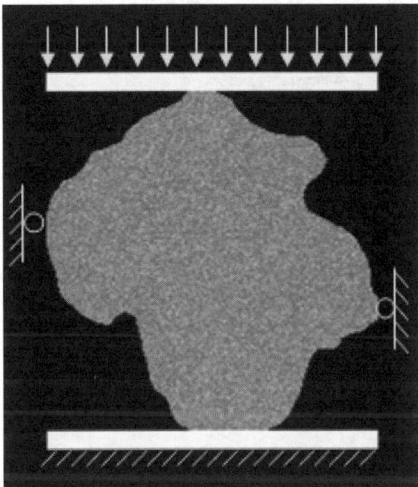

Figure 19.7 2-D irregularly shaped particle model and the loading and confining conditions.

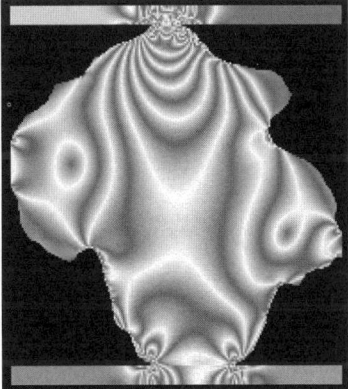

Figure 19.8 The fringe (shear stress) contours in the particle under diametral loading with confinement.

The response in the horizontal direction is also included in this Figure. Initially, the material is elastic and exhibits a stiff and stable response. After the maximum load is reached, a load drop with a high energy release is observed. The curve of the plateau load shows a strong saw-tooth undulation which indicates a complex stress response induced by the micro-fractures. Much more energy is released in the later crushing process.

Figures 19.11 and 19.12 show a sequence of deformed configurations corresponding to the steps shown in Figure 19.10, with the biaxially loaded particle now exhibiting more crushing failure. Subsequently, the fractures are mostly initiated by progressive destabilisation in the vicinity of the upper loading point and then in a wide

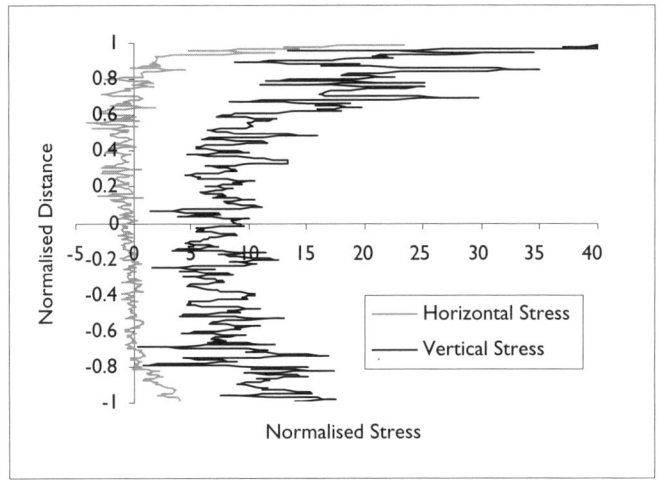

Figure 19.9 Normalised stress distributions, σ_x and σ_y, along the loading axis inside the irregularly shaped particle under diametral loading with confinement.

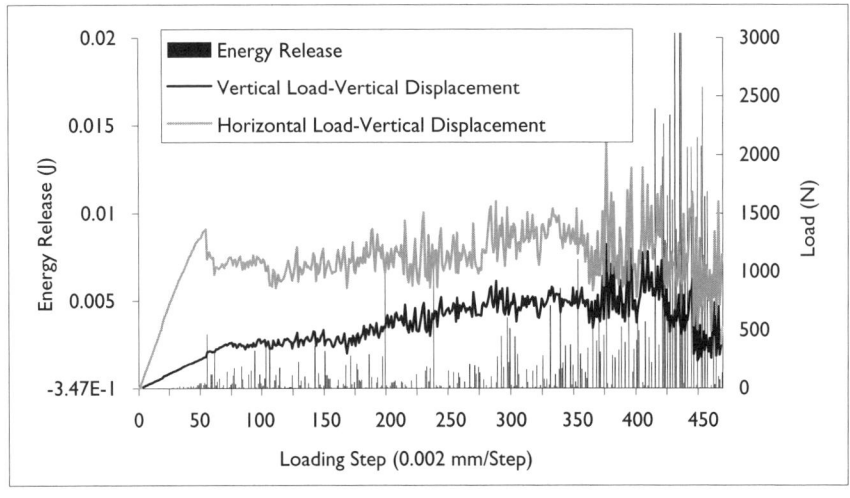

Figure 19.10 Load–displacement and energy–displacement curves for the particle under diametral loading with confinement.

Particle breakage and comminution 281

shear band between the upper loading point and the bottom left loading point. The failure is affected by the presence of the confinement and, as a result, the failure mechanisms are more or less in shear mode. Also, there are three larger fragments formed at a later stage of crushing which result from the side cracks between the neighbouring loading points, as seen in Steps 300 to 438 of Figures 19.11 and 19.12.

Figure 19.11 Failure mode and associated *major principal stress* field in the particle under diametral loading with confinement.

Figure 19.12 Failure mode and associated *minimum principal stress* field in the particle under diametral loading with confinement.

Because of the difference between the loading conditions, the compressive response under diametral loading with confinement differs considerably from the one without confinement. As shown in Figure 19.10, the lateral constraint increases the initial stiffness of the particle a little (5%) compared to the particle without confinement as shown in Figure 19.4. Also, the load–displacement curve of the particle with confinement is more ductile than the one without confinement (as shown in Figures 19.4 and Figure 19.10). The maximum load needed to break the particle that failed in a more ductile mode is 14% higher than that required for the unconfined particle which fails in a more brittle manner with a larger load drop. Due to the confinement, the plateau load level is about 60% higher than that of the corresponding unconstrained compression test, and the energy release is even hundreds of times higher (as shown in Figures 19.4 and 19.10). In addition, Figures 19.11 and 19.12 indicate that, with confinement, highly stressed zones evolve at the centre and near the contact areas, leading to a large volume of crushed fine particles which consumes most of the failure induced energy release.

If we define a confinement index as $I = P_x/P_y$, where P_y and P_x are the vertical load and the confinement induced horizontal load, respectively, we find that the confinement index I varies with the loading steps and the particle breakage process. Figure 19.13 shows this variation with respect to the loading steps. The variation of the index includes three stages: 1) before the maximum vertical load is reached, the index stays relatively constant; 2) the index has a sudden increase as soon as the maximum load is reached, and maintains the increasing tendency until the collapse of the particle through the formation of three chips due to the side crack development; and 3) the index decreases rapidly indicating the final breakage of the particle. It is seen clearly from Figure 19.13 that the energy release increases significantly during the third stage.

In the model, material heterogeneity inside the particle is taken into account but, because of the randomly distributed mechanical parameters, the particle material is statistically isotropic. Despite this, the crushing response is sensitive to local variations in the mechanical properties. The spreading of fractures is affected by the heterogeneity and the load plateau that is traced during the spread of the collapse is also

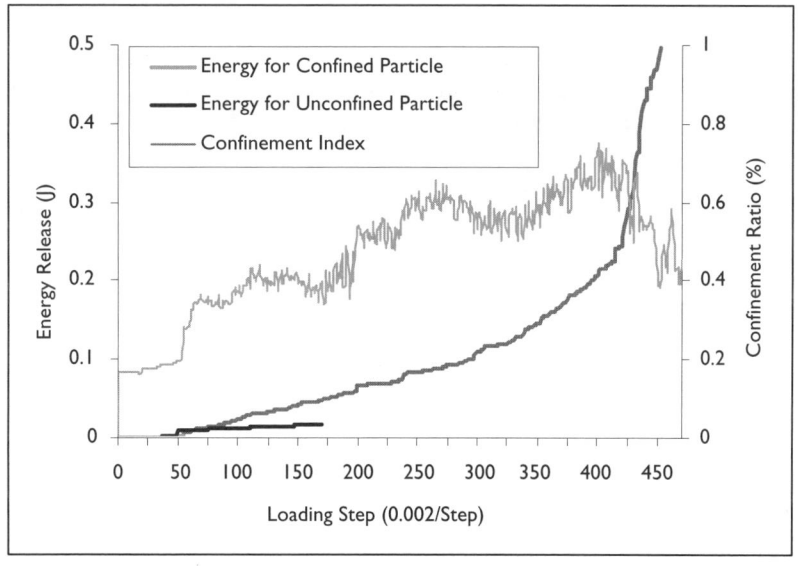

Figure 19.13 Cumulative energy release and the confinement index.

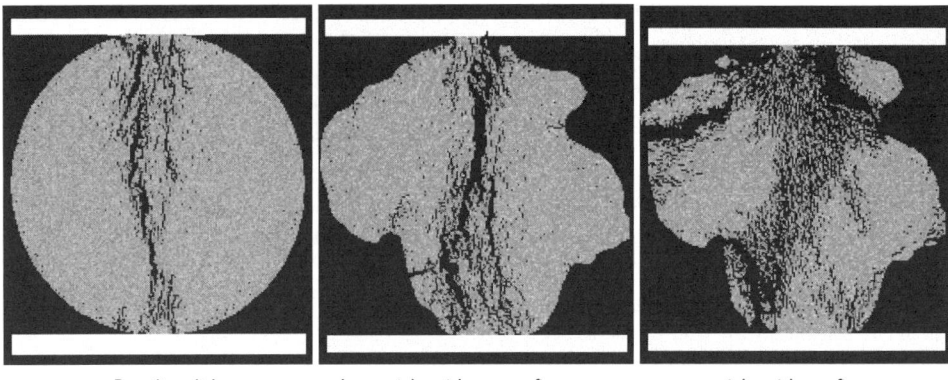

a. Brazilian disk   b. particle without confinement   c. particle with confinement

Figure 19.14 Failure modes of the three cases with different geometry and loading conditions.

affected by the heterogeneity characteristics of the particle. In fact, when a particle is broken by compression or crushing, the products fall into two distinct size ranges: coarse particles are products resulting from the induced tensile failure; and fines from compressive or shear failure near the points of loading.

It is found that the single particle breakage may be categorized into two limiting cases. In the first case, purely elastic deformation occurs. Under diametral loading without confinement, curved fracture trajectories are observed to form between contact zones (as shown in Figure 19.14a and b). These fractures approximately follow the principal stress trajectories that are perpendicular to the maximum tensile stresses. The fractures separate the disc or particle into two main parts with some finer fragments formed along the fracture zones as shown in Figure 19.14a and b. In the other limiting case, inelastic deformation, i.e. micro-fracture processes, is the primary mode—in this case, cone-shaped regions at the points of loading deform with high compressive stress. With increasing external load, localised shear mode failures occur in this cone-shaped zone, forming the so-called crushed zone (as shown in Figure 19.14c).

Although all the samples show a pronounced increase in the intensity of shear stresses immediately adjacent to the regions of the point contact areas, the diametral loading without confinement produces more coarse particles (as shown in Figure 19.14a and b); whereas, the diametral loading with confinement generates many more fines, particularly in the vicinity of the contact points (as shown in Figure 19.14c). It is also observed that, even though under confinement loading conditions, tensile fractures may also occur. The three big chips formed between the neighbouring loading points by side cracks are examples of tensile fracture.

19.3 MULTIPLE PARTICLE BREAKAGE

When a group of particles is being crushed, inter-particle breakage occurs when a particle has contact points shared with other surrounding particles. In this section,

Particle breakage and comminution 285

the inter-particle breakage process under confined conditions in mechanical crushing is investigated and discussed in terms of two loading geometries: quasi-uniaxial compression and quasi-triaxial compression.

19.3.1 Fragmentation process of a rock particle assemblage in a container

The model shown in Figure 19.15, is used to investigate the inter-particle breakage process in a particle assembly. In practice, the shape of the working surface may be a plane, cylindrical or be spherical. Any combination of these shapes is possible for investigating a particular configuration in a mill but a parallel plane loading surface is used, Figure 19.15. The test corresponds to the compressing component of the machine cycle of the crusher, when the liners move towards each other. The material is then locked between the chamber walls and can only deform elastically or break into smaller particles. Here we treat the breakage process as quasi-static.

The model (Figure 19.15) consists of a crushing chamber and 27 randomly placed rock particles with radii following a Weibull distribution within the chamber, where the individual particles are subjected to a set of contact forces. The model consists of a steel container measuring 180 mm in width and height. The thickness of the container walls is 5 mm. A steel platen measuring 170 mm in width is used as a cover on the top for transferring a compressive load down to the rock particles in the vertical direction. The particle bed is loaded under uniform conditions with the assumption of plane strain; the axial load is increased by moving the upper loading platen downwards, step by step in a displacement controlled mode. In the model, the walls, having the same modulus as the loading platen, impose a horizontal constraint against the particles inside. This provides the necessary confined condition for inter-particle breakage.

The rock elements are heterogeneous with homogeneity index $m = 2$ and mean elastic modulus $E = 60$ GPa, mean $UCS = 200$ MPa. However, to begin, the particles are firstly assumed to be homogeneous and the fringe patterns (shear stresses) in each particle and in the wall of the container are shown in Figure 19.16a. The major and minor principal stress fields are shown in Figures 19b and c and indicate how the load is

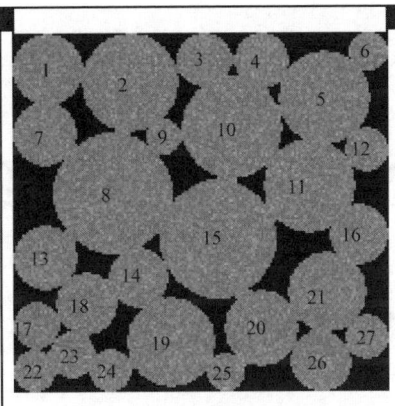

Figure 19.15 Model of a crushing chamber containing 27 rock particles.

transferred from one particle to another through the contacts. The compressive stresses in the particles attenuate from the top to the bottom, as can be seen from Figure 19.16b; this is because part of the load acting on the top cover is borne by the container wall.

Figure 19.17 shows the inter-particle progressive breakage process under compression in confined conditions. At the beginning of the process, grain fragmentations are located first in the small grains, such as the grains numbered 1, 4, 6, 7, 12, 13, 16, 18, 19, 20, 21, 23, 24 and 25 (refer to Figure 19.15 for the particle numbers) in Figure 19.17 A, B, C and D. In these grains, splitting macroscopic cracks are initiated and propagated along the lines between the two highest stressed contact points. For the large grains, the fragmentation is more difficult, because their large number of surrounding contacts effectively create a hydrostatic effect around the grains, i.e. quasi-triaxial compression, for example the grains numbered 2, 5, 8, 10, 11 and 15 in Figure 19.17 A, B, C and D.

As the loading displacement increases, with more and more small-sized grains fragmenting, the large grains also undergo failure. Although the splitting cracks are still initiated and propagated along the lines between the two most highly stressed contact points (for example the grains numbered 8 and 15 in Figure 19.17 E, F, G, H and I), because the previous failures release the confinement, grain crushing has also become an important failure mechanism (for example the grains numbered 2, 5, 10 and 11 in Figure 19.17 G, H and I). A large number of Hertzian cracks are initiated from the highly stressed contact points to form chips.

The development of the major principal and minor principal stress fields that initiate material failure are visually shown in Figures 19.18 and 19.19, corresponding to loading displacements of 0.015, 0.08, 0.145, 0.16, 0.27, 0.36, 0.405, 0.47, 0.48 and 0.53 mm. As shown in Figures 19.18 and 19.19, the contact forces are displayed through the contact points with a high stress field. The externally applied load is carried by the particles, but it is not uniformly distributed on all the particles. At the beginning of the process, one can observe that the particle breakage starts first with the small-sized particles because of the small number of contact points with their neighbouring particles or the walls of the crushing chamber. After the small-sized particles have fractured, the large particles begin to break in the middle of the assembly.

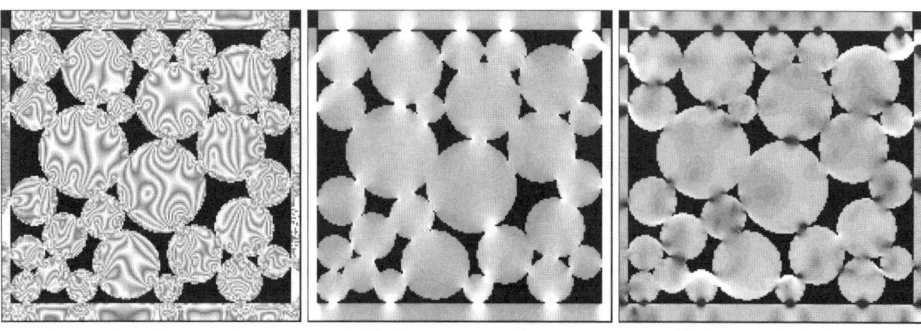

a. Shear stress fringe pattern   b. Maximum principal stress   c. Minimum principal stress

Figure 19.16 Elastic stress distributions in homogeneous particles in a crushing chamber before breakage.

Particle breakage and comminution 287

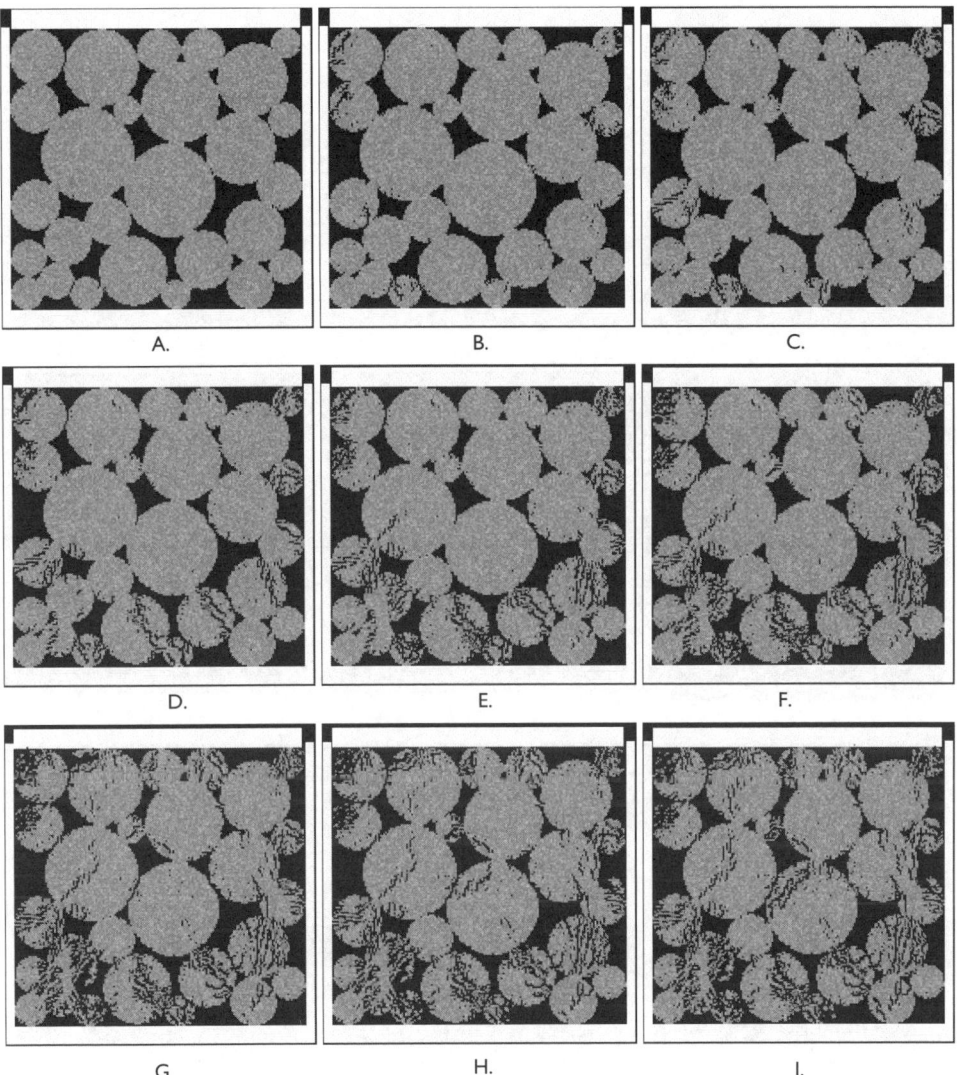

Figure 19.17 Progressive fragmentation process for rock particles inside a crushing chamber.

19.3.2 Force and displacement relation during the breakage process

Figure 19.20 records the total crushing force (F) and the displacement (S) curve during the breakage process. The sequence of failure in Figures 19.17, 19.18 and 19.19 corresponds to the equilibrium states identified on the response curve by the letters A, B etc. Initially, the response is relatively stiff and nearly linear (AB in Figure 19.20). Each particle deforms elastically. At a load of 1364 N (point B in Figure 19.20), the response begins to soften, mainly due to the breakage of particles 1, 6, 7, 12, 13, 16

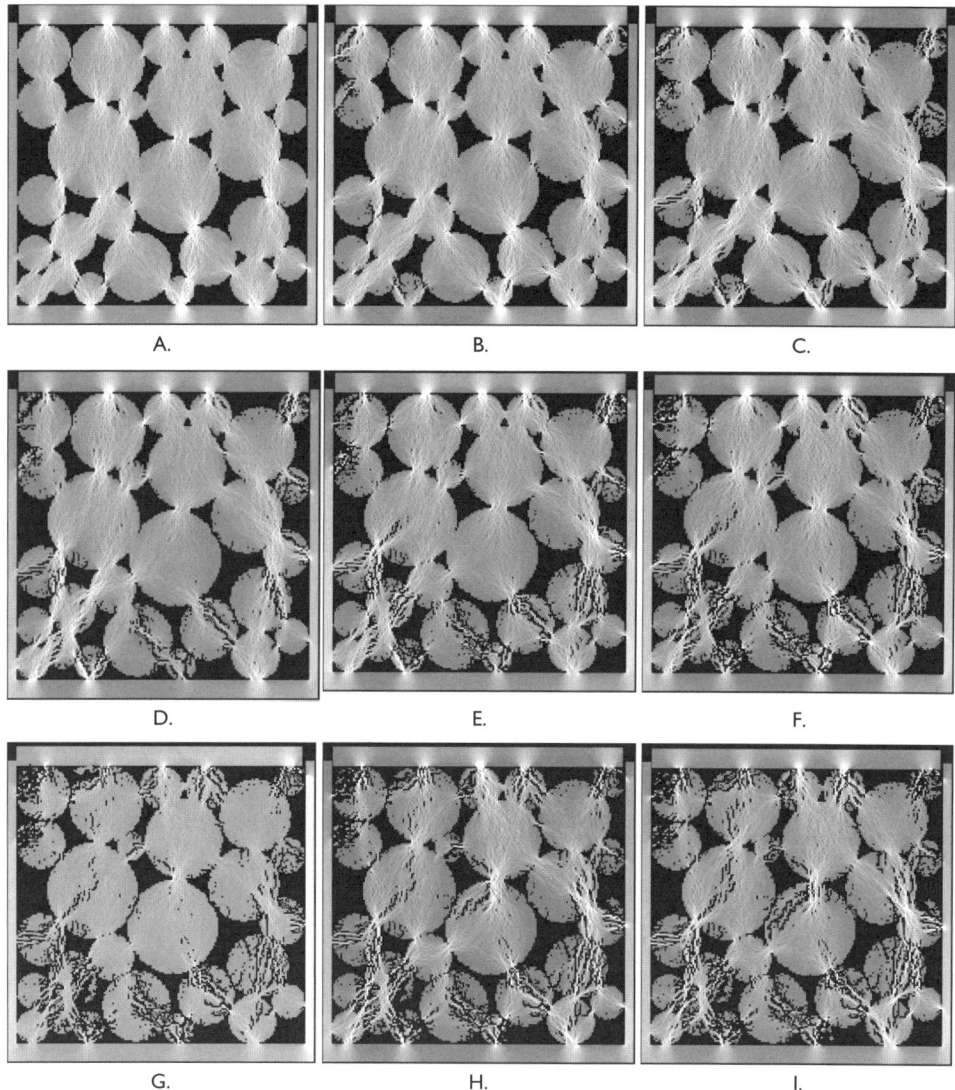

Figure 19.18 The major principal stress distribution within rock particles inside a crushing chamber during the inter-particle breakage process.

and 24 (Figures 19.17, 19.18 and 19.19B), and eventually a limit load develops at 1858 N (point C in Figure 19.20). The fracture is localised in particles 1, 4, 6, 7, 12, 13, 16, 18, 19, 20, 21, 23, 24 and 25 (Figures 19.17, 19.18 and 19.19C and D) beyond the limit load (point D in Figure 19.20). These particles collapse in a tensile-type mode under quasi-uniaxial compression, while other particles next to them remain more or less elastically deformed because of the confinement from neighbouring particles, covers or container walls. As the collapse of these particles progresses, the failures spread to the neighbouring particles. Eventually the whole assembly collapses.

Particle breakage and comminution 289

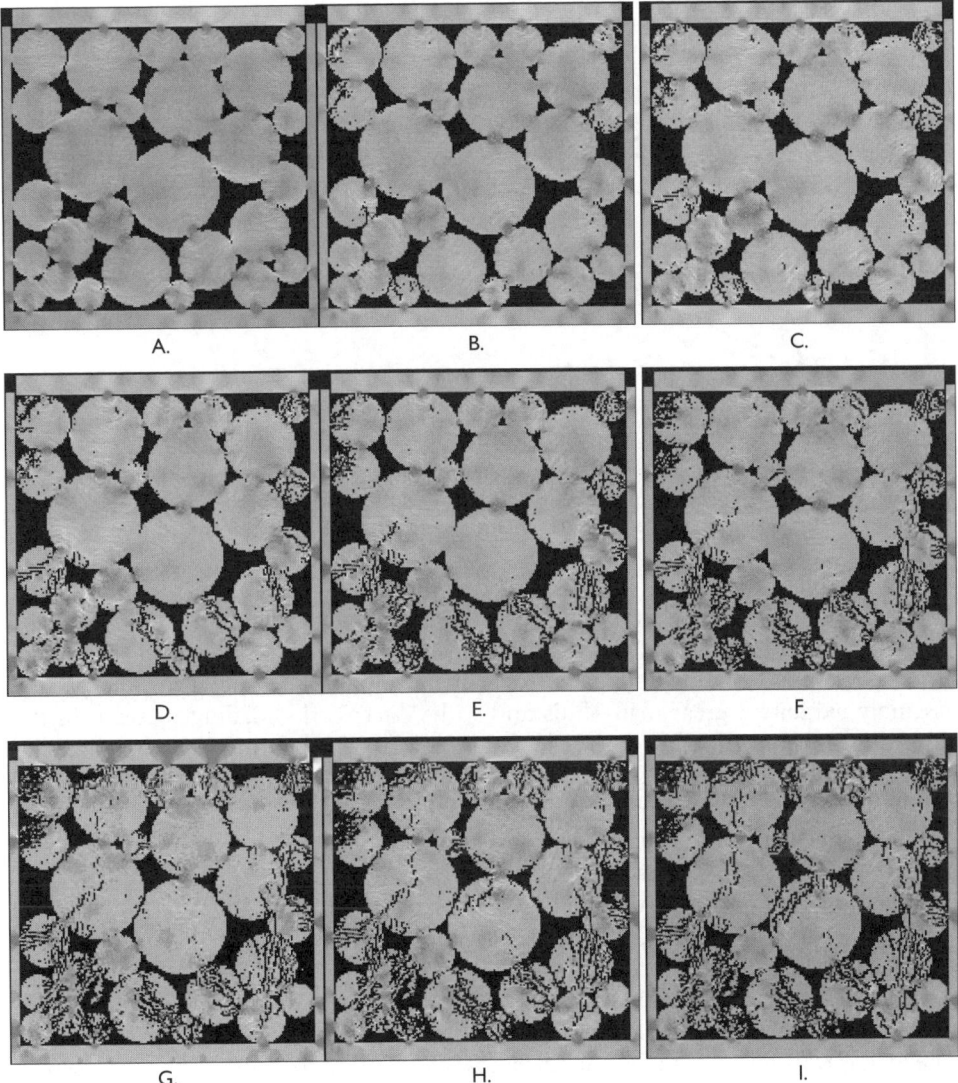

Figure 19.19 The minor principal stress distribution in rock particles inside a crushing chamber during the inter-particle breakage process.

The spreading of the collapse from particle to particle continues, creating a spiky load plateau, as shown in Figure 19.20 after point C. One can observe the fragmentation regime, characterised by the irregular saw-toothed curve (points D, E, F, G and H) of the load. At a loading displacement of 0.36 mm, corresponding to point F in Figure 19.20, almost all the particles have collapsed (Figures 19.17, 19.18 and 19.19 F), except a few larger ones in the middle of the particle bed. After this point, the load starts to increase again (point G in Figure 19.20), which implies a material-hardening characteristic. The propagation of the collapse involves the shear-type

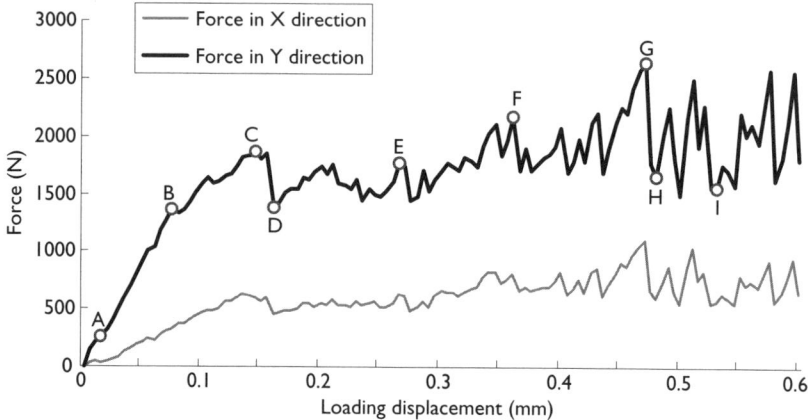

Figure 19.20 Resultant force–displacement curve during the inter-particle breakage process. (Points on the curve labelled alphabetically A, B, etc. correspond to the images in Figures 19.17, 19.18 and 19.19.)

mode as well as the tensile-type mode in the large grains (Figure 19.17, 19.18 and 19.19 G, H and I), which causes the initial instability. Furthermore, each load undulation corresponds to the collapse of one or more particles. Therefore, the particle assembly exhibits a great many hills and valleys across the loading plateau. In this case, the maximum load (point G in Figure 19.20) extends much higher than the assembly strength at point C in Figure 19.20. The mean load of the plateau, which will be called the propagation load, is about 2000 N.

Thus, the inter-particle breakage consists of two phases: Phase I, consisting of stiff particles, behaves as an elastic structure that dissipates little energy and imposes local deformation on Phase II particles; Phase II acts like a brittle-and-plastic material that dissipates energy at the contacts and serves as a restraint on Phase I. As the applied displacement increases, certain changes take place in the material structure in that some Phase-I particles become Phase-II particles; i.e. more and more particles progressively fail, and the particle assembly, therefore, becomes less able to carry the load. The load is transferred from new load chains that develop from other particles nearby. An anisotropic fabric is established corresponding to the direction of the applied loading. The resultant force–displacement response on the side-walls shows the same features as that in the loading platens (the lower curve in Figure 19.20).

19.3.3 Energy considerations

The energy is supplied to the particle bed by moving the loading platen. By integration of the crushing force (F) over the displacement of the loading platen (S) in Figure 19.20, the energy consumption (W) can be obtained

$$W = \int_0^{S_{max}} FdS = 1.003\ J$$

where S_{max} is the maximum displacement of the loading platen when loading the particle bed.

Figures 19.21 and 19.22 show the fracture event rate and the elastic energy release (ENR) inside the particle during the inter-particle breakage process. The failure event rate shows the following expected features: 1) during the initial deformation or linear elastic phase (line AB in Figure 19.20), little elastic energy (Figure 19.22) was released, although some fracture events (Figure 19.21) occurred; 2) an increasing rate of fracture events (Figure 19.21) accompanied the inelastic phase and the load plateau (point C in Figure 19.20). This agrees with the understanding that the fracture events are generated by micro-fractures that result in non-linear deformation behaviour. Although nearly 60% of the fracture events (Figure 19.21) occur before point F

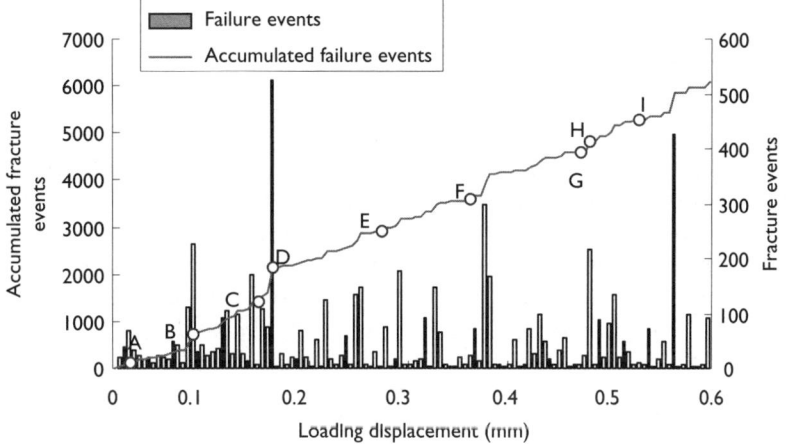

Figure 19.21 Relation between the failure event rate and the cumulative failure event rate during the inter-particle breakage process.

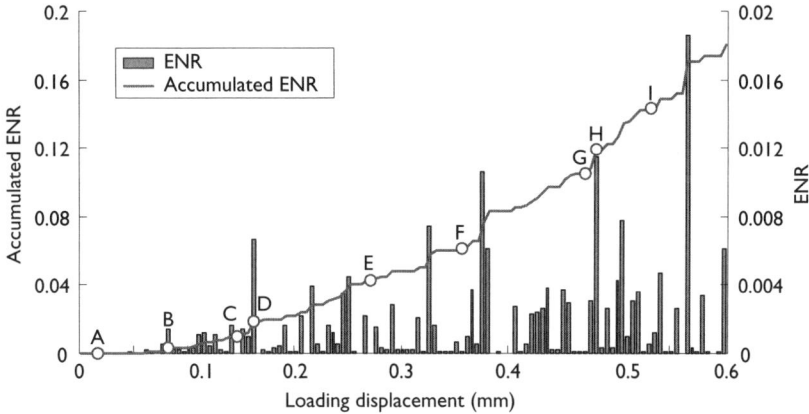

Figure 19.22 Relation between the elastic energy release (ENR) and the cumulative ENR during the inter-particle breakage process.

in Figure 19.20, less than 35% of the elastic energy is dissipated during this stage, as shown in Figure 19.22. A comparison between Figures 19.20, 19.21 and 19.22 shows a good relation between the load curve, failure event rate and energy release. Note that most of the large load drop on the load curves shown in Figure 19.20 corresponds to a high failure event rate (Figure 19.21) and a large energy release (Figure 19.22).

The applied work equals the energy consumed in breakage of the particles and energy losses. According to the cumulative energy release shown in Figure 19.22 and the applied work shown in Figure 19.20, about 18% of the applied work is consumed in breaking particles. Compared with the single-particle breakage, inter-particle breakage has a lower energy utilisation ratio because of local crushing at contact points. The applied energy is mainly consumed by a) acoustic emission, b) the formation of new surfaces, and c) local crushing at contact points.

19.3.4 Size distribution

The size distribution is another important consideration in the inter-particle breakage process during comminution. A quantitative description of the size reduction would be helpful for a better understanding of comminution in mechanical crushing and especially for modelling the crushing performance. The distribution of fragment size, corresponding to the inter-particle breakage process, is shown in Figure 19.23, in which the letters correspond to those in Figure 19.17. An image analysis program has been used to measure the fragment size with the Diameter of the Equivalent Circle Area (DECA) of a particle being used to measure the size of the particle.

In Figure 19.23, the abscissa values are the DECAs of the particles (mm) and the ordinate is the cumulated weight distribution (%). Before crushing (Step A), less than 8% of the DECAs of the particles are smaller than 12 mm. With the loading displacement increasing, the size reduction effect increases rapidly before step F (the onset of the material-hardening regime) in Figure 19.23—from less than 8% of the DECAs of the

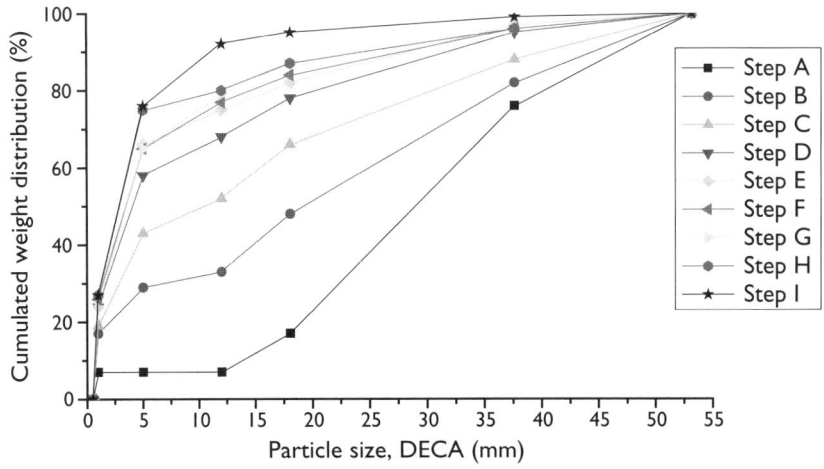

Figure 19.23 The fragment size distributions corresponding to the images in Figure 19.17.

particles being smaller than 12 mm to more than 70% of the DECAs of the particles being smaller than 12 mm, as shown in Figure 19.23 (Steps B, C, D and E).

After Step F, local crushing at contact points becomes the important failure mechanism and the size reduction effect increases slowly, as shown in Figure 19.23 (Steps F, G, H and I). In practice, in order to reduce the fines or to control the micro-cracks within the reduced particles, a careful design of the normal stroke is important. In the case of this particular model, with the assumed mechanical properties and the size and shape of the assembled rock particles, as well as the height of the container, a normal stroke between 0.3 and 0.4 mm (Figure 19.20F) is a good choice to avoid over-breakage (Step F in Figure 19.23). After crushing (Step I), over 90% of the DECAs of the particles are below 12 mm but the size distribution does depend on parameters such as the bed height, the stroke and the initial size of the crushed particle.

19.3.5 Influence of particle shape

In order to investigate the influence of the particle shape on the inter-particle breakage process, a model similar to that illustrated in Figure 19.15 was constructed. The particle bed consists of irregularly shaped particles, Figure 19.24, but they are 'mono-dispersed', which means that they have an approximately equal area. The irregular shape determines the number of contacts among neighbouring particles and the walls of the crushing chamber.

Figure 19.24 shows the elastic stress distribution of irregularly shaped particles in a steel container. Similarly to Figure 19.16a, Figure 19.24a indicates that the overall load produces contact forces between the particles. It is seen that in the current particle bed arrangement, large forces are carried by chains of particles that are more or less aligned in the direction of the major compression. These chains become shorter and more kinked as an increasing load is applied. Figure 19.25 illustrates the progressive inter-particle breakage process of the irregularly shaped but mono-dispersed particles. In contrast to the disc particle breakage process, where the fragmentation starts in the small-sized particles, here the grain fragmentation starts first with a few of the grains contacting the two assumed rigid walls, as shown in the particles numbered 8

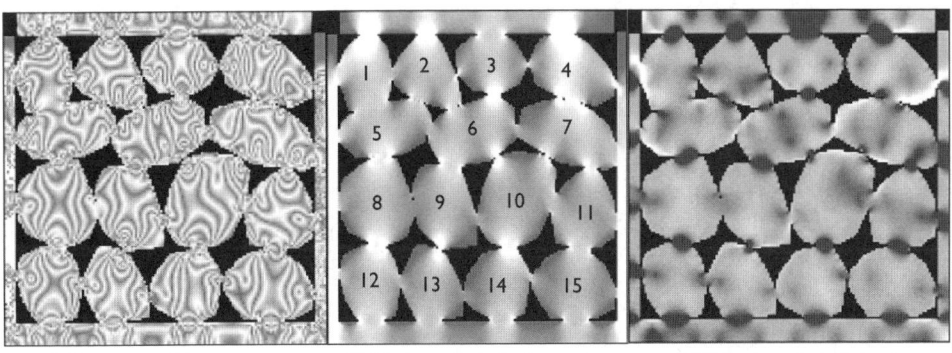

a) Shear stress fringe pattern   b) Maximum principal stress   c) Minimum principal stress

Figure 19.24 Elastic stress distributions of irregular 'mono-dispersed' particles in a crushing chamber before breakage.

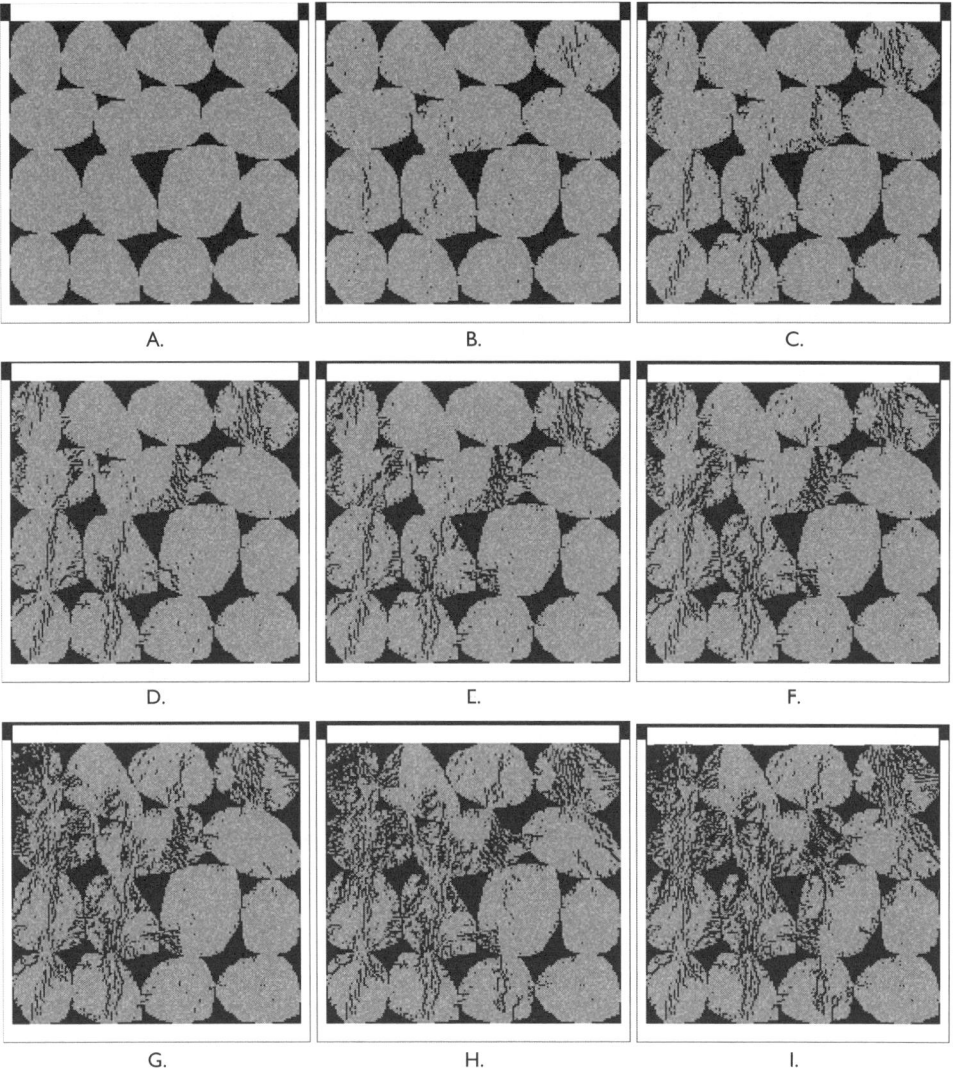

Figure 19.25 Simulated progressive fragmentation process for irregularly shaped but mono-dispersed rock particles inside a crushing chamber.

and 4 (referring to Figure 19.24 for the particle numbers in Figure 19.25B. The grains in contact with the walls present geometric configurations close to those of grains submitted to two opposed forces, i.e. quasi-uniaxial compression, which favours their fragmentation.

Previously, we have shown in the poly-dispersed circular particle bed that the fragmentation starts with the small particles. In fact, irrespective of whether it is a small-sized circular particle or an irregular particle contacting the rigid walls, a particle is selected to break first if two criteria have been fulfilled: 1) the particle is located in such a way that quasi-uniaxial compression can be achieved, i.e. in such a way that splitting failure easily occurs; and 2) the stress level of the particle has reached a critical value. After the first few grains have been fractured, one begins to observe that the grain fragmentations are mainly located on the grains under an almost quasi-uniaxial compression, as shown in the particles numbered 4, 6, 8, 9, 12 and 13 in Figure 19.25C, D and E. At the same time, because of the failure releasing the confinement, the particles loaded at first under quasi-triaxial compression begin to fail also, as shown in the particles numbered 1, 5 and 10 in Figure 19.25 C, D and E. After that, local grain crushing at the contact points becomes an important breakage mechanism, as shown in the particles numbered 1, 4, 5, 7, 8, 9, 12 and 13 in Figure 19.25 F, G, H and I, although a splitting failure of the quasi-uniaxial compression type (particle 14 in Figure 19.25) still occurs.

It was noted that the particle breakage strength depends on the particle shape. The results suggest that the more spherical the particle becomes, the higher is the breakage strength that may be expected (particle 7, 10, 11, 14 and 15 in Figure 19.25). Presumably, a disc particle has a more regular stress distribution than an irregular particle. In fact, in order to achieve the same propagation load as is shown in Figure 19.20, a higher percentage (~90%) of irregularly shaped particles (Figure 19.25) is put into the crushing chamber compared with the percentage (~85%) of circular particles (Figure 19.17).

Chapter 20

3-D modelling and 'turtle crack formation' in rock

20.1 INTRODUCTION

When natural and engineered systems are subjected to shrinkage, driven by cooling or drying, the resulting stresses may lead to the formation of fractures. Experimental studies of this phenomenon and fracture spacing theory show that the spacing between fractures initially decreases as extensional strain increases in the direction perpendicular to the fractures. At a certain ratio of spacing to layer thickness, however, no new fractures form and the additional strain is accommodated by the further opening of existing fractures: the spacing then simply scales with layer thickness. In field and laboratory observations, two kinds of fracture patterns are commonly observed—parallel fractures (Figure 20.1a) and polygonal fractures (Figure 20.1b).

A parallel fracture pattern occurs in layered materials usually under a mechanical layer-parallel stretching force, i.e., directional extension, or biaxial stretch, with one of the principal stresses much greater than the other. A polygonal fracture pattern is often observed in surface layered materials under cooling or shrinking induced isotropic stretch in all layer-parallel directions. Examples of the polygonal fracture pattern are desiccation fractures in dried-out mud flats or in a 'turtle cracking' rock (Figure 20.1b) and permafrost.

An interesting area of research in engineered systems is to consider the fracture patterns in composite materials. When thin glass strips are exposed to uniaxial tensile strain, such as that caused by a thermal gradient or drying of thin colloidal suspensions, uniformly spaced fractures form parallel to the direction of the temperature or moisture gradient. Alternatively, for a brittle film that is attached to a substrate subjected to biaxial tension, the fractures formed divide the film into a series of polygonal shaped islands in a pattern that is often seen in drying mud. The surface cooling fracturing, an ancient technique used to decorate china with a surface fracture network, is another classic example of this polygonal fracture phenomenon.

In an experimental study, Shorlin et al. (2000) observed another mode of fracture pattern that they refer to as 'laddering'. In their investigation, two types of experiments were performed: 'isotropic drying', in which the entire layer was dried uniformly; and 'directional drying', in which the layer was dried from one end. They observed a change in the way in which fragmentation occurred as the drying method in the system was varied. For isotropic drying, more uniform polygons formed. For directional drying, polygons did not form in a regular shape, but rather by 'laddering'. In this process,

298 Rock failure mechanisms

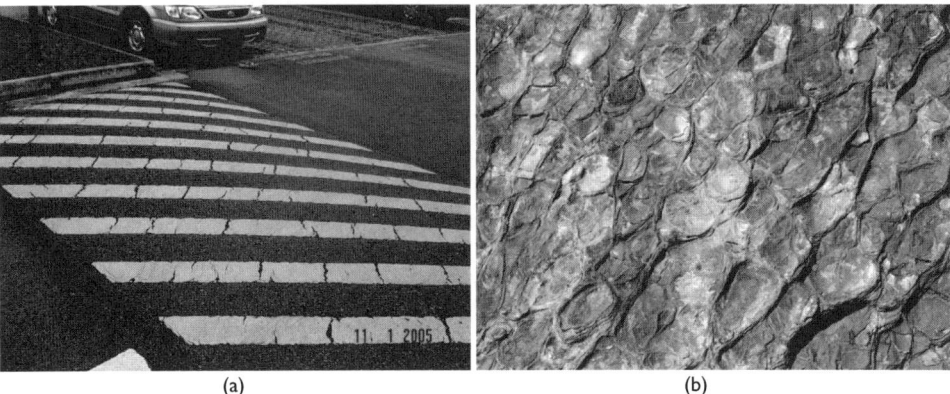

Figure 20.1 Examples of a) a parallel fracture pattern in a road surface and b) a polygonal rock fracture pattern in 'turtle cracking' mode, from the Chinese Sinian period (600 million years ago).

fractures propagate more or less parallel to each other. As the layer behind the moving fracture tips dries further, perpendicular fractures form, joining two of the parallel fractures like the rungs of a ladder. This process is dominant in the directional drying case, and clearly involves a local directionality of the drying process.

In this Chapter, we explain and illustrate these different fracture patterns. We find that the modelled fracture patterns show a continuous pattern transition from parallel fractures, laddering fractures, to polygonal fractures, which depend on the far-field loading conditions in terms of the principal stress ratio ($\lambda = \sigma_2/\sigma_1$), from uniaxial ($\lambda = 0$), biaxial ($0 < \lambda < 1$) to isotropic stretch ($\lambda = 1$). Note that the λ parameter used here is equivalent to the k parameter used in Chapter 16.

20.2 THE THREE-LAYER MODEL

Here we consider a three-layered model (Figure 20.2) that fails under a quasi-static, slowly increasing biaxial strain (induced, for example, by temperature changes, desiccation, or mechanical deformations). The 3-D finite element model consists of 1.6 million elements. In order to limit the fractures that occur in the central layer, the strength of the upper and lower layers are set to be many times higher than the central layer. The stretching of the layered model then corresponds to a gradual, homogeneous change of co-ordinates. Of interest are the ensuing pattern of fractures and the dependence of the pattern on the stretching in terms of the principal stress ratio, λ.

We use a constant displacement increment in the x-direction along the left and right boundaries, and another increment in the y-direction along the top and bottom boundaries. The principal stress ratio, λ, is then the loading ratio of the displacement in the x-direction to the displacement in the y-direction. The λ values selected are 0, 0.125, 0.25, 0.375, 0.5, 0.625, 0.75, 0.875 and 1, thus representing loading from directional to isotropic conditions.

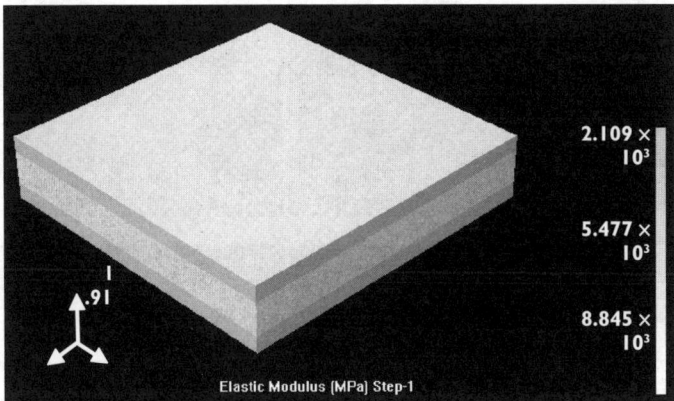

Figure 20.2 FEM model with a heterogeneous central embedded layer bonded to the top and bottom layers and consisting of 1.6 million elements (200 × 200 × 40). The horizontal plane is defined as the x-y plane and z is the vertical direction. The whole lower boundary is fixed in the z-direction, and the top boundary is free to displace as required. A constant displacement increment is implemented in the x-direction along the left and right boundaries, and another increment in the y-direction along front and back boundaries. λ is the loading ratio of the displacements in the x and y directions.

Figure 20.3 shows the model images that demonstrate the complete process of the fracture pattern evolution for isotropic loading ($\lambda = 1$). It can be seen from Figure 20.3a that, in the first stage of fracture pattern development, fractures nucleate at a small number of points. Fractures nucleate at the weaker sites and in many cases do not propagate long distances across the layer, but rather move in small steps from one weak site to the next, occasionally meeting another fracture moving in a similar fashion (Figure 20.3b-c). After the initiation of a few longer fractures, most of the new fractures then start at the vicinity of existing fractures and propagate away from their parent fracture, approximately at right angles. Finally, successive generations of fractures form, mostly joining older fractures and forming a complicated array of polygons (Figure 20.3c-d).

As shown in Figure 20.3, a new fracture propagates away from an existing fracture and eventually stops when it runs into another fracture. The region is then fragmented into two parts. Although randomly distributed, the size of the two polygonally fractured areas are found to be approximately the same. But, after reaching a certain strain, the number of polygons formed in the fractured layer no longer increases (Figure 20.3e). Instead, interface debonding is found to dominate the post-failure processes. This suggests that 'fracture saturation' exists in such polygonal fracture patterns.

In order to understand the role of the stress distribution in the fracture saturation phenomenon, we plot the minimum principal stress distribution in the right-hand columns of Figure 20.3. This indicates that the network fractures behave as free surfaces and, as a result, the normal tensile stress in the vicinity of a fracture is greatly reduced and, particularly, the minimum tensile stresses in the centre of the polygonal shaped islands are found to be small when fracture saturation is well-developed

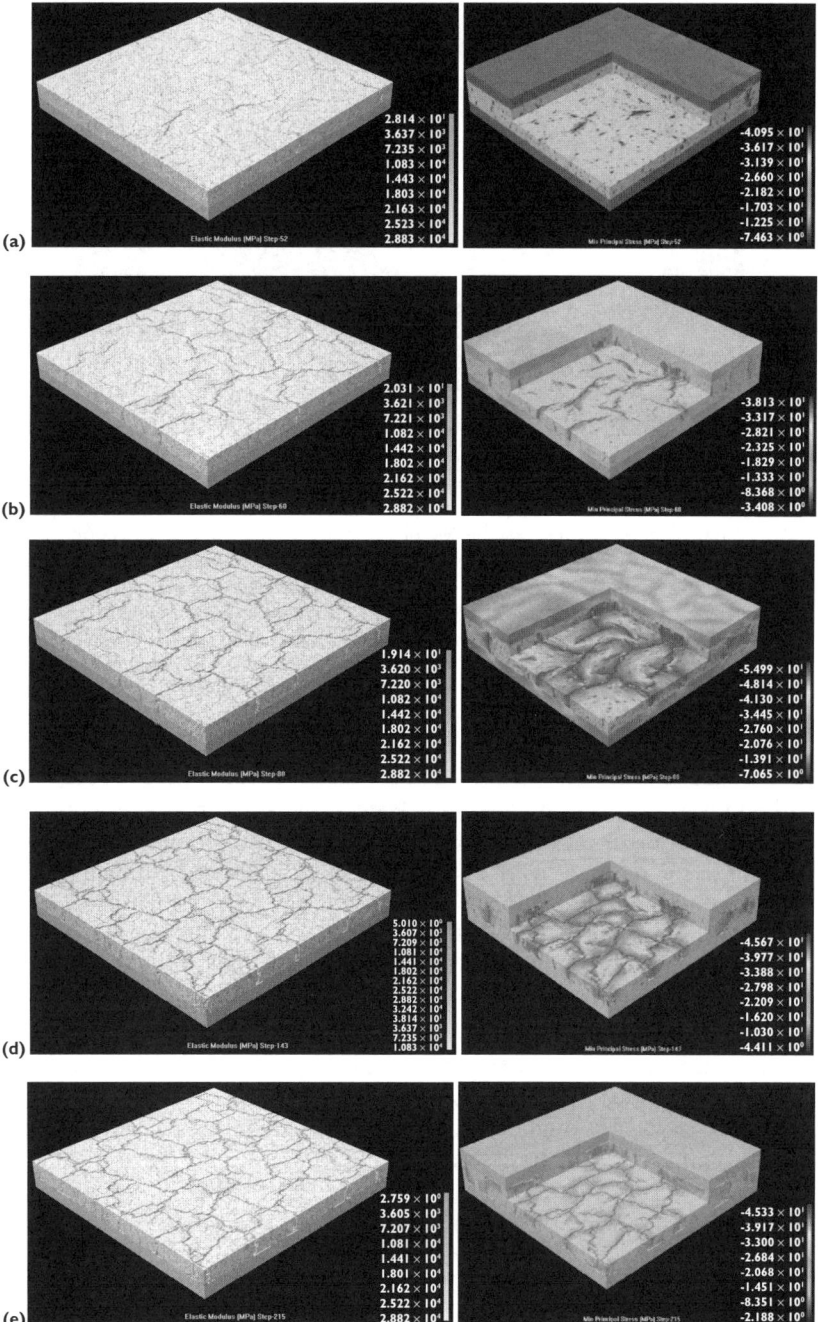

Figure 20.3 Fracture evolution (left column, with top layer not shown) and induced stress redistribution (right column) for model with principal stress ratio $\lambda = 1$ (isotropic stretch). The cross-sections are taken from the central plane in the embedded layer. The stress is expressed as minimum principal stress. The stages a to e represent the different characteristic stages of the fracture process.

(Figure 20.3e), and therefore the formation of new fractures is inhibited. This zone of reduced stress, referred to as the 'stress reduction shadow', with the traction-free islands, scales with layer thickness and is responsible for the observed correlation between fracture spacing and layer thickness.

Figure 20.4 shows the fracture patterns well-developed in the central cross-section of the embedded fractured layer for the nine models, corresponding to the loading conditions for $\lambda = 0$, 0.125, 0.25, 0.375, 0.5, 0.625, 0.75, 0.875 and 1. These modelling results indicate a continuous pattern transition from parallel fractures to polygonal fractures, depending on the loading conditions in terms of the principal stress ratio. In contrast to the results obtained for $\lambda = 1$ (Figure 20.3), Figure 20.4a for the model subjected to directional loading ($\lambda = 0$) shows a clear parallel fracture pattern. Due to the effect of the introduced heterogeneity, the obtained fracture propagations demonstrate realistic pathways. Most of the fractures terminate at certain lengths because they overlapped other fractures propagating from the opposite edge or interior. Even for those that appear to extend from one edge to the other, they are actually composed of several closely-spaced *en echelon* segments.

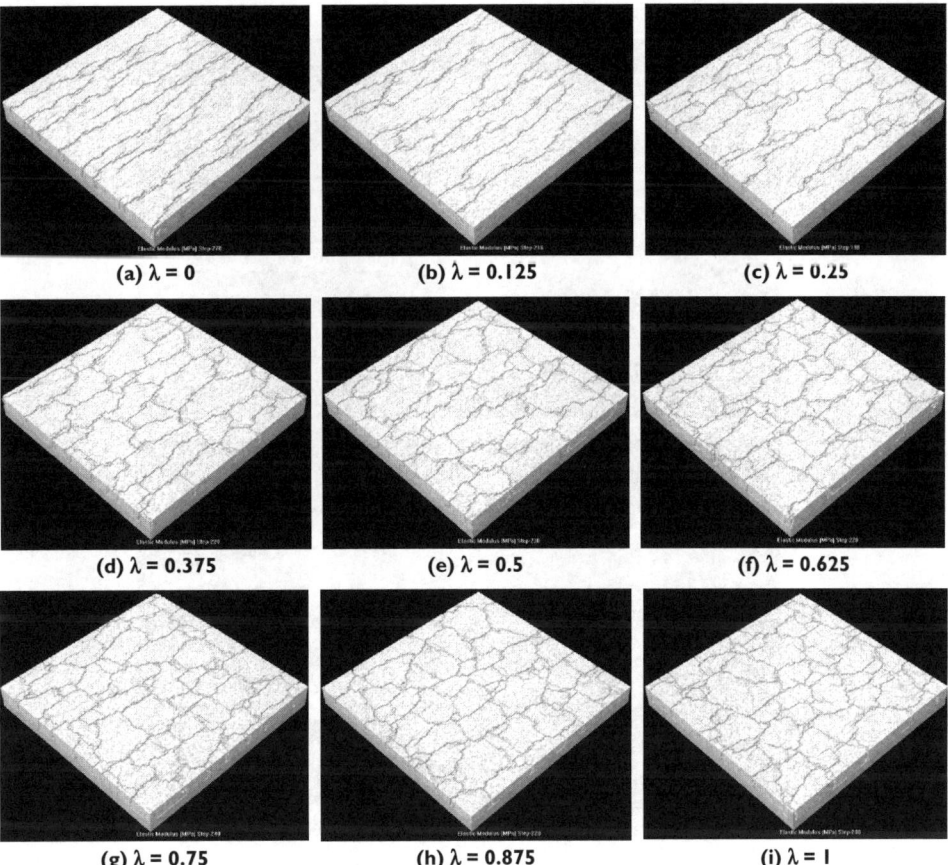

Figure 20.4 Models with different principal stress ratios varying from uniaxial tension, $\lambda = 0$, to equal biaxial tension, $\lambda = 1$.

302 Rock failure mechanisms

In biaxial loading with the principal stress ratio λ ranging from 0.125 to 0.5, another dominant mode of pattern formation is observed, as shown in Figure 20.4c-e, particularly for $\lambda = 0.375$. Figure 20.4d shows that, during the fracture forming process, fractures that are perpendicular to the parallel fractures were formed and they are like the rungs of a ladder. The model indicates that, under isotropic stretch conditions, $\lambda = 1$, isolated fractures initially appear by nucleation at a few points. Triple junctions, at which three fractures meet with junction angles of 120°, are formed by such nucleation in the early stages of pattern development. At later development times, however, fractures meet predominately at 90° junctions. This is due to the fact that fractures propagate in the direction which most efficiently relieves the stress. Since the stress near a given fracture is parallel to its surface, other fractures will tend to approach and meet it at right angles. But this situation is only observed in the model under isotropic stretch, i.e. with the principal stress ratio, λ, close to 1.

For the results with different principal stress ratios, λ, a transition in the distribution of junction angles was observed as the ratio, λ, decreases: for a λ less than 0.5, a marked increase in the number of junctions larger than 120° or less than 60°

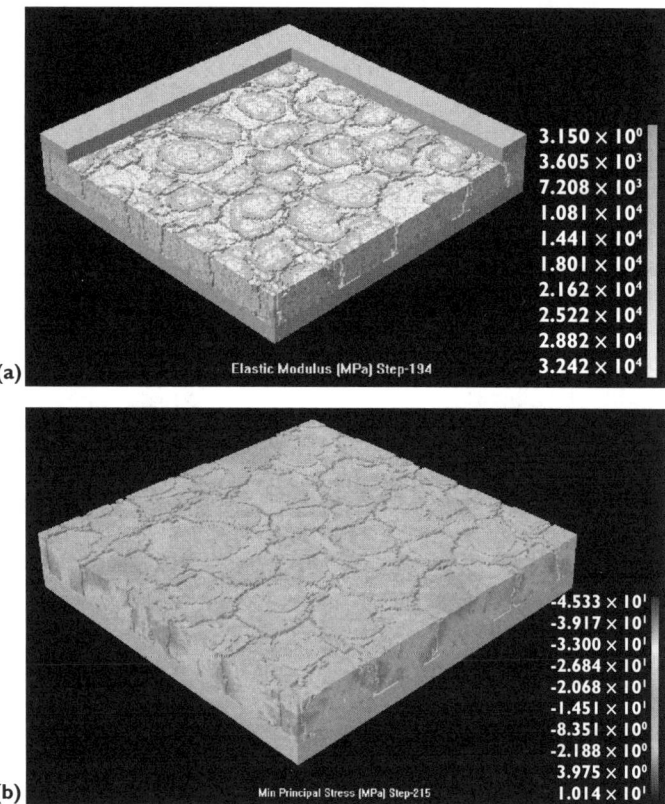

Figure 20.5 Interface fracturing induced stress redistribution for model with principal stress ratio $\lambda = 1$. The cross-sections are taken from the interface plane: a) the interface debonding; b) minimum principal stress redistribution.

was observed. In the extreme situation, that is, directional stretch, the junction angle becomes 180° or zero, and the parallel fractures dominate.

Interface debonding is found to dominate the fracture development when reaching the fracture saturation stage. Before that, mechanically the fractures arrest at the layer boundary during sequential infilling. Only after fracture saturation is reached does the interface debonding dominate fracture occurrence. As an example, Figure 20.5a shows the modelling results of interface debonding for a model under isotropic loading ($\lambda = 1$). During the fracture infilling process (as shown in Figure 20.3a-e), the polygonal fracture network forms gradually until fracture saturation is reached (Figure 20.3e). After that, as shown in Figure 20.5a, the interface debonding starts from the fracture network, which forms many 'islands' (the areas bonded to the top layer) and 'lakes' (the area of interface debonding) surrounding them. When the external loads continue to increase, the debonding area increases. Consequently, the 'islands' become smaller and smaller, and the 'lakes' around the islands become correspondingly larger and larger. So the strain accommodation is through fracture propagation along the layer interface. Consideration of the fracture induced stress redistribution in the interface area after fracture saturation helps to illustrate the interface debonding mechanism. As shown in Figure 20.5b, comparing with other areas, the tensile stresses in the 'islands' may be high enough to cause further debonding of the interface, and, therefore, the fractures propagating horizontally in the layer interfaces dominate the fracture pattern development.

20.3 FRACTURE SPACING MEASUREMENTS

This type of modelling is helpful in guiding the method of measuring fractures during site investigation for engineering projects. The most common method for measuring fracture spacing is the borehole or scanline method. As illustrated in Figure 20.6,

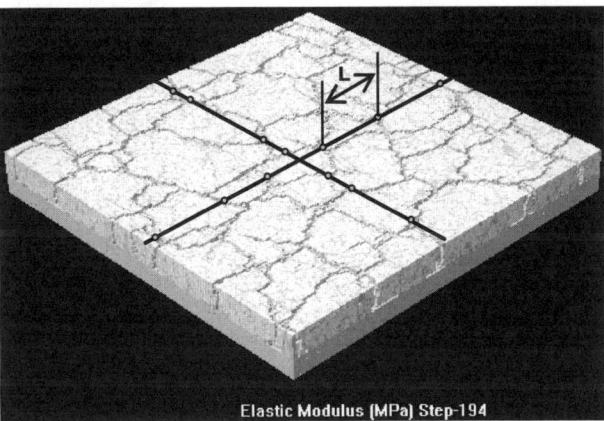

Figure 20.6 Measurement of fracture spacing using scanlines across the central point of the numerical model and parallel to the two principal stress directions. The points along the scanlines are the intersection points with fractures. The distance L between two neighbouring points represents its fracture spacing.

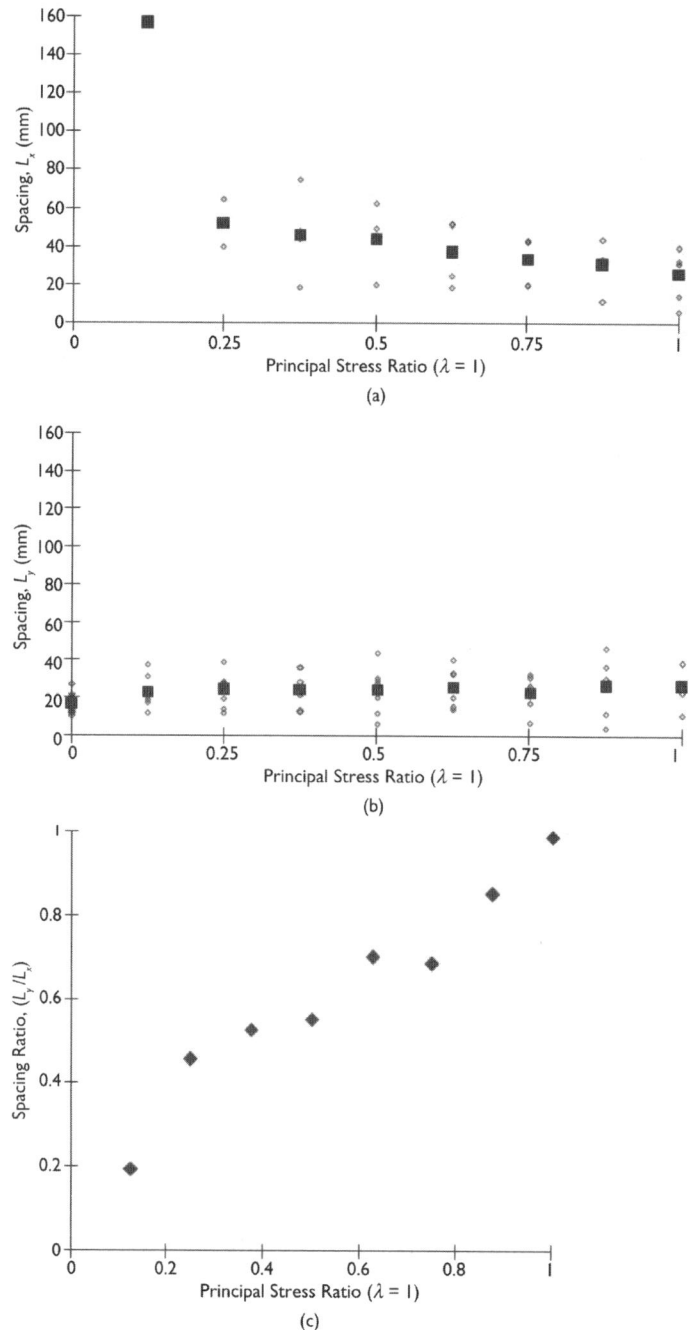

Figure 20.7 Modelling results for fracture spacing, L, vs. principal stress ratio, λ. a) and b) Spacing L_x and L_y vs. principal stress ratio λ; c) Spacing L_y/L_x vs. the principal stress ratio, λ.

scanlines are often taken perpendicular to the mean strike of the fracture set, either along cross-sections of a layer or along the surface of a layer, to estimate the distances, L, between two neighbouring fractures.

We might expect a consistent measure of L for a well-developed fracture pattern (saturated). However, this is true only when the layer is loaded uniaxially, in which parallel fractures form. For a fractured layer under biaxial loading conditions, the spacing of the fractures can be widely scattered even for a well-developed fracture pattern (Figure 20.4). As shown in Figure 20.6, even for isotropic loading ($\lambda = 1$), due to the 'island' fracture network developed in the x-y plane, the spacing observed along two scanlines perpendicular to each other across the central point will exhibit different fracture distributions.

On plotting the spacings, L, versus the principal stress ratio, λ, for data obtained using the scanline method (Figure 20.7a,b), we find that the fracture spacings measured along lines in different directions have a different relation of fracture spacing to principal stress ratio. When the ratio increases from 0 to 1, the spacings L_y measured along the scanline parallel to the minimum principal stress decrease; whereas, the spacings L_x measured along the scanline parallel to the maximum principal stress show a just a little increase. In both situations, scatter is found.

Bearing this in mind, and by extension to other geological formations, it is clear that care must be taken in establishing scanlines and in interpreting the resultant data. There is a tendency to assume that, if three mutually perpendicular scanlines are used, then the bias associated with directionality will be eliminated. However, and as can be deduced from the preceding discussion, the situation can be more subtle than this and the engineering measurements of fractures along scanlines should be also guided by geological understanding.

Chapter 21

Concluding remarks

In the previous twenty Chapters, we have explored a variety of circumstances in which rock failure occurs—with examples from structural geology, laboratory experiments and numerical modelling. In the Explanatory Notes at the beginning of the book, we explained the concept of the generic force–displacement curve and it is clear that all the cases of rock failure are variations on the same theme: the rock is loaded by some means in some configuration; at first, there is elastic displacement; then there is the initial development of cracks before the peak force; and finally the progressive breakdown of the rock.

We concentrated on the outputs of the numerical simulation of rock failure because these enable one to study aspects of the rock failure process that cannot be observed in laboratory experiments, e.g. the detailed stress field within a rock specimen or rock mass. Some readers may ask why we are so confident that these simulations represent the rock reality—and the answer is that experimental evidence confirms the trends of the results. Using the term 'verification' to mean that the numerical results are confirmed by a different computer program and the term 'validation' to mean absolute confirmation of the numerical results by comparison with physical observations, we can verify the computer results but we will never be able to achieve full validation because it is not possible to obtain the detailed physical evidence of the internal rock responses.

Another related aspect is that we have studied relatively simple loading configurations and circumstances in order to be able to explain and illustrate the rock failure. In practice, an *in situ* rock mass will be simultaneously subjected to a combination of variables such as heat, water, stress, chemical effects and time. In this case, the rock mass response must be evaluated through fully-coupled thermo-hydro-mechanical-chemical numerical models, as is being progressed, *inter alia*, by the DECOVALEX programme mentioned in the introductory section of Chapter 12. There is even less chance of being able to validate such models, especially at the full scale of rock engineering, and so the concept of technical auditing of the codes and their operation is critically important. Does the code contain all the necessary variables? Does it contain all the interactions between the variables? Have all the data been input correctly? These and many other questions are crucial if the design of a rock engineering project is to be based on the output from such fully-coupled codes.

The approach strategy to the technical auditing subject requires the following example 'high level' questions to be addressed. How are the key components of a rock engineering problem to be decided upon *a priori*? How are the mechanisms necessary to be incorporated in any particular code to be determined? How is the adequacy of a specific code to be established? How is confidence in the use of a numerical code to be ascertained?

308 Rock failure mechanisms

The International Society for Rock Mechanics (ISRM) currently has a Commission on "Rock Engineering Design Methodology" which will be presenting its report in a Taylor and Francis book by Feng and Hudson in 2011. This book will address these questions and outline the full technical auditing procedure for computer codes supporting rock engineering design.

Hopefully, through the examples in this book, we have demonstrated the advantages of numerical simulation, not just to illustrate different rock loading configurations but also as a research tool. There is no question that the development of computing capability and the coupled simulation codes will continue to develop rapidly, and so the future of this subject is assured. For physical testing of rocks, the ISRM produces Suggested Methods recommending appropriate testing procedures. This raises the intriguing question of whether virtual rock testing procedures are required in order to be able to directly compare the capabilities of different codes.

Just a glance at Figure 21.1 illustrating the Itasca codes supporting the numerical compilation and analysis of a synthetic rock mass confirms that we are at a truly exciting time in rock mechanics research. In the early days of rock mechanics from the 1960s onwards, there was emphasis on physical rock testing. This was followed over the last thirty years by the ever-increasing development of numerical simulation methods. From now on, we are moving in the direction of even more enhanced computer capabilities, 3-D transparent earth visualisation, virtual rock laboratories, more use of the Internet, etc. These developments are not only directly exciting for the new generation of rock mechanics researchers and engineers but will also lead to safer rock engineering structures, improved methods of mining, less loss of life through structural failures, and more efficient use of the Earth's limited resources.

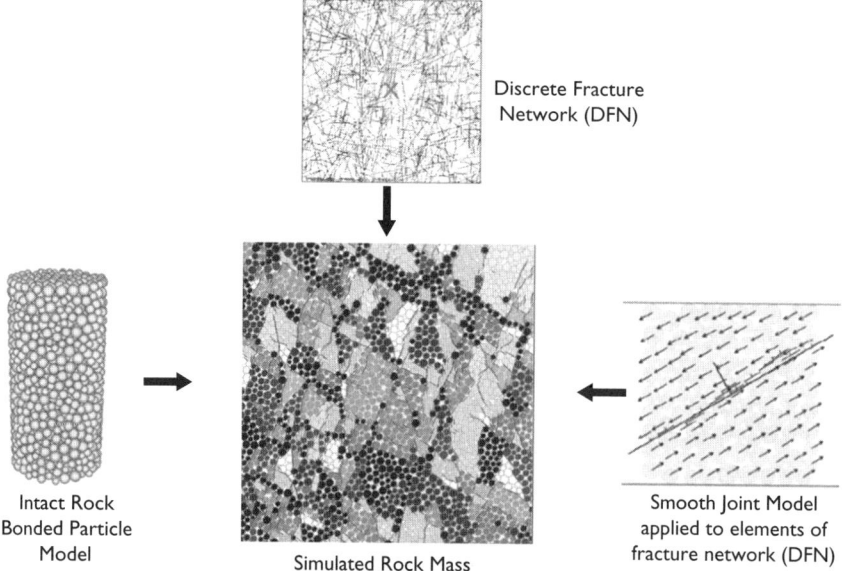

Figure 21.1 Numerical compilation and analysis of a synthetic fractured rock mass (from Mas Ivars, 2010).

References and bibliography

Aaltonen, I., Heikkinen, E., Paulamäki, S., Säävuori, H., Vuoriainen, S. & Öhman, I.: *Summary of Petrophysical Analysis of Olkiluoto Core Samples 1990–2008*. Working Report 2009-11, Posiva, Finland, 2009. (Download available from www.posiva.fi).
Addinall, E. & Hackett, P.: Tensile Failure in Rock-Like Materials. In: E.M. Spokes and C.R. Christiansen (Eds): *Proc. 6th US Rock Mech. Symp.*, Rolla, Missouri, 1964, pp. 515–38.
Alehossein, H. & Hood, M.: State-of-the-Art Review of Rock Models for Disc Roller Cutters. In: M. Aubertin, F. Hassani, & H. Mitri (Eds): *Rock Mechanics*. Balkema, Rotterdam, 1996, 693p.
Alehossein, H. & Poulsen, B.A.: Stress Analysis of Longwall Top Coal Caving. *Int. J. Rock Mech. Min. Sci.*, 47 (2010), pp. 30–41.
Amadei, B. & Stephansson, S.: *Rock Stress and Its Measurement*. Chapman & Hall, London, 1997, 490p.
Anderson, E.M.: *The Dynamics of Faulting and Dyke Formation with Applications to Britain*. Oliver & Boyd, Edinburgh, 1942, 191p.
Andersson, J., Ahokas, H., Hudson, J., Koskinen, L., Luukkonen, A., Löfman, J., Keto, V., Pitkänen, P., Mattila, J., Ikonen, A.T.K. & Ylä-Mella, M.: *Olkiluoto Site Description 2006. Part 1*. Report 2007-3. (Download available from www.posiva.fi).
Andersson, J.C.: *Rock Mass Response to Coupled Mechanical Thermal Loading. Äspö Pillar Stability Experiment, Sweden*. PhD thesis, Royal Institute of Technology (KTH), Stockholm, 2007. (Download available from www.kth.se).
Ashby, M.F. & Hallam, S.D.: The Failure of Brittle Solids Containing Small Cracks under Compressive Stress States. *Acta. Metall.* 34 (1986), pp. 497–510.
Ashurst, J. & Dimes, F.: *Conservation of Building and Decorative Stone*. Butterworth-Heinemann Series in Conservation and Museology, 1998, 468p.
Atkinson, B.K.: Sub-Critical Crack Growth in Geological Materials. *J. Geophysics* 89 (1984), pp. 4077–114.
Atkinson, B.K. & Meredith, P.G.: The Theory of Sub-Critical Crack Growth with Applications to Minerals and Rocks. In: B.K. Atkinson (Ed): *Fracture Mechanics of Rock*. Academic Press, London, Orlando, 1987, pp. 111–66.
Bäckström, A.: *Rock Damage Caused by Underground Excavation and Meteorite Impacts*. PhD thesis, Royal Institute of Technology, Sweden, 2008. (Download available from www.kth.se).
Baecher, G.B., Lanney, N.A. & Einstein, H.H.: Statistical Description of Rock Properties and Sampling. In: *Proc. 18th US Symp. Rock. Mech.* Am. Inst. Mining Eng., Ref: 5C1-8, Johnson Publishing, Keystone, Co., 1977.
Bahat, D.: *Tectonofractography*. Springer-Verlag, Berlin, 1991, 354p.

Bahat, D., Rabinovitch, A. & Frid, V.: *Tensile Fracturing in Rocks.* Springer-Verlag, Berlin, 2005, 569p.

Barton, N.: *Rock Quality, Seismic Velocity, Attenuation and Anisotropy.* Taylor & Francis, London, 2007, 729p.

Beamish, B.B. & Crosdale, P.J.: Instantaneous Outbursts in Underground Coal Mines: An Overview and Association with Coal Type. *Int. J. Coal. Geol* 35 (1998), pp. 27–35.

Bieniawski, Z.T.: Mechanism of Brittle Fracture of Rock. *Int. J. Rock Mech. Min. Sci.* 1967, 4 (1967), pp. 407–23.

Biot, M.A.: General Theory of Three-Dimensional Consolidation. *J. Appl. Phys.* 12 (1941), pp. 155–64.

Blair, S.C. & Cook, N.G.W.: Analysis of Compressive Fracture in Rock using Statistical Techniques: Part I. A Non-Linear Rule-Based Model. *Int. J. Rock Mech. Min. Sci.* 35 (1998), pp. 837–48.

Blair, S.C. & Cook, N.G.W.: Analysis of Compressive Fracture in Rock using Statistical Techniques: Part II. Effect of Microscale Heterogeneity on Macroscopic Deformation. *Int. J. Rock Mech. Min. Sci.*, 35 (1998), pp. 849–61.

Bobet, A.: Modeling of Crack Initiation, Propagation and Coalescence in Uniaxial Compression. *Rock Mech. Rock Eng.* 33 (2000), pp. 119–39.

Bobet, A. & Einstein, H.H.: Fracture Coalescence in Rock-Type Materials Under Uniaxial and Biaxial Compression. *Int. J. Rock Mech. Min. Sci.* 35 (1998), pp. 863–88.

Bobet, A. & Einstein, H.H.: Numerical Modeling of Fracture Coalescence in a Model Rock Material. *Int. J. Fracture* 92 (1998), pp. 221–252.

Borecki, M. & Kwasniewski, M.A.: Experimental and Analytical Studies on Compressive Strength of Anisotropic Rocks. In: Proc. Seventh Plenary Scientific Session of the International Bureau of Rock Mechanics, Katowice, Poland, 1981, 23p, 26p.

Bracc, W.F., Walsh, J.B. & Frangos, W.T.: Permeability of Granite under High Pressure. *J. Geophys. Res.* 73 (1968), pp. 2225–36.

Brady, B.H.G. & Brown, E.T.: *Rock Mechanics.* 2nd edition. Chapman & Hall, London, 1993.

Brooks, Z., Ulm, F-J., Einstein, H.H. & Abousleiman, Y.: A Nanomechanical Investigation of the Crack Tip Process Zone. In: Proc. 44th US Rock Mechanics Symposium and 5th U.S.-Canada Rock Mechanics Symposium. Salt Lake City, UT, June 27–30, 2010. Paper on CD.

Brown, E.T.: Fracture of Rock under Uniform Biaxial Compression. In: *Proc. 3rd Congress of International Society for Rock Mechanics.* Denver, CO, USA, pp. 111–17, 1976.

Brown, S., Caprihan, A. & Hardy, R.: Experimental Observation of Fluid Flow Channels in a Single Fracture. *J. Geophys. Res.* 103 (1998), pp. 5125–32.

Cao, Y.X., He, D.D. & Glick, D.C.: Coal and Gas Outbursts in Footwalls of Reverse Faults. *Int. J. Coal Geol.* 48 (2001), pp. 47–63.

Cazacu, O., Cristescu, N.D., Shao, J.F. & Henry, J.P.: A New Failure Criterion for Transversely Isotropic Rocks. *Int. J. Rock Mech. Min. Sci.* 35 (1998), 130p, 421p.

Chan, L-Y. & Goodman, R.E.: Predicting the Number and Dimensions of Key Blocks of an Excavation using Block Theory and Joint Statistics. In: I.W. Farmer, J.J.K. Daemen, C.S. Desai, C.E. Glass & S.P. Neuman (Eds): *Proceeding of the 28th US Symposium on Rock Mechanics*, University of Arizona, Tuscon, 1987, pp. 81–7.

Chau, K.T., Zhu, W.C., Tang, C.A. & Wu, S.Z.: Numerical Simulations of Failure of Brittle Solids under Dynamic Impact using a New Computer Program—DIFAR. *Key Eng. Mat.* 261-63 (2004), pp. 239–44.

Chen, S., Yue, Z.Q. & Tham, L.G.: Digital Image-Based Numerical Modeling Method for Prediction of Inhomogeneous Rock Failure. *Int. J. Rock Mech. Min. Sci.* 41 (2004), pp. 939–57.

Cho, S.H. & Kaneko, K.: Influence of the Applied Pressure Waveform on The Dynamic Fracture Processes in Rock. *Int. J. Rock Mech. Min. Sci.* 41 (2004), pp. 771–84.

Cook, N.G.W.: The Failure of Rock. *Int. J. Rock Mech. Min. Sci.* 2 (1965), pp. 389–403.

Copur, H.: Linear Stone Cutting Tests with Chisel Tools for Identification of Cutting Principles and Predicted Performance of Chain Saw Machines. *Int. J. Rock Mech. Min. Sci.* 47 (2010), pp. 104–20.

Cox, S.J.D. & Meredith, P.G.: Microfracture Formation and Material Softening in Rock Measured by Monitoring Acoustic Emissions. *Int. J. Rock Mech. Min. Sci.* 30 (1993), pp. 11–24.

Cristescu, N.D.: *Rock Rheology*. Kluwer Academic Publishers Group, 1989.

Cristescu, N.D. & Hunsche, U.: *Time Effects in Rock Mechanics*. Wiley-Interscience-Europe, 1998, 342p.

Crouch, S.L. & Starfield, A.M.: *Boundary Element Methods in Solid Mechanics*. Allen & Unwin, London, 1983.

Cuisiat, F.D.E. & Hudson, J.A.: The Influence of Rock Anisotropy on Borehole Breakouts: A Microstatistical Approach. *Int. J. Rock Mech. Min. Sci.* 30 (1993), pp. 1077–83.

Cundall, P.A.: Formulation of a Three-Dimensional Distinct Element Model—Part I: A Scheme to Detect and Represent Contacts in a System Composed of Many Polyhedral Blocks. *Int. J. Rock Mech. Min. Sci.* 25 (1988), pp. 107–16.

Dahou, A., Shao, J.F. & Bederiat, M.: Experimental and Numerical Investigations on Transient Creep of Porous Chalk. *Mech. Mater.* 21 (1995), pp. 147–58.

Davis, G.H. & Reynolds, S.J.: *Structural Geology of Rocks and Regions–2nd Edition*. John Wiley & Sons, Inc., 1996, 757p.

DECOVALEX: Special DECOVALEX Issue. *Environ. Geol.* 57 (2009), pp. 1217–389.

Donath, F.A.: Strength Variation and Deformational Behaviour in Anisotropic Rock. In: W.R. Judd (Ed): *State of Stress in the Earth's Crust*. Elsevier, New York, 1964.

Duveau, G. & Shao, J.F.: Single Discontinuity Theory for the Failure of Highly Stratified Rocks. *Int. J. Rock Mech. Min. Sci.* 35 (1998), pp. 807–13.

Dyskin, A.V.: On the Role of Stress Fluctuations in Brittle Fracture. *Int. J. Fracture* 100 (1999), pp. 29–53.

Eberhardt, E., Stead, D. & Coggan, J.S.: Numerical Analysis of Initiation and Progressive Failure in Natural Rock Slopes—The 1991 Randa Rockslide. *Int. J. Rock Mech. Min. Sci.* 41 (2004), pp. 69–87.

Eberhardt, E., Stimpson, B. & Stead, D.: Effect of Grain Size on the Initiation and Propagation Thresholds of Stress-Induced Brittle Fractures. *Rock Mech. Rock Eng.* 32 (1999), pp. 81–99.

Eberhardt, M.E.: *Why Things Break*. Three Rivers Press, Random House, Inc., New York, 2003, 257p.

Evans, H. & Brown, K.M.: Discussion on "Coal Structures in Outbursts of Coal and Firedamp Conditions". *Min. Eng.* 137 (1973), pp. 457–60.

Fairhurst, C.: On the Validity of the Brazilian Test for Brittle Materials. *Int. J. Rock Mech. Min. Sci.* 1 (1964), pp. 535–46.

Fairhurst, C. & Cook, N.G.W.: The Phenomenon of Rock Splitting Parallel to the Direction of Maximum Compression in the Neighbourhood of a Surface. In: *Proc.1st. Congress of the International Society for Rock Mechanics*, Lisbon, 1966, pp. 687–92.

Fang, Z. & Harrison, J.P.: Development of a Local Degradation Approach to the Modelling of Brittle Fracture in Heterogeneous Rocks. *Int. J. Rock Mech. Min. Sci.* 39 (2002), pp. 443–57.

Farmer, I.W. & Pooley, F.D.: A Hypothesis to Explain the Occurrence of Outbursts in Coal Based on a Study of West Wales Outburst Coal. *Int. J. Rock Mech. Min. Sci.* 4 (1967), pp. 189–93.

Fielden, B.M.: *Conservation of Historic Buildings*. Architectural Press, Elsevier, 2003, 392p.

Folias, E.S., Hohn, M. & Nicholas, T.: Predicting Crack Initiation in Composite Material Systems due to a Thermal Expansion Mismatch. *Int. J. Fracture* 93 (1998), pp. 335–49.

Fourney, W.L.: Mechanisms of Rock Fragmentation by Blasting. In: J.A Hudson (Ed) *Comprehensive Rock Engineering, Principles, Practice and Projects, Volume 4*. Pergamon Press, Oxford, 1993, pp. 39–69.

Franciss, F.O.: *Fractured Rock Hydraulics*. CRC Press, Taylor & Francis, London, 2010, 179p.

Frocht, M.M.: *Photoelasticity*. Wiley, New York, 1941, 411p.

Gale, J.E.: *A Numerical, Field and Laboratory Study of Flow in Rocks with Deformable Fractures*. PhD thesis. University of Berkeley, California, USA. 1975.

Gibowicz, S.J. & Kijko, A.: *An Introduction to Mining Seismology*. Academic Press, London, 1994. 399p.

Goodman, R.E.: *Methods of Geological Engineering in Discontinuous Rocks*. West Publishing Co., St. Paul, MN., 1976.

Goodman, R.E.: *Introduction to Rock Mechanics*. John Wiley & Sons, New York, 1980.

Goodman, R.E. & Shi, G-h.: *Block Theory and its Application to Rock Engineering*. Prentice-Hall, Englewood Cliffs, N.J., 1985, 338p.

Gray, I.: The Mechanism of, and Energy Release Associated with Outbursts. In: *Proc. Symposium on the Occurrence, Prediction and Control of Outbursts in Coal Mines*. The Australasian Inst. Min. Metall., Melbourne, 1980, pp. 111–25.

Griffiths, D.V. & Kidger, D.J.: Enhanced Visualization of Failure Mechanisms in Finite Elements. *Comput. Struct.* 55 (1995), pp. 265–69.

Hagros, A., Johansson E. & Hudson, J.A.: *Time Dependency in the Mechanical Properties of Crystalline Rocks: A Literature Survey*. Posiva Working Report 2008-68, 2008. (Download available from www.posiva.fi).

Hajiabdolmajid, V. & Kaiser, P.K.: Modelling Slopes in Brittle Rock. In: *Proceedings of the Fifth North American Rock Mechanics Symposium and the 17th Tunneling Association of Canada Conference*: NARMS-TAC 2002, pp. 331–39.

Hao, Y.H. & Azzam, R.: The Plastic Zones and Displacements around Underground Openings in Rock Masses Containing a Fault. *Tunn. Undergr. Sp. Tech.* 20 (2005), pp. 49–61.

Hargraves, A.J.: Instantaneous Outbursts of Coal and Gas: A Review. *Proc. Aust. Inst. Min. Metall.* 285 (1983), pp. 1–37.

Harrison, J.P. & Hudson, J.A.: *Engineering Rock Mechanics: Illustrative Worked Examples*. Elsevier, Oxford, 2000, 506p. (Also published in Chinese by Science Press of Beijing, 2009).

Hart, R., Cundall, P.A. & Lemos, J.: Formulation of a Three-Dimensional Distinct Element Model—Part II: Mechanical Calculations for Motion and Interaction of a System Composed of Many Polyhedral Blocks. *Int. J. Rock Mech. Min. Sci.* 25 (1988), pp. 117–25.

Hazzard, J., Young, R. & Maxwell, S.: Micromechanical Modeling of Cracking and Failure in Brittle Rocks. *J. Geophys. Res.* 105(B7) (2000), pp. 16683–97.

Hermann, H.J. & Roux, S.: *Statistical Models for the Fracture of Disordered Media*. Elsevier Science, Amsterdam, 1990.

Hobbs, B.E., Means, W.D. & Williams, P.F.: *An Outline of Structural Geology*. Wiley, Chichester. 1975, 571p.

Hobbs, D.W.: An Assessment of a Technique for Determining the Tensile Strength of Rock. *Brit. J. Appl. Phys.* 16 (1965), pp. 259–68.

Hoek, E.: Fracture of Anisotropic Rock. *J S Afr. Inst. Min Metall.* 64 (1964), pp. 510–8.

Hoek, E., Kaiser, P.K. & Bawden, W.F.: *Support of Underground Excavations in Hard Rock*. Balkema. Rotterdam, 1995, 215p.

Hoerger, S.F. & Young, D.S.: Probabilistic Prediction of Key Block Occurrences. In: *Proc. 31st US Symp. Rock Mech.*, Balkema, Rotterdam, 1990, pp. 229–36.

Horii, H., & Nemat-Nasser, S.: Compression-Induced Microcrack Growth in Brittle Solids: Axial Splitting and Shear Failure. *J. Geophys Res.* 90 (B4) (1985), pp. 3105–25.

Horii, H. & Nemat-Nasser, S.: Brittle Failure in Compression: Splitting, Faulting and Brittle–Ductile Transition. *Phil. Trans. Roy. Soc. London* A319 (1986), pp. 337–74.
Hudson, J.A.: Tensile Strength and the Ring Test. *Int. J. Rock Mech. Min. Sci.* 6 (1969), pp. 91–7.
Hudson, J.A. & Fairhurst, C.: Tensile Strength, Weibull's Theory and a General Statistical Approach to Rock Failure. In: M. Te'eni. (Ed): *Proc. Int. Conf. on Structure, Solid Mechanics and Engineering Design in Civil Engineering Materials*, Part II. Wiley, London, 1969, pp. 901–14.
Hudson, J.A. & Harrison, J.P.: *Engineering Rock Mechanics: An Introduction to the Principles.* Elsevier, Oxford, 1997, 444p. (Also published in Chinese by Science Press of Beijing, 2009).
Hudson, J.A., Crouch, S.L. & Fairhurst, C.: Soft, Stiff and Servo-Controlled Testing Machines: A Review with Reference to Rock Failure. *Eng. Geol.* 6 (1972), pp. 155–89.
Hudson, J.A., Brown, E.T. & Rummel, F.: Controlled Failure of Rock Discs and Rings Loaded in Diametral Compression. *Int. J. Rock Mech. Min. Sci.* 9 (1972), pp. 241–48.
Hudson, J.A., Bäckström, A., Rutqvist, J., Jing, L., Backers, T., Chijimatsu, M., Christiansson, R., Feng, X.-T., Kobayashi, A., Koyama, T., Lee, H-S., Neretnieks, I., Pan, P.Z., Rinne, M. & Shen, B. T.: Characterising and Modelling the Excavation Damaged Zone in Crystalline Rock in the Context of Radioactive Waste Disposal. *Environ. Geol.* 57 (2009), pp. 1275–97.
Hunt, S.P., Meyers, A.G. & Louchnikov, V.: Modelling the Kaiser Effect and Deformation Rate Analysis in Sandstone using the Discrete Element Method. *Comput. Geotech.* 30 (2003), pp. 611–21.
ISRM: Rock Dynamics Papers. *ISRM News Journal* (2009), pp. 72–95. (Download available from www.isrm.net).
Jaeger, J.C.: Shear Failure of Anisotropic Rocks. *Geol. Mag.* 97 (1960), pp. 65–72.
Jaeger, J.C., Cook, N.G.W. & Zimmerman, R.W.: *Fundamentals of Rock Mechanics.* Blackwell, Oxford, 2007, 475p.
Jagiello, J., Lason, M. & Nodzenski, A.: Thermodynamic Description of the Process of Gas Liberation from a Coal Bed. *Fuel* 71 (1992), pp. 431–5.
Jeon, S., Kim, J., Seo, Y. & Hong, C.: Effect of a Fault and Weak Plane on the Stability of a Tunnel in Rock-Ascaled Model Test and Numerical Analysis. *Int. J. Rock Mech. Min. Sci.* 41 (2004), pp. 658–63.
Jespersen, C.D., Spence, R.J., MacLaughlin, M.M., Parkhurst, J. & Hudyma, N.: Strength and Failure Modes of Macroporous Rock: Results from Laboratory Testing and 3D Numerical Models. In: *Proc. 42nd US Rock Mech. Symp. & 2nd US-Canada Rock Mech. Symp*, San Francisco 2008.
Jiang, C.L. & Yu, Q.X.: *Spherical Shell Losing Stability Mechanism of Coal and Gas Outbursts.* China University of Mining and Technology Press, Xuzhou, 1998.
Jiang, Y., Tanabashi, Y., Li, B. & Xiao, J.: Influence of Geometrical Distribution of Rock Joints on Deformational Behavior of Underground Opening. *Tunn. Undergr. Sp Tech.* 21 (2006), pp. 485–91.
Jiang, Z.Q. & Ji, L.J.: The Laboratory Study on Behaviour of Permeability of Rock along the Complete Stress–Strain Path. *Chin. J. Geotech. Eng.* 23 (2001), pp. 153–6.
Jing, L. & Stephansson, O.: *Fundamentals of Discrete Element Methods for Rock Engineering—Theory and Applications.* Elsevier, Amsterdam, 2007, 545p.
Jonsson, M., Bäckström, A., Feng, Q.H., Berglund, J., Johansson, M., Mas Ivars, D. & Olsson M.: *Äspö Hard Rock Laboratory: Studies of Factors that Affect and Control the Excavation Damaged/Disturbed Zone.* Report R-09-17, Swedish Nuclear Fuel & Waste Management Co., Stockholm, 2009, 312p. (Download available from www.skb.se).
Kaiser, J.: *Untersuchung über das Auftreten von Geräuschen beim Zugversuch.* Dr.-Ing. Dissertation, Fakultät für Maschinenwesen und Elektrotechnik der Technischen Universität München (TUM), 1950.

Kaiser, J.: *Arch. Für das Eisenhüttenwesen.* 24 (1953) (1/2), pp. 43–5.

Kidybinski, A.: Significance of In-Situ Strength Measurements for Prediction of Outburst Hazard in Coal Mines of Lower Silesia. In: *Proc. Symposium on the Occurrence, Prediction and Control of Outbursts in Coal Mines.* The Australasian Inst. Min. Metall., Melbourne, 1980, pp. 193–201.

Kingery, W.D.: Factors Affecting Thermal Stress Resistance of Ceramic Materials. *J. Am. Ceram. Soc.* 38 (1955), pp. 3–7.

Kolsky, H.: *Stress Waves in Solids.* Dover, Mineola, New York, 2003, 213p.

Korinets, A.R., Chen, L., Alehossein, H., Lim, W.: DIANA Modeling of a Rolling Disc Cutter and Rock Indentation. In: M. Aubertin, F. Hassani, H. Mitri, (Eds): *Proc. 2nd NARMS, Montreal, Canada—Rock mechanics.* Rotterdam: Balkema, 1996, 647p.

Kou, S., Tan, X. & Lindqvist, P-A.: *Modelling of Excavation Depth and Fractures in Rock caused by Tool Indentation.* SKB Report R-99-11, 1999. (Download available from www.skb.se).

Koyama, T.: *Stress, Flow and Particle Transport in Rock Fractures.* PhD thesis, Royal Institute of Technology, Stockholm, 2007. (Download available by from www.kth.se).

Kupfer, H.B. & Gerstle, K.H.: Behavior of Concrete under Biaxial Stresses. *J. Eng. Mech.-ASCE* 99(EM4), (1973), pp. 852–66.

Lama, R.D. & Bodziony, J.: Management of Outburst in Underground Coal Mines. *Int. J. Coal. Geol.* 35 (1998), pp. 83–115.

Lavrov, A.: Theoretical Investigation of the Kaiser Effect Manifestation in Rocks after True Triaxial Pre-Loading. *Arch. Min. Sci.* 46 (2001), pp. 47–65.

Lavrov, A.: The Kaiser Effect in Rocks: Principles and Stress Estimation Techniques. *Int. J. Rock Mech. Min Sci.* 40 (2003), pp. 151–71.

Lehtonen, A.: *Kaiser Effect-Based Stress Measurements in Olkiluoto Drillholes OL-KR28 and OL-KR29.* Posiva Working Report 2008-76, 2008, 88p. (Download available from www.posiva.fi).

Li, C. & Norlund, E.: Experimental Verification of Kaiser Effect in Rocks. *Rock Mech. Rock Eng.* 26 (1993), pp. 333–351.

Li, H.: Application of the Deformation Coefficients of Fold in Forecasting Outburst of Coal and Gas. *Henan Geol.* 13(1995), pp. 304–08. (In Chinese with English abstract).

Li, H.Y.: Major and Minor Structural Features of a Bedding Shear Zone Along a Coal Seam and Related Gas Outburst, Pingdingshan Coalfield, Northern China. *Int. J. Coal Geol.* 47 (2001), pp. 101–13.

Li, L.C., Tang, C.A., Li, C.W. & Zhu, W.C.: Slope Stability Analysis by SRM-Based Rock Failure Process Analysis (RFPA). *Geomech. Geoeng.* 1 (2006), pp. 1–12.

Li, Y.J. & Zimmermann, T.: Numerical Simulation of Fracture Propagation in an Anisotropic Layered Medium. In: *Proceedings of the Ninth International Conference on Computing Methods and Advances in Geometry.* Wuhan, China, 1997, pp. 319–24.

Lin, S.D.: Numerical Simulation Process Analysis of Upper Layered Rock's Failure Mechanism. *Chin.J. Rock Mech. Eng.* 18 (1999), pp. 392–96.

Lindqvist, P.A.: *Rock Fragmentation by Indentation and Disc Cutting.* PhD thesis, Luleå University of Technology, 1982.

Litwiniszyn, J.: A Model for the Initiation of Coal-Gas Outbursts. *Int. J. Rock Mech. Min. Sci.* 22 (1985), pp. 39–46.

Liu, H.Y.: *Virtual Research of Overburden Rock Strata Failure Process in Mining Engineering.* Master Degree Dissertation, Northeast University, Shenyang, 2000.

Liu, Z., Myer, L.R. & Cook, N.G.W.: Numerical Simulation of The Effect of Heterogeneities on Macro-Behavior of Granular Materials. In: H.J. Siriwardane & M.M. Zaman (Eds): *Computer Methods and Advances in Geomechanics,.* Proceedings of 8th Int. Conf. on Computer Methods and Advances, Vol. I, Balkema, Rotterdam, 1994, pp. 611–16.

Lockner, D.A., Byerlee, J.D., Kuksenko, V., Ponomarev, A. & Sidorin, A.: Quasi-Static Fault Growth and Shear Fracture Energy in Granite. *Nature* 350 (1991), pp. 39–42.

Lockner, D.A., Byerlee, J.D., Kuksenko, V., Ponomarev, A. & Sidorin, A.: Observations of Quasi-static Fault Growth from Acoustic Emissions. In: B. Evans & T.F. Wong (Eds): *Fault Mechanics and Transport Properties of Rocks*, Academic Press, Harcourt Brace Jovanovich, Publishers, New York, 1992, pp. 1–20.

Malan, D.F.: Simulating the Time-dependent Behaviour of Excavations in Hard Rock. *Rock Mech. Rock Eng.* 35 (2002), pp. 225–54.

Mandl, G.: *Faulting in Brittle Rocks*. Springer-Verlag, Berlin, 2000, 434p.

Mas Ivars, D.: *Bonded Particle Model for Jointed Rock Mass*. PhD thesis, Royal Institute of Technology, Stockholm, Sweden, 2010, 94p. (Download available from www.kth.se).

Matsui, T. & San, K.C.: Finite Element Slope Stability Analysis by Shear Strength Reduction Technique. *Soils Found.* 32 (1992), pp. 59–70.

Mellor, M. & Hawkes, I.: Measurement of Tensile Strength by Diametral Compression of Discs and Annuli. *Eng. Geol.* 5 (1971), pp. 173–225.

Mishnaevsky, L.J.: Physical Mechanisms of Hard Rock Fragmentation under Mechanical Loading: A Review. *Int. J. Rock Mech. Min. Sci.* 32 (1995), pp. 763–66.

Mishaevsky Jr., L.: *Damage and Fracture of Heterogeneous Materials: Modelling and Application to*. Balkema, Rotterdam, 1998, 214p.

Mogi, K.: *Earthquake Predication*. Academic Press, Harcourt Brace Jovanovich, Tokyo, 1985, 376p.

Mogi, K.: *Experimental Rock Mechanics*. Taylor & Francis, 2007, 361p.

Myer, L.R., Kemeny, J.M., Zheng, Z., Suarez, R., Ewy, R.T. & Cook, N.G.W.: Extensile Cracking in Porous Rock under Different Compressive Stress. *Appl. Mech. Rev.* 45 (1992), pp. 263–80.

Nature debate: Is the Reliable Prediction of Individual Earthquakes a Realistic Scientific Goal? http://helix.nature.Com/debates/earthquake/, April 1999.

Nemat-Nasser, S. & Horii, H.: Compression-Induced Nonplanar Crack Extension with Application to Splitting, Exfoliation, and Rockburst. *J. Geophys. Res.* 87(B8) (1982), pp. 6805–21.

Niandou, H., Shao, J.F., Henry, J.P. & Fourmaintraux D.: Laboratory Investigation of the Mechanical Behavior of Tournemire Shale. *Int. J. Rock Mech. Min. Sci.* 34 (1997), pp. 3–16.

Okubo, S. & Fukui, K.: Complete Stress–Strain Curves for Various Rock Types in Uniaxial Tension. *Int. J. Rock Mech. Min. Sci.* 33 (1996), pp. 549–556.

Ord, A. & Hobbs, B.E.: Fracture Pattern Formation in Frictional, Cohesive, Granular Material. *Phil. Trans. R. Soc. A* 368 (2010), pp. 95–118.

Ortlepp, W.D.: *Rock Fracture and Rockbursts—an Illustrative Study*. S. Afr. Inst. Min. Metall., Johannesburg, 1997, 98p.

Pan, P.-Z., Feng, X.-T. & Hudson, J.A.: Numerical Simulations of Class I and Class II Uniaxial Compression Curves using an Elasto-Plastic Cellular Automaton and a Linear Combination of Stress and Strain as the Control Method. *Int. J. Rock Mech. Min. Sci.* 43 (2006), pp. 1109–17.

Pan, P.-Z., Feng, X.-T. & Hudson, J.A.: Study of Failure and Scale Effects in Rocks under Uniaxial Compression using 3D Cellular Automata. *Int. J. Rock Mech. Min. Sci.* 46 (2009), pp. 674–85.

Pariseau, W.G.: *Design Analysis in Rock Mechancis*. Taylor & Francis, London, 2007, 560p.

Paterson, L.: A Model for Outburst in Coal. *Int. J. Rock Mech. Min. Sci.* 23 (1986), pp. 327–32.

Paterson, M.S. & Wong, T.-f.: *Experimental Rock Deformation—The Brittle Field*. 2nd edn. Springer-Verlag, Berlin, 2005, 347p.

Peng, J., Tang, C.A., Yang, T.H. & Li, L.C. Application of Strength Reduction Method on the Stability Study of Rock Tunnel. *Chinese Journal of Mechanics in Engineering* 29 (2007), pp. 50–55.

Peng, L.S.: *Introduction to Gas-Geology*. China Coal Industry Publishing House, Beijing, 1990, 250p. (In Chinese).

Peng, S.S.: A Note on the Fracture Propagation and Time-Dependent Behavior of Rocks in Uniaxial Tension. *Int. J. Rock Mech. Min. Sci.* 12 (1975), pp. 125–7.

Peng, S.S.: *Surface Subsidence Engineering*. Soc. Mining Metallurgy & Exploration Inc., USA, 1992.

Pietruszczak, S., Lydzba, D. & Shao, J.F.: Modeling of Inherent Anisotropy in Sedimentary Rocks. *Int. J. Solids Struct.* 39 (2002), pp. 637–48.

Posiva: *Olkiluoto Site Description 2008*. Posiva Report 2009-01, 2009. 714p. (Download available from www.posiva.fi).

Price, N.J. & Cosgrove, J.W.: *Analysis of Geological Structures*. Cambridge University Press, Cambridge, UK, 1990, 502p.

Qian, M.G., Miao, X.X. & Xu, J.L.: Key Stratum Theory in Stratum Controlling. *J. China Coal Soc.* 3 (1996), pp. 225–30.

Qing, S.Q. & Li, Z.D.: *Acoustic Emission Technology on Rock*. Xinan Jiaotong University Press, Chengdu, China, 1993.

Ramamurthy, T.: Strength and Modulus Responses of Anisotropic Rocks. In: J.A. Hudson (Ed.) *Comprehensive Rock Engineering, Vol. I. Fundamentals*. Pergamon Press, Oxford. (1993) pp. 313–29.

Sammis, C.G. & Ashby, M.F.: The Failure of Brittle Porous Solids under Compressive Stress States. *Acta Metall.* 34 (1986), pp. 511–26.

Schaffer, R.J.: *The Weathering of Natural Building Stones*. Department of Scientific and Industrial Research, Building Research Special Report 18, 1932, 149p.

Shao, J.F., Zhu, Q.Z. & Su, K.: Modeling of Creep in Rock Materials in Terms of Material Degradation. *Comput. Geotech.* 30 (2003), pp. 549–55.

Sheorey, P.R.: *Empirical Rock Failure Criteria*. A.A. Balkema, Rotterdam, 1997.

Shepherd, J., Rixon, L.K. & Griffiths, L.: Outbursts and Geological Structures in Coal Mines: A Review. *Int. J. Rock Mech. Min. Sci.* 18 (1981), pp. 267–83.

Shorlin, K.A., de Bruyn, J.R., Graham, M. & Morris, S.W.: Development and Geometry of Isotropic and Directional Shrinkage-Crack Patterns. *Phys. Rev. E* 61 (2000), 6950.

Siegesmund, S., Ullemeyer, K., Weiss, T. & Tschegg, E.K.: Physical Weathering of Marbles Caused by Anisotropic Thermal Expansion. *Int. J. Earth Sci.* 89 (2000), pp. 170–82.

Skoczylas, F. & Henry, J.P.: A Study of the Intrinsic Permeability of Granite to Gas. *Int. J. Rock Mech. Min. Sci.* 32 (1995), pp. 171–9.

Smith, C.H.: Observations on Stone used for Building. *Transactions of the Royal Institute of British Architects* 1 (1842).

Song, J.Q.: *Practical Support Stress Control in Mining Engineering*. Coal Industry Publishing House, Beijing, 1992.

Song, J.-J., Lee, C-I. & Seto, M.: Stability Analysis of Rock Blocks Around a Tunnel using a Statistical Joint Modeling Technique. *Tunn. Undergr. Sp. Tech.* 16 (2001), pp. 341–51.

Song, Y. & Song, Z.Q.: Laws of Uphold Pressure in Mining and its Relation with Overburden Strata Movement. *J. China Coal Soc.* 9 (1991), pp. 47–56.

Souley, M., Homand, F. & Thoraval, A.: The Effect of Joint Constitutive Laws on the Modelling of an Underground Excavation and Comparison with In-Situ Measurements. *Int. J. Rock Mech. Min. Sci,* 34 (1997), pp. 97–115.

Souley, M., Homand, F., Pepa, S. & Hoxha, D.: Damage-Induced Permeability Changes in Granite: A Case Example at the URL in Canada. *Int. J. Rock Mech. Min. Sci.* 38 (2001), pp. 297–310.

SP Swedish National Testing and Research Institute.: *Testing and Assessment of Marble and Limestone.* TEAM Report under EU Contract N°: G5RD-CT-2000-00233, Project N°: GRD1-1999-10735, 2005, 133p.

Stavrogin, A.N. & Tarasov, B.G.: *Experimental Physics and Rock Mechanics.* Balkema, Abingdon, 2001. 356p.

Stephansson, O., Hudson, J.A. & Jing, L. (Eds): *Coupled Thermo-Hydro-Mechanical-Chemical Processes in Geo-Systems.* Elsevier, Oxford, 2004, 832p.

Stephansson, O., Jing, L. & Tsang, C.-F. (Eds): *Coupled Thermo-Hydro-Mechanical Processes of Fractured Media.* Elsevier, Oxford, 1996, 575p.

Sundberg, J., Wrafter, J., Back, P.-E. & Rosén, L.: *Thermal Properties Laxemar.* SKB report R-08-61, 2008. (Download available from www.skb.se).

Tan, X.C., Kou, S.Q. & Lindqvist, P-A.: Simulation of Rock Fragmentation by Indenters using DDM and Fracture Mechanics. In: M. Aubertin, F. Hassani, H. Mitri, (Eds).: *Proc. 2nd NARMS, Montreal, Canada—Rock mechanics.* Rotterdam: Balkema, 1996.

Tang, C.A.: *Catastrophe in Rock Failure.* PhD thesis, Northeastern University, Shenyang, China, 1988.

Tang, C.A.: Numerical Simulation of Progressive Rock Failure and Associated Seismicity. *Int. J. Rock Mech. Min. Sci.* 34 (1997), pp. 249–62.

Tang, C.A., Chen, Z.H., Xu, X.H. & Li, C.: A Theoretical Model for Kaiser Effect in Rock. *Pure Appl. Geophys.* 150 (1997), pp. 203–15.

Tang, C.A. & Kaiser, P.K.: Numerical Simulation of Cumulative Damage and Seismic Energy Release in Unstable Failure of Brittle Rock—Part I. Fundamentals. *Int. J. Rock Mech. Min. Sci.* 35 (1998), pp. 113–21.

Tang, C.A., Lin, P., Wong, R.H.C. & Chau, K.T.: Analysis of Crack Coalescence in Rock-Like Materials Containing Three Flaws–Part II:Numerical Approach. *Int. J. Rock Mech. Min. Sci.* 38 (2001), pp. 925–39.

Tang, C.A., Liu, H., Lee, P.K.K., Tsui, Y. & Tham, L.G.: Numerical Tests on Micro-Macro Relationship of Rock Failure under Uniaxial Compression, Part I: Effect of Heterogeneity. *Int. J. Rock Mech. Min. Sci.* 37 (2000), pp. 555–69.

Tang, C.A., Liu, H.Y., Zhu, W.C., Yang, T.H., Li, W.H., Song, L. & Lin, P.: Numerical Approach to Particle Breakage under Different Loading Conditions. *Powder Technol.* 143-4 (2004), pp. 130–43.

Tang, C.A., Tham, L.G., Lee, P.K.K., Tsui, Y. & Liu, H.: Numerical Tests on Micro-Macro Relationship of Rock Failure under Uniaxial Compression, Part II: Constraint, Slenderness and Size Effects. *Int. J. Rock Mech. Min. Sci.* 37 (2000), pp. 570–7.

Tang, C.A., Tham, L.G., Wang, S.H., Liu, H. & Li, W.H.: A Numerical Study of the Influence of Heterogeneity on the Strength Characterization of Rock under Uniaxial Tension. *Mech. Mater.* 39 (2007), pp. 326–39.

Tang, C.A., Wong, R.H.C., Chau, K.T. & Lin, P.: Modeling of Compression-Induced Splitting Failure in Heterogeneous Brittle Porous Solids. *Eng. Fract. Mech.* 72 (2005), pp. 597–615.

Tang, C.A., Yang, T.H., Tham, L.G., Lee, P.K.K. & Li, L.C.: Coupled Analysis of Flow, Stress and Damage (FSD) in Rock Failure. *Int. J. Rock Mech. Min. Sci.* 39 (2002), pp. 477–89.

Tang, C.A., Zhang, Y.B, Liang, Z.Z., Xu, T., Tham, L.G., Lindqvist, P.-A., Kou, S.Q. & Liu, H.Y.: Fracture Spacing in Layered Materials and Pattern Transition from Parallel to Polygonal Fractures. *Phys. Rev. E*, 73, 5 (2006).

Terzaghi, K.: Stability of Steep Slopes on Hard Unweathered Rock. *Geotechnique* 12 (1962), pp. 251–70.

Tien, Y.M. & Kuo, M.C.: A Failure Criterion for Transversely Isotropic Rocks. *Int. J. Rock Mech. Min. Sci.* 38 (2001), pp. 399–412.

Tien, Y.M, Kuo, M.C. & Juang, C.H.: An Experimental Investigation of the Failure Mechanism of Simulated Transversely Isotropic Rocks. *Int. J. Rock Mech. Min Sci.* 43 (2006), pp. 1163–81.

Tien, Y.M. & Tsao, P.F.: Preparation and Mechanical Properties of Artificial Transversely Isotropic Rock. *Int. J. Rock Mech. Min Sci.* 37 (2000), pp. 1001–12.

Tsang, Y.W. & Tsang, C.F.: Channel Model of Flow Through Fractured Media. *Water Resour. Res.* 23 (1987), pp. 467–79.

Ulusay, R. & Hudson, J.A. (Eds): *The Complete ISRM Suggested Methods for Rock Characterisation, Testing and Monitoring: 1974–2006.* International Society for Rock Mechanics, Lisbon, Portugal. 2007, 628p.

Van Mier, J.G.M.: *Fracture Process of Concrete.* CRC Press, New York, 1997, 253p.

Villaescusa, E., Seto, M. & Baird, G.: Stress Measurements from Oriented Core. *Int. J. Rock Mech. Min Sci.* 39 (2002), pp. 603–15.

Wang, C., Tannant, D.D. & Lilly, P.A.: Numerical Analysis of The Stability of Heavily Jointed Rock Slopes using PFC2D. *Int. J. Rock Mech. Min. Sci.* 40(2003), pp. 415–24.

Wang, J.K. & Lehnhoff, T.F.: Bit Penetration into Rock—A Finite Element Study. *Int. J. Rock Mech. Min. Sci.* 13 (1976), pp. 11–6.

Wawersik, W.R. & Brace, W.F.: Post-Failure Behavior of a Granite and Diabase. *Rock Mech.* 3 (1971), pp. 61–85.

Wawersik, W.R. & Fairhurst, C.: A Study of Brittle Rock Failure in Laboratory Compression Experiments. *Int. J. Rock Mech. Min. Sci.* 7 (1970), pp. 561–75.

Whittaker, B.N. & Reddish, D.J.: *Subsidence, Occurrence, Prediction and Control.* Elsevier Science, Amsterdam, 1989.

Williams, D.B.: *Stories in Stone.* Walker, New York, 2009, 260p.

Williams, R.J. & Weissmann, J.J.: Gas Emission and Outburst Assessment in Mixed CO_2 and CH_4 Environments. In: *Proc. ACIRL Underground Mining Seminar.* Australian Coal Industry Res. Lab., North Ryde, Brisbane, 1995,12p.

Wong, L.N.Y. & Einstein, H.H.: Systematic Evaluation of Cracking Behavior in Specimens Containing Single Flaws under Uniaxial Compression. *Int. J. Rock Mech. Min. Sci.* 46 (2009), pp. 239–249.

Wong, R.H.C. & Chau, K.T.: Crack Coalescence in a Rock-Like Material Containing Two Cracks. *Int. J. Rock Mech. Min. Sci.* 35 (1998), pp. 147–64.

Wong, R.H.C., Chau, K.T., Lin, P. & Tang, C.A.: Analysis of Crack Coalescence in Rock-Like Materials Containing Three Flaws. I. Experimental Approach. *Int. J. Rock Mech. Min. Sci.* 38 (2001), pp. 909–24.

Wong, R.H.C., Chau, K.T., Tsoi, P.M. & Tang, C.A.: Pattern of Coalescence of Rock Bridge between Two Joints under Shear Testing. In: G. Vouile & P. Berest (Eds): *Proceedings, 9th International Congress on Rock Mechanics, Paris.* Vol. 1, Balkema, Rotterdam, 1999, pp. 735–8.

Wong, R.H.C., Lau, K.W., Lin, P., Chau, K.T. & Tang, C.A.: A Numerical and Experimental Study of the Effects of Non-Persistent Joints on Slope Stability. In: *Proceedings Symposium on Slope Hazards and their Prevention.* HKU, Hong Kong, 2000, pp. 126–31.

Wong, R.H.C., Tang, C.A., Chau, K.T. & Lin, P.: Splitting Failure in Brittle Rocks Containing Pre-Existing Flaws under Uniaxial Compression. *Eng. Fract. Mech.* 69 (2002), pp. 1853–1871.

Worsey, P.: *Geotechnical Factors Affecting the Application of Pre-Split Blasting to Rock Slopes.* PhD thesis, University of Newcastle, UK, 1981.

Wylie, D.C.: *Foundations on Rock.* Taylor & Francis, London, 1999, 401p.

Yanagidani, T., Sano, O., Terada, M. & Ito, I.: The Observation of Cracks Propagating in Diametrically-Compressed Rock Discs. *Int. J. Rock Mech. Min. Sci.* 15 (1978), pp. 225–35.

Yang, T.H., Tang, C.A., Zhu, W.C. & Feng, Q.Y.: Coupling Analysis of Seepage and Stress in Rock Failure Process. *Chin. J. Rock Soil Eng.* 23 (2001), pp. 489–93.

Yang, Z.Y., Chen, J.M. & Huang, T.H.: Effect of Joint Sets on the Strength and Deformation of Rock Mass Models. *Int. J. Rock Mech. Min. Sci*, 35 (1998), pp. 75–84.

Yavuz, H.: An Estimation Method for Cover Pressure Re-Establishment Distance and Pressure Distribution in the Goaf of Longwall Coal Mines. *Int. J. Rock Mech. Min. Sci.* 41 (2004), pp. 193–205.

Yeung, M.R. & Leong, L.L.: Effects of Joint Attributes on Tunnel Stability. *Int. J. Rock Mech. Min. Sci.* 34 (1997), pp. 3–4.

Yoshinaka, R. & Yamabe, T.: Joint Stiffness and the Deformation Behaviour of Discontinuous Rock. *Int. J. Rock Mech. Min. Sci.* 23 (1986), pp. 19–28.

Yu, G.M.: *Nonlinear Theory and Practice of Mining Subsidence*. Coal Industry Publishing House, Beijing, 1998.

Yu, M., Xia, G. & Kolupaev, V.A .: Basic Characteristics and Development of Yield Criteria for Geo-materials. *J. Rock Mech. Geotech. Eng.* 1 (2009), pp. 71–88.

Yu, M.H.: *Double Twin Strength Criterion and its Application*. Science Press, Beijing, 1998. (in Chinese).

Yu, Q.X.: *Prevention and Control of Methane in Coal Mines*. China University of Mining and Technology Press, Xuzhou, 1993.

Yu, R.C., Ruiz, G. & Pandolfi, A.: Numerical Investigation on the Dynamic Behaviour of Advanced Ceramics. *Eng. Fract. Mech.* 71 (2004), pp. 897–911.

Yuan, S.C. & Harrison, J.P.: Development of a Hydro-Mechanical Local Degradation Approach and its Application to Modelling Fluid Flow during Progressive Fracturing of Heterogeneous Rocks. *Int. J. Rock Mech. Min. Sci.* 42 (2005), pp. 961–84.

Yuan, S.C. & Harrison, J.P.: A Review of the State of the Art in Modelling Progressive Mechanical Breakdown and Associated Fluid Flow in Intact Heterogeneous Rocks. *Int. J. Rock Mech. Min. Sci.* 43 (2006), pp. 1001–22.

Yue, Z.Q., Chen, S. & Tham, L.G.: Finite Element Modeling of Geomaterials using Digital Image Processing. *Comput. Geotech.* 30 (2003), pp. 375–97.

Zeuch, D.H., Swenson, D.V. & Finger JT.: Subsurface Damage Development in Rock During Drag-Bit Cutting: Observations and Model Predictions. In: *Proc. 24th US Symposium on Rock Mechanics*, 1983, pp. 733–42.

Zhang, A. & Stephansson, O.: *Stress Field of the Earth's Crust*. Springer, Netherlands, 2010, 322p.

Zhang, X. & Sanderson, D.J.: *Numerical Modelling and Analysis of Fluid Flow and Deformation of Fractured Rock Masses*. Elsevier, Oxford, 2002, 288p.

Zhao, J.: Applicability of Mohr–Coulomb and Hoek–Brown Strength Criteria to the Dynamic Strength of Brittle Rock. *Int. J. Rock Mech. Min. Sci.* 37 (2000), pp. 1115–21.

Zhao, Y.S.: Coupled Mathematical Model on Coal Mass-Gas and its Numerical Method. *Chin. J. Getech. Eng.* 13 (1994), pp. 229–39. [in Chinese].

Zhao, Y.S.: *Fluid Mechanics of Mining Rock*. Publishing House of the Chinese Coal Mining Industry, Beijing, (1994).

Zhao, J., Zhou, Y.X., Hefny, A.M., Cai, J.G., Chen, S.G., Li, H.B., Liu, J.F., Jain, M., Foo, S.T. & Seah, C.C.: Rock Dynamics Research Related to Cavern Development for Ammunition Storage. *Tunn. Undergr. Sp. Tech.* 14 (1999), pp. 513–26.

Zhou, S.N. & Lin, B.Q.: *Theory of Gas Flow and Storage in Coal Seams*. China Coal Industry Publishing House, Beijing, 1998.

Zhu, W.C., Liu, J., Tang, C.A., Zhao, X.D. & Brady, B.H.: Simulation of Progressive Fracturing Processes around Underground Excavations under Biaxial Compression. *Tunn. Undergr. Sp. Tech.* 20 (2005), pp. 231–47.

Zhu, W.C. & Tang, C.A.: Micromechanical Model for Simulating the Fracture Process of Rock. *Rock Mech. Rock Eng.* 37 (2004), pp. 25–56.

Zhu, W.C., Tang, C.A., Huang, Z.P. & Liu, J.S.: A Numerical Study of the Effect of Loading Conditions on the Dynamic Failure of Rock. *Int. J. Rock Mech. Min. Sci.* 41 (2004), p. 424. Full paper available in Vol. 41, Supplement 1 on CD, Proc. ISRM SINOROCK 2004 Symposium, pp. 348–53.

Zhu, Z.H. & Meguid, S.A.: On the Thermoelastic Stresses of Multiple Interacting Inhomogeneities. *Int. J. Solids Struct.* 37 (2000), pp. 2313–30.

Index

anisotropy 8, 121
   types 122
   simulation 123

biaxial loading 97
Brazilian test 32
   cushion materials 38
   specimen size 40
building stone degradation 144

Caen stone 145
Carrara marble quarry 211
Class I and Class II in uniaxial
   compression 82
comminution 273 (see particle
   breakage)
concluding remarks 307
confinement and shear 89
crack propagation 109
creep 136
cutting rock 229
   chipping process 233
   crushed zone 236

Daliuta coal mine in China 255
DECOVALEX 189
DIANE 7
dogbone tensile specimen 14
dynamic loading 181
   fracture influence on pre-
     split blasting 182
   simulation 183

elastic-plastic cellular automaton 86
Ellora cave temples in India 4

Finlandia Hall in Helsinki, Finland 145, 211
flaws 148
   array 158
   physical tests 179
   pore-like 165, 172
   propagation 149
   wing cracks 162
fracture coalescence 147
fracture formation 92
   acoustic emission 95
   interaction matrix 93

gas outbursts 263
generic force–displacement curve xxxix

indirect tension 29
   stress distributions 33
inelasticity 10
inhomogeneity 7, 101
   stress distribution 105, 106

JinPing II hydroelectric project
   in China 6

Kaiser Effect 129

longwall coal mining 249

micro-structure simulation 116

Norwich Cathedral 145

particle breakage 273
  energy 290
  multiple particles 284
  shape 293
  single particle 274
  size distribution 292
pillar stresses 258
plate test 41
purpose of the book 1
P-wave velocities 9

ring test 42
  secondary fractures 50
runestone 4

slope failure 219
  jointed rock mass 225
  layered rock mass 223
  Loch Lomond 220
  safety factor 220
  Tectonic Bore Mine in Australia 220
stress relaxation 136, 141
structural geology 3
subsidence 260
synonyms used in the book xl
synthetic rock mass 308

thermal stress 209
  Äspö HRL 210
  disc–ring model 213
  Gobi desert 210
  inclusions 215
  stress variation 212
  thermal conductivity modelling 217
time dependency 135
  micro-structural damage 138
tunnels 237
  ϕ_j 245
  lateral stress 243
  progressive failure 238
  South Africa 238
'turtle' crack formation 297
  fracture spacing 303
  three-layer model 298

uniaxial compression 55
  AE generation 62
  complete stress–strain curve 56
  end constraint 73
  fracturing 63
  size effect 81
  slenderness 78
  variability 70
uniaxial tension 13
  complete load–deformation curves 25
  displacement vectors 28
  irregularity of stress distribution 23

water flow 189
  experimental configuration 191
  flow paths 196
  influence of inhomogeneity 193
  Yuan and Harrison (2005) 200
Weibull distribution xli
why do things break? 1